AF704168

WITHDRAWN

RECEPTORS AND HUMAN DISEASES

RECENT MACY PUBLICATIONS

Aspects of the History of Medicine in Latin America, edited by John Z. Bowers and Elizabeth F. Purcell

Protein-Energy Malnutrition in Barbados: The Role of Continuity of Care, by Frank C. Ramsey

The Impact of Health Services on Medical Education: A Global View, edited by John Z. Bowers and Elizabeth F. Purcell

New Medical Schools at Home and Abroad, edited by John Z. Bowers and Elizabeth F. Purcell

Care of the Aging, edited by Carolyn Spieler

Renal Function, edited by Gerhard H. Giebisch and Elizabeth F. Purcell

* * * * *

A complete catalogue of books
in print will be sent on request

All publications may be ordered directly from
Independent Publishers Group
14 Vanderventer Avenue
Port Washington, New York 11050

Edited by
Alexander G. Bearn
and
Purnell W. Choppin

RECEPTORS AND HUMAN DISEASES

Report of a Macy Conference

JOSIAH MACY, JR. FOUNDATION
One Rockefeller Plaza, New York, New York 10020

QH
603
.C43
R4

© 1979 Josiah Macy, Jr. Foundation

All rights reserved

Library of Congress Catalog Card No.: 79-90867

Bowers, John Z. and Elizabeth F. Purcell
Receptors and Human Diseases.
Josiah Macy, Jr. Foundation, The
333 p.
7912 790914

ISBN: 0-914362-30-5
Manufactured by Port City Press, Baltimore, Maryland
Distributed by the Independent Publishers Group
14 Vanderventer Avenue, Port Washington, New York 11050

CONTENTS

FOREWORD

From its inception during World War II until 1960 the principal thrust of the Macy Conference Program focused on developments in the biomedical sciences. When the program was reactivated in 1965 the emphasis shifted to aspects of medical education and medical care in the United States and elsewhere in the world.

Recently, however, primarily through the foundation's Faculty Scholar Award Program, we have become conscious of the striking progress being made in cellular biology and the structure and function of cell membranes. Therefore, when Alexander G. Bearn and Purnell W. Choppin suggested that we sponsor a conference on receptors and human diseases, we agreed to do so with the understanding that they would plan and direct the meeting in cooperation with our conference director and editor, Elizabeth F. Purcell, as well as act as the scientific editors of the published volume.

The conference was held in New Orleans on 4 to 6 December 1978. This report of the proceedings is an excellent testimonial to the success of that meeting and to the skills of its organizers.

John Z. Bowers, M.D.
President
Josiah Macy, Jr. Foundation

9 November 1979

INTRODUCTION

The scientific exploration of cellular structure, particularly during the last decade, has led to a remarkable increase in our understanding of the chemistry and molecular configuration of cell surfaces. The molecular architecture of cell membranes has taken on a deeper significance as the biological properties of specialized sites, in the form of specific receptors, have come into sharp focus. The role of receptors in the regulation of complex biological activities has ramifications for physicians, immunologists, virologists, pharmacologists, and parasitologists, to name but a few of the disciplines for whom keeping abreast of advances in receptor biology and chemistry is becoming increasingly important.

This conference was convened in the belief that the usual stratification of science by field does not provide sufficient opportunities for fruitful interactions among scientists who, while working in different disciplines, have a common interest in receptor biology. The meeting began with a discussion of virus-receptor interactions, including the importance of receptors as determinants of host range and cellular tropism, and the role of the incorporation of viral proteins into cell membranes in virus production and in alerting host defense mechanisms. The finding that other host factors affecting virus-cell membrane interactions, that is, cellular proteases that activate viral activities, are important determinants of virus multiplication and pathogenesis emphasizes the need to focus on the host as well as the virus.

In perhaps no other field has the molecular dissection of receptor molecules been so successful and so evident as in immunology. The investigation of the histocompatibility antigens and the rich diversity of B lymphocyte receptors has provided new insights into the fundamental nature of immunological processes, which have been found to be interwoven in a wide variety of disease states, from virus diseases to myasthenia gravis, for example, as discussed in this volume.

The field of endocrinology has provided many examples of the fruitful elucidation of interactions between small molecules, such as polypeptide hormones, with the receptors, and in the control of metabolic processes by such interactions. Studies on diabetes mellitus and familial hypercholesterolemia indicate how our understanding of important human metabolic diseases can be greatly enhanced by studying receptor molecules on cell surfaces. The study of the interactions of another effector molecule, acetylcholine, with receptors has illuminated our understanding of both normal functions and disease processes.

Toxins and drugs often achieve their biological effects by their interaction with specific receptors, and the field of endorphins has exploded with the discovery of receptors for opiates and related compounds. The striking failure of merozoites of *P. knowlesi* to invade red blood cells lacking Duffy blood-group determinants has not only provided a fascinating model in which to study the biological relationship of man and his parasites at the level of receptor interaction, but has provided a probable explanation for the resistance of Duffy-negative individuals to *vivax* malaria.

The bringing together in one place of investigators from different disciplines provided a most stimulating experience for those present. The interrelationships of these disciplines and the similarity of principles in receptor interactions was strikingly demonstrated, and the cross-fertilization of ideas was extensive. We hope this publication will bring to a larger audience a summary of many exciting aspects of the rapidly growing field of receptors and their relationship to human biology and disease.

Alexander G. Bearn, M.D.
Purnell W. Choppin, M.D.
Editors

1 November 1979

I. INTERACTIONS OF VIRUSES AND RECEPTORS

RECEPTORS AS DETERMINANTS OF CELLULAR TROPISM IN PICORNAVIRUS INFECTIONS *

Richard L. Crowell
and Burton J. Landau †

The large number of diverse diseases of man and animals caused by picornaviruses (Table 1) suggests that the different viruses in this family are more heterogeneous in disease potential than their biochemical and biophysical properties would suggest.[1, 2] Although polioviruses have been considered the prototypes of this group, it has become clear that differences exist between the cellular receptors for polioviruses and members of the other virus subgroups. These differences presumably reflect the variety of tissue tropisms identified with these viruses. The major tissue tropism of picornaviruses in humans are tabulated in Table 2 according to the embryological origin of the tissues.

The primary purpose of this chapter is to explore the concept that specific receptors for picornaviruses are major determinants of cellular tropism influencing the course of events leading to disease. Since many studies have been performed with cells grown in culture, considerations of their applicability to events occurring in the in vivo environment will be attempted. Finally, because the answers to many questions about viral pathogenesis remain open, suggestions will be made as to how to evaluate further the

---

* The work reported here was supported in part by research grant AI-03771 from the National Institutes of Health.

† Dr. Landau, of the Department of Microbiology and Immunology, Hahnemann Medical College, did not attend the conference.

TABLE 1. PICORNAVIRUS CLASSIFICATION AND ASSOCIATED DISEASE SYNDROMES

Picornaviridae	Major Disease Syndromes
Enterovirus	
Polio T1-3	Poliomyelitis, meningitis
Coxsackie A1-24	Herpangina, hand-foot-mouth
Coxsackie B1-6	Pleurodynia, meningitis Myocarditis, pancreatitis, etc.
Echo 1-34	Meningitis, rash
Swine Vesicular Disease	Hand-foot-mouth, pleurodynia
Bovine	Enteritis
Rhinovirus	
Rhinovirus > 100	Rhinitis
Cardiovirus	
Mengo, EMC, ME	Encephalomyocarditis
Aphthovirus	
FMDV	Foot-and-mouth disease

Enteroviruses: [1] 68—pneumonia; 70—conjunctivitis; 71—encephalitis

NOTE: 1) More recently classified human enteroviruses.

TABLE 2. MAJOR PICORNAVIRUS TISSUE TROPISM IN HUMANS GROUPED ACCORDING TO THE EMBRYOLOGIC ORIGIN OF INVOLVED TISSUES [1]

Germ Layer and Tissue	Virus				
	Polio	Cox A	Cox B	ECHO	RHINO
Endoderm					
Oropharynx [2]	+	+	+	+	+
Liver			+		
Pancreas			+		
Intestines	+	+	+	+	
Mesoderm					
RES [3]	+	+	+	+	+
Heart			+		
Skeletal muscle		+	+		
Kidney			(+)		
Ectoderm					
Central nervous system	+	+	+	+	
Buccal mucosa		+			
Nasal mucosa					+
Skin		+		(+)	

NOTES: 1) + denotes tissue involved; (+) denotes tissue reaction most likely due to viral antigen-antibody complexes; absence of a symbol denotes no significant involvement. 2) Oropharynx and the upper respiratory tract. 3) RES = reticuloendothelial system.

degree of influence that the early events in virus-cell interaction have on pathogenesis and disease production.

PATHOGENESIS OF PICORNAVIRUSES

A complete study of the pathogenesis of a viral infection would include a description of the attributes of the virus that permit infection to occur, and the course of events that follows at the cellular and tissue levels leading to the disease syndrome, as well as the host responses to the infection. Despite the extensive number of studies that have been performed to identify the essential factors associated with virulent strains of poliovirus that cause paralytic disease, these factors remain largely undefined.[3] Nevertheless the classic studies of D. Bodian convincingly showed the sequence of events and the distribution of poliovirus in monkeys at intervals following infection (Figure 1),[4] upon which much of our present understanding of picornavirus pathogenesis is based.[5-7]

Enterovirus infections of humans occur by mouth as well as by inhalation. Following ingestion, the virus becomes implanted and multiplies in cells comprising the lymphoid tissue of the oropharynx. It has been suggested that cells in the oropharyngeal mucosa may also be infected. Penetration of the virus to underlying lymphoid tissues may occur as a result of the injury of the epithelium by mechanical acts such as chewing food. Early in disease virus may be isolated from oropharyngeal secretions.

Poliovirus and, presumably, the other enteroviruses that multiply in the oropharynx are transmitted by lymphatic vessels to the regional cervical lymph nodes, where additional viral replication occurs. For poliovirus, at least, another major primary replication site includes Peyer's patches located in the lamina propria of the ileum. Drainage to regional mesenteric lymph nodes provides an additional secondary site for replication. Polioviruses, echoviruses, and the coxsackieviruses usually can be isolated from the feces for weeks to months postinfection. In contrast, the rhinoviruses are restricted to replication in selected cells of the mucosa and submucosa of the upper respiratory tract. These viruses, being acid-labile, are unlikely to survive the low pH of the gastric fluid.

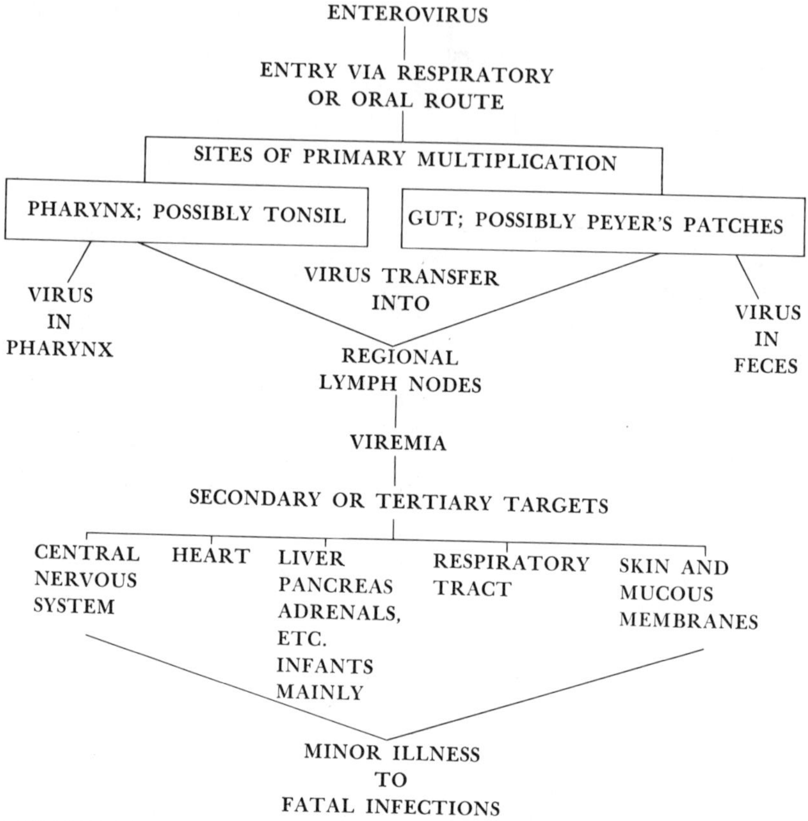

Figure 1. Schematic diagram depicting essential features of the replication and spread of enteroviruses in the body.

SOURCE: Reproduced, with permission, from D. Bodian, "Emerging Concept of Poliomyelitis Infection," *Science* 122 (1955): 105-08.

Prior to the onset of major illness caused by poliovirus, a transient viremia occurs. The amount of virus in the blood reflects the number of infected susceptible cells, as well as the number of replicative growth cycles of virus occurring in the primary cellular sites prior to viremia. Since the virus is freely dispersed in the plasma, escape of virus to extravascular sites may occur by either ingestion by macrophages or by capillary endothelial cells. A more detailed description of the precise mechanism for this virus to reach the extravascular areas does not exist.

For poliovirus, dissemination of virus to the central nervous system (CNS) occurs after a viremia, although, experimentally, virus may travel along nerve pathways to the CNS.

The severity of clinical neurological disease depends on quantitative phenomena associated with the number and distribution of neurons infected and destroyed. In certain instances infected neurons may be destroyed during the preparalytic phase of the disease; in others, the destruction of nerve cells is accompanied by an inflammatory response involving polymorphonuclear leukocytes, mononuclear cells, and macrophages. While the entire CNS may be involved, lesions are more pronounced in the spinal cord and brain stem. In reactions involving cell death followed by an inflammatory response, there is nuclear necrosis and dissolution of Nissl bodies (cytoplasmic reticulum degeneration).

The spectrum of nerve cell susceptibility to poliovirus infections as evidenced by the location of lesions is striking. In the cerebellum, for example, the hemispheres and the corticopontine system from the cerebral cortex to pontine nuclei have been reported as being spared, whereas the cerebellar vermis and cerebellar nuclei are severely damaged. Virus infection of the spinal cord is mainly restricted to the anterior horn cells, although in some instances cells of the intermediate and posterior columns may be involved. There is a relationship between the extent of damage of cells in the spinal cord and those found in the spinal sensory ganglia.

Muscular involvement in poliovirus infection is due to infection of anterior horn cells rather than of muscle fibers, as in coxsackieviruses. Thus a single infected neuron may be responsible for atrophy of many skeletal fibers. This detailed description of the pathogenesis of polioviruses is presented to illustrate the tissue and cell specificity these viruses exhibit. Although not as well documented in human infections, the other picornaviruses also show marked tissue and cellular tropisms, as evidenced by the diversity of disease syndromes associated with the respective viruses.

One of the major unanswered questions about poliovirus pathogenesis concerns the reason(s) why certain individuals develop paralysis and others do not. Factors presumed to influence the occurrence of paralytic disease in a given individual include: the selective permeability of the blood-brain barrier of certain persons to poliovirus, which increases after exhaustive exercise to potentiate the probability of paralytic disease.[8] Additional factors involve the regulatory role of interferon,[9] and the immune response to infection that may suppress the level of virus titer below the critical threshhold necessary for production of the disease syndrome.[10] It is also recognized that certain strains of virus are virulent, whereas others are avirulent for a given species. Development of vaccine strains of poliovirus that produce immunity in the absence of disease provides a clear demonstration of the rationale for considering the role played by viral genetic information in determining disease potential. Since the preceding have not fully explained virus pathogenesis and disease production, other essential factors must be involved,[11,12] for example, the relationship of capsid polypeptides, and the capacity of cellular receptors to bind and process virions leading to uncoating. The absence of information in defining virulence for picornaviruses is due, in part, to difficulties in segregating genes and obtaining recombinant virus strains for analysis of gene products.

Although an exhaustive list of factors involved in determining severity of infection cannot be made at this time, it is of importance to recognize that individuals respond differently to a given virus infection because their quantitative responses may differ widely (Table 3). As will be discussed later, these differences

TABLE 3. QUANTITATIVE ASPECTS OF PICORNAVIRUS PATHOGENESIS 1

1. Location of primary sites of replication relative to portal of entry
2. Amount of inoculum relative to number of available host cells
3. Yields of virus per host cell per unit time (permissiveness)
4. Number of host cells per gram of tissue
5. Attainment of excess virions to permit their translocation to secondary targets:
   a) Hemotogenous, lymphogenous, neurogenic
   b) Sufficient numbers of virions to exceed natural and acquired inhibitors
   c) Availability of secondary target cells and their concentration
6. Relative importance of destroyed host cells to normal tissue function
7. Immunopathology and protective immune response

NOTE: 1) The severity of the disease is influenced by the virulence of the virus strain and the genetic predisposition of the host.

may involve variations among individuals in the specific virus receptors on selected cells. In this case a more rapid rate for binding and processing of variant virions initiating infection of certain cells provides the parasite with a selective advantage over the immune response of the host.

VIRION CAPSID STRUCTURE AND ASSIGNED POLYPEPTIDE FUNCTIONS

In brief, the picornaviruses are nonenveloped virions containing nonsegmented, single-stranded RNA surrounded by capsids composed of four polypeptides.[13] While it is clear that one or more of the capsid polypeptides of each of the picornaviruses is involved in the specific binding to cellular receptors on susceptible cells, to date no single polypeptide has been shown unequivocally to be the common virus attachment protein (VAP). For example, with foot-and-mouth disease viruses (FMDV), the virion polypeptide VP1 serves as the VAP,[14] whereas VP4 has been proposed to serve this function for poliovirus.[15] A number of studies suggest, however, that VP4 is an internal virion polypeptide for poliovirus, rhinovirus, bovine enterovirus, and mengovirus, and is not likely to serve as the VAP. In lieu of a single polypeptide for the poliovirus VAP, there is evidence for cooperation among the capsid polypeptides to serve this function.[16]

The assignment of VPI [17] or VP3 [18] as the neutralizing immunogen for FMDV is not in concert with that found for coxsackievirus B3 [19] or poliovirus.[20] It is regrettable that the data at hand do not permit extrapolation of structure-function relationships from one picornavirus to another.

It becomes apparent that instead of focusing more attention on similarities among the picornaviruses, comparative analyses of virulent and avirulent variants of the same immunotype should be continued in order to identify the biochemical basis for virulence.[21,22] Since the polypeptide structure for the neutralizing immunogen would be shared between these variants, it might be predicted that major differences would reside in the VAP and/or the intercapsomeric regions proximal to the VAP. To this end, analyses of tryptic digests and amino acid sequencing of the respective virion polypeptides are appropriate. Assuming that cellular receptors not only bind, but also process virions prior to translation of the viral genome, studies of uncoating may prove to be crucial to discerning the virulence attribute.

CELLULAR RECEPTOR SITES FOR BINDING AND PROCESSING PICORNAVIRUSES

The successful extraction of infectious RNA from purified poliovirus proved to be a valuable tool for J.J. Holland and coworkers in verifying their hypothesis that the presence of receptors on the surface of primate cells was essential for infection.[23-25] Nonprimate cells that lacked poliovirus receptors were refractory to infection by native virions, but not by infectious RNA, which bypassed the specific virion-receptor interaction. These classic studies provided the impetus for others to consider the role of receptors as one of the major factors that make cells permissive or nonpermissive to virus infection. Earlier work with bacteriophages and hemagglutinating influenza viruses had provided the rationale that gave these ideas broad acceptance. The interaction of picornaviruses with cellular receptor sites on a variety of cultured mammalian cells has been reviewed.[26,27]

Cellular receptor sites for the human enteroviruses are limited to the external surface of the plasma membrane, and since they are not easily dissociated [28,29] they are considered to be integral membrane components. In these studies demonstrable receptor activity could not be found on any subcellular components following fractionation of cells that had had their receptor sites inactivated by proteolytic enzymes. It was found that trypsin inactivated the poliovirus receptors on HeLa cells, while chymotrypsin inactivated the receptors for group B coxsackieviruses without killing the cells.[30] The receptors for coxsackievirus B3 were found to be located on both the microvilli and the body of HeLa cells.[31] The cellular location of the neuraminidase-sensitive type of receptors utilized by the cardioviruses, equine rhinovirus, and FMDV has not been determined.

The evidence that cellular receptors serve to bind virions specifically as the initial event in infection is widely accepted.

Undoubtedly, without this specificity there would be little interest in further defining the nature of the different receptor sites. Considerable debate has taken place, however, over the sequence of events that occurs subsequent to attachment of virions to receptors, which involves presentation of the viral genome for translation. The issues are whether: a) the virion must be in a specific conformational state for processing by receptors; [32] b) the receptors coalesce around the virion to disrupt the capsid, causing virus eclipse in a specific manner,[33] and whether this eclipse occurs at the periphery of the plasma membrane, within vesicles surrounded by plasma membrane, or in both places; and c) a specific receptor-mediated event is required for the final stage in viral uncoating, or whether this step is mediated by endogenous cellular proteolytic enzymes or thermal disruption.[34-37] Whether the native virion crosses the plasma membrane intact [38, 39] or as a subviral particle is not clear.[40]

The location of receptor sites on the cell surface is of considerable importance, since it may determine the fate of the attached virions. Depending on whether the virions attach to receptors on microvilli or the body of the cell, the virions might be processed in different ways. For example, those virions bound to the microvilli might be eluted, whereas those bound to the body of the cell may lead to productive infection. The converse could be suggested equally well. These alternative possibilities are subject to experimental testing. In the case of those virions that remain intact, being neither eluted nor uncoated,[41,42] it is possible that they are bound to receptors on adjacent microvilli and are unable to move in either direction to be processed (Figure 2). It is possible that virus variants relatively resistant to urea and thermal inactivation [43] are also more resistant to uncoating by susceptible cells. HeLa cell variants, which are relatively resistant to poliovirus infection, are considered to be more resistant as a consequence of a defect in their capacity to productively uncoat virus.[44] A careful examination of this defect has not been made, however.

It has been recognized for many years that there exists a rather high ratio of virions to infectious units [45] in all purified

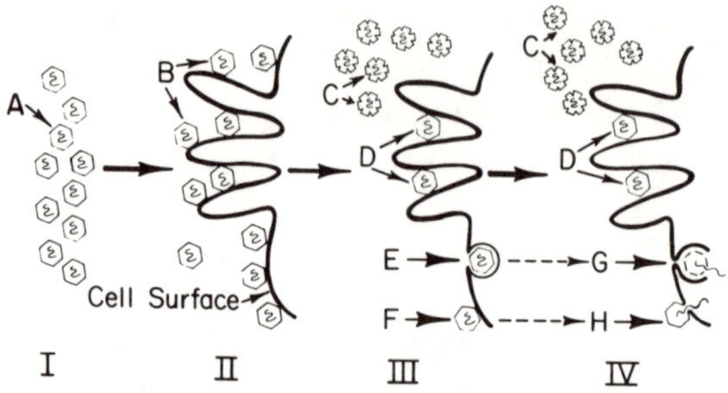

Figure 2. A schematic portrayal of the early events in picornavirus-cell interaction at 37°C. A population of virions containing a full complement of capsid polypeptides (VP1-4) enclosing the RNA is shown in I. Following attachment to the cell surface (II), populations of the virions are known to be eluted, remain stationary at the cell surface, be engulfed, or be firmly bound (III). The final sequence (IV) suggests that virions may lose all or as few as one of their pentameric polypeptides from within a vacuole or at the cell surface, to permit emergence of the virion RNA (uncoating) initiating productive infection. The percentages indicated in the following parentheses give the approximate distribution of the attached virions: (A) Picornavirions (VP1-4, RNA); (B) Attached (90%) (VP1-4, RNA); (C) Eluted (50-80%) (VP1-3, RNA); (D) Static (20%) (VP1-4, RNA); (E) Penetrated (10%) or (F) Firmly bound (10%) (VP1-4, RNA); (G) Uncoated (10%) (VP1-3 ?) or (H) Uncoated (10%) (Loss of 1 pentamer?).

SOURCE: Reproduced, with permission, from R.L. Crowell, "Comparative Generic Characteristics of Picornavirus-Receptor Interactions," in *Cell Membrane Receptors for Viruses, Antigens, and Antibodies, Polypeptide Hormones and Small Molecules,* ed. R.F. Beers, Jr. and E.G. Bassett (New York: Raven Press, 1976): 179-202.

picornavirus preparations. Ratios range from a low of about 30:1 to several thousand:1 for poliovirus. The difficulty this presents for the interpretation of results of studies of the early events in virus infection cannot be overemphasized. For example, to elucidate the sequential steps in the uncoating of virions by electron microscopy requires a very high input of virions/cell (in the range of 10^4-10^5/cell) to locate viral particles in the preparation. If only about 5 percent or less [46] of attached virions become productively uncoated, how can one know which particles are on the pathway to infection and which are being rejected? One can easily make a similar criticism of studies utilizing radioactively labeled virions, where comparable problems exist in terms of obtaining labeled virion preparations with better than 1 CPM/10^5 virions.

These facts are emphasized because for polioviruses,[47,48] echoviruses,[49] and coxsackieviruses[50] it is well established that most of the virions that become cell-associated are eluted or sloughed from the cell as eclipsed particles ("A" particles) that have lost VP4[51] and are labilized for further degradation by proteolytic enzymes, high salt concentrations, or a temperature of 37°C. Although it is logical to consider that any alteration of the capsid following virion-receptor interaction at 37°C is an event that must precede infection, one is confronted with the problem of separating out those particles that are going to become uncoated in a specific way from those that may undergo thermal degradation. On the other hand, the initial event of labilization of these compact capsids by receptors may be the only specific event that is needed to allow subsequent release of the RNA within the cytoplasm. J. Enders[52] and T. Wilson[53] and their coworkers showed that poliovirus could be entrapped in insusceptible cells fused with Sendai virus or liposomes, respectively. Replication occurred at a very low level, however, probably reflecting an inefficient processing of virions in the absence of receptors. It is now well known that isolated viral RNA or viral RNA packaged in liposomes[54] initiates productive infection (Table 4). In the latter case the relatively high efficiency of infection focuses attention on the importance of uncoating events to initiate infection by native virions.

A better understanding is needed of the genes that control virion capsid protein synthesis, their assembly,[55] and the production of specific functional receptors on cells.[56] Thus analysis of both virus and cell variants defective in their respective functions

TABLE 4. NONRECEPTOR-MEDIATED INITIATION OF PICORNAVIRUS INFECTION OF INSUSCEPTIBLE CELLS

Initiator	Conditions
Viral RNA	DEAE-D, DMSO, protamine, $MgCl_2$
Viral RNA	Liposomes
Virions	Liposomes
Virions	Cell fusion by Sendai
Virions	Very high titers ($>10^{10}$ PFU)

for binding and processing virions should help resolve the foregoing issues.

FACTORS INFLUENCING THE QUANTITATIVE DETERMINATION OF CELLULAR RECEPTOR SITES

The quantitative measurement of cellular receptor sites on cells or tissues for binding viruses is more difficult than one might expect from a cursory examination of this subject. Determination of receptor activity for binding virions is influenced by all the environmental factors that regulate the attachment of enzymes to substrates. These factors must be experimentally determined for each virus-receptor system.

Attachment kinetics of virions is followed by measurement of the amount of free virus that remains unattached at intervals of incubation following removal of cells by centrifugation. Dilution of the virus-cell system by 100-fold effectively stops the interaction of virus with receptors. It is recommended that the amount of attached virus also be determined so the total amount of virus in the system can be measured, thereby avoiding the possibility of artifactual adsorption of virus to the walls of the vessel.

It is important to determine whether loose or firm binding has occurred. Recovery of infectious (unaltered) virus from cells by simple dilution in buffer at physiological pH is considered to be a consequence of loose binding. Firm binding of virions to receptors most likely results from interaction of adjacent receptor sites with the repeating subunits at the vertices of the virion. Dissociation of infectious virus in this case may be accomplished only by extensive alteration in the ionic environment, that is, by low pH [57] or by detergents.[58] It is likely that receptors separated by large distances in the plasma membrane will bind virions loosely, especially at low temperatures (rhinoviruses), where lateral mobility of receptors is low.[59] Dissociation of virions from receptors by ETDA also is considered to be a consequence of loose binding. On the other hand, firm or tight binding of virions by receptors, even at 4°C, may reflect the presence of multiple receptor sites in close proximity, where lateral mobility in the lipid matrix is not essential to securely anchor the virions—group B coxsackieviruses, polioviruses.

The quantitative determination of cellular receptor sites can be made either by measuring the comparative virus attachment rates under optimal conditions or by assaying the maximum number of virions required to saturate all the available receptors. As given in Table 5 a large number of factors influence the attachment rate of virions to receptors. In this regard let us consider examples of each of the major factors. Thus changes in cell concentration and, by extension, the number of receptors per cell affect attachment kinetics of picornaviruses in a directly proportional manner. On the other hand, the virion concentration over a wide range (Mi 0.01 to 50) has relatively little effect on attachment rate. As the upper limit is approached, saturation of receptors begins to occur, and the reduced rate of attachment reflects competition for receptor sites.

Attachment of virions to their receptors is mainly an electrostatic interaction that is strongly influenced by the charge distribution on both the ligand and the receptor.[60] The latter concept is inferred from studies demonstrating that binding of virus does not occur in the absence of ions,[61] and that some rhinoviruses, coxsackievirus A9, and FMDV require divalent cations,[62-64] while some Group B coxsackieviruses and echoviruses are strongly influenced by the pH of the environment.[65] It is also likely that other forces, such as Van der Waal's forces and hydrophobic bonding, play a role in firm binding of viruses to receptors. Attachment has been considered to be independent of temperature, although in studies with selected picornaviruses,

TABLE 5. FACTORS INFLUENCING ATTACHMENT RATE OF VIRIONS TO RECEPTORS[1]

1. Cell concentration	7. Ion species and concentration
2. Number of receptors per cell	8. Viscosity
3. Virion concentration	9. Selected sulfhydryl cpds.
4. Charge distribution	10. Proteolytic enzymes
5. pH	11. Inhibitors
6. Temperature	

NOTE: 1) See text for an explanation of terms and references.

for example, rhinoviruses and coxsackievirus B3, temperature has been shown to be an important factor. The rate of attachment of cardioviruses and FMDV are, however, influenced much less by temperature. These differences undoubtedly reflect the differences in receptor composition. The addition of reagents that increase viscosity, sucrose, for example, decreases attachment rates.[66] It is known that sulfhydryl groups on the surface of some picornaviruses have an influence on virus attachment.[67] Polioviruses, for example, are less influenced by the presence of reagents that block sulfhydryl groups, whereas ECHO viruses are profoundly affected. Coxsackieviruses form an intermediate group in this regard. No correlation exists between the O/R potential produced by selected sulfhydryl reagents and their capacity to enhance attachment.*

Proteolytic enzymes are commonly used to dissociate tissues and to detach cultured cells from glass or plastic substrates. It has been found that low concentrations of enzymes such as trypsin will enhance attachment of virus to cells,[68] perhaps by removal of sialic acid residues, which reduces the negative charge at the cell surface.[69] At higher concentrations, trypsin inactivates poliovirus receptors selectively.[70]

Finally, caution must be exercised in ascribing the disappearance of virus infectivity to attachment to receptors in systems that may contain antiviral antibody. In this instance, animal tissue that may possess antibodies or virus inhibitors may be mistakenly thought to possess receptors, when in actuality they are absent. Anticellular serum, on the other hand, can block receptors.[71,72]

Unless optimal conditions are utilized for virus-receptor interaction, the kinetic results may not be very meaningful. For example, the kinetics of attachment of coxsackieviruses B2, B4, and B6 to HeLa cells were 100 times slower than for coxsackieviruses B1, B3, and B5. These results were interpreted initially to reflect about 100 times fewer receptor sites per cell for the even-numbered group B viruses.[73] The finding that these viruses would nevertheless saturate the receptors for the odd-numbered viruses and block their attachment to HeLa cells suggested that

* R. L. Crowell and L. Philipson: unpublished observations.

the number of receptor sites per cell was not different. Attention then was focused on the nature of the virions. Since the different group B viruses migrated in an electric field at different rates, it was assumed that the different attachment kinetics reflected a different distribution of charges on the virion-attaching proteins.[74] Further studies showed that the attachment kinetics of B4 and B6 viruses were rate-limiting for an efficient plaque assay, and that the relatively slow kinetics for these viruses could be increased by selecting the optimal pH and increasing the cell concentration relative to the volume of the virus inoculum.[75]

Thus the environmental factors surrounding virion-receptor interaction are numerous. Unless attention is given to defining all the optimal conditions for each virus-cell system, the quantitative comparison of receptor activity by virus attachment kinetics may be invalid.

Additional problems to be considered in estimating the number of receptor sites per cell in tissues derived from humans or animals include the observations that: 1) most tissues are composed of heterogeneous cell types of which only selected ones possess receptors; 2) the methods utilized to dissociate tissues may destroy the receptors; and 3) if the tissues are not dissociated, the receptors will remain cryptic and undetected. When the number of virions in the virus inoculum is known, the most definitive technique for estimation of the number of receptor sites per cell is that of virus saturation. In this regard, utilization of preparations of purified virions is recommended, since crude virus preparations would contain VAP not assembled into virions, which would compete with virions for the receptor sites. Under optimal conditions the results of such experiments yield a minimal value for the number of receptor sites per cell.[76] In similar fashion the specificity or relatedness between two potentially similar viral receptors on cells may be determined by reciprocal competition experiments using receptor-saturating amounts of each virus.[77-79]

Following attachment of virions to their specific receptors, the virions are considered to be processed in a manner that will release the genome to initiate translation. It is likely that multiple receptor sites act in concert to disrupt the capsid. The operational

TABLE 6. OPERATIONAL CRITERIA USED TO FOLLOW EARLY EVENTS OF PICORNAVIRUS INFECTION

Event	Temperature (C)	Characteristic of Event
Attachment	0-37°	Decrease in unattached virions
a) Loose	0°	Reversible with EDTA
b) Tight	0-20°	Reversible at pH2
Penetration	20-37°	Virions inaccesible to neut. ab., but recoverable by SDS
Eclipse ("eluted")	37°	Loss of infectivity ("A" particles), not recoverable by pH2 or SDS, loss of photosensitivity
Uncoating	37°	Viral RNA hydrolyzed by RNase

criteria used to follow these early events are outlined in Table 6. Details of these events are not fully understood and may differ for the different subgroups of picornaviruses. These differences will undoubtedly reflect the different intercapsomeric bonds that hold the capsid together. It is believed that the capsid of the acid-stable picornaviruses is disrupted sequentially, with VP4 first being released to form an "A" particle that subsequently releases VP2 and the RNA (Figure 3). On the other hand, other picorna-

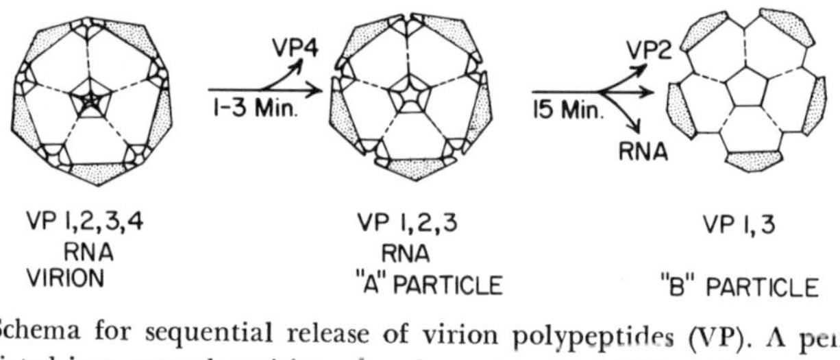

Figure 3. Schema for sequential release of virion polypeptides (VP). A pentamer of VP4 is depicted in a central position of each pentamer of VP2, which are considered to be located at each of the twelve vertices of the icosahedron-shaped virion. It is suggested that interaction of the virion with specific receptors dissociates the pentameric polypeptides VP4 and VP2 to permit release (uncoating) of viral RNA in a manner similar to that observed with urea (3 M, pH 9, 37°C).

SOURCE: Reproduced, with permission, from R.L. Crowell and J-S. Siak, "Receptor for Group B Coxsackieviruses: Characterization and Extraction from HeLa Cell Membranes," in *Perspectives in Virology,* vol. 10, ed. M. Pollard (New York: Raven Press, 1978): 39-53.

viruses, such as FMDV and the cardioviruses, are disrupted into 12S or 14S substructures containing VP1,2,3, and do not appear to go through an "A" particle stage.[80,81] The evolutionary significance of these differences is unknown. Nevertheless we believe that the receptor complex in the plasma membrane brings about the capsid disruption in a rather specific way. It is likely that a defect in the receptor structure will limit the amount of uncoating that may occur, and the cell will be designated nonpermissive, or relatively so. Further studies of these events for the different picornaviruses are needed in order to better understand and quantitate receptor-mediated virus uncoating.

CORRELATION BETWEEN RECEPTOR ACTIVITY AND CELLULAR TROPISM

A number of workers have postulated that tropism of viruses for selected cell types during infection is determined by the presence or absence of specific receptor sites on the cell surface that attract certain viruses toward them.[82-85] This hypothesis found support in the results of studies that showed there was a correlation between the presence of receptors on cells of only those animal species that supported virus replication; they were absent from nonpermissive hosts.[86,87] Host-range studies of picornaviruses that have proved useful in their identification [88] also raise questions as to the reason for the observed host-range specificity. Why, for example, are polioviruses limited to humans and primates, the human rhinoviruses and echoviruses devoid of suitable animal hosts? And why do the coxsackieviruses and FMDV have a high predilection for the newborn mouse? Although it is common knowledge that the picornaviruses of humans and animals have been observed to have a rather restricted host range in nature and under experimental conditions, the reason for this is not clear.

Histopathological examination of tissues removed from animals or humans infected with assorted picornaviruses has revealed selective cell damage. Among the well-recognized features of this examination are the destructive effects of polioviruses on neurons, coxsackieviruses of group A causing generalized myositis, and group B viruses destroying the pancreas, fat cells, cardiac, and brain tissues. It is obvious that the different picornaviruses display a specific cellular tropism. The reason for this tropism, however, is not so obvious. Another well-known phenomenon is the selection of virus variants with altered cellular tropism by serial passage in selected tissues.[89-93] The ability to obtain these variants has been fundamental to the selection of virus strains for use in vaccines and for experimental studies of pathogenesis.

Studies of the selection of variant viruses by susceptible cells are accomplished most readily by using cells in culture, since they are more homogeneous than those in tissues.[94] For receptor sites to select a variant virus from the parental virus population, one could assume that the selecting cells, for example, RD cells,[95,96] had a receptor site configuration different from that of the standard cells (HeLa cells). In this example RD cells selected a hemagglutinating variant (HA$^+$) of coxsackievirus B3, after serial passage in RD cells, from a nonhemagglutinating (HA$^-$) parental population grown in HeLa cells.[97] Both the HA$^+$ variant and the HA$^-$ virus grow in HeLa cells, whereas only the HA$^+$ variant grows in RD cells. If the HA$^+$ variant had a monocomponent virion-attaching protein designated (a) and the RD cell had a monocomponent receptor designated (A), then the RD cell would selectively bind the HA$^+$ variant from a population of virions with different virion-attaching proteins. Since the HeLa cell could have a multicomponent receptor (A,B,C), etc. (Figure 4), it would bind the HA$^+$ variant in addition to the other variants in the viral population: HA$^-$, etc. It is also possible that the binding of HA$^+$ virions would block or compete for the same receptor on HeLa cells as HA$^-$ virions, since the components A, B, and C could be blocked sterically by virions binding to any one or combination of components. It is considered that this schema could accommodate selection of many kinds of virus variants if the configuration of each receptor site component is subject to modification through mutation and evolution on different animal species and on different cell types within the host. The determination of receptor specificity of the different picornaviruses, and their host range variants, would help to extend the virus grouping based on receptor competition studies [98,99] and provide

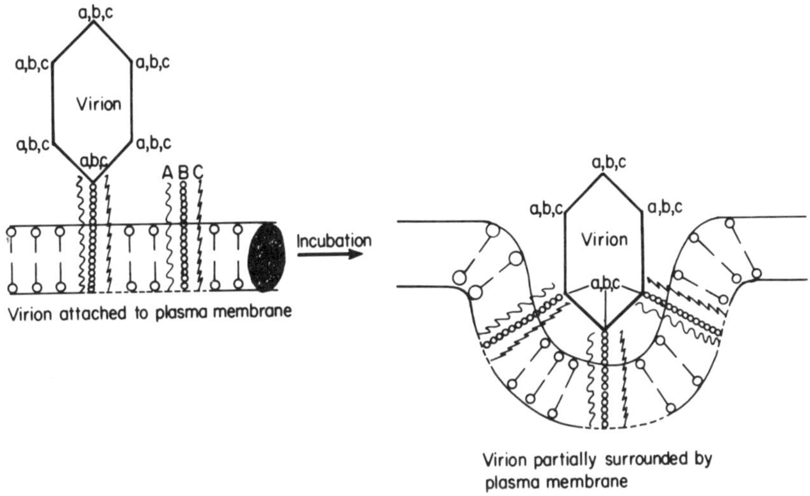

Figure 4. Schematic diagram of a hypothetical multicomponent receptor site for picornaviruses. A tricomponent receptor site is shown protruding from the plasma membrane with a virion attached. Incubation results in the virion becoming surrounded by the plasma membrane. Additional interactions of the adjacent sites with the corresponding vertices of the virion constitute a preliminary step to virus eclipse and uncoating. The hypothesis also includes the possibility that mono- or dicomponent receptor sites may exist for selecting variant virions possessing complementary attaching proteins (VAP). See text for further discussion.

a basis for testing the proposed hypothetical multicomponent receptor site.

The potentially high degree of complexity of picornavirus receptor sites, if multicomponent, is further exaggerated by the recognized differences that might occur in the different lateral mobilities these components may display. Perhaps those picornaviruses that display loose binding at 4°C (rhinoviruses) have receptor components that are spatially distant from each other; unless they diffuse together their binding strength is low. Other picornaviruses, such as coxsackieviruses of group B and polioviruses, demonstrate tight binding even at low temperatures, suggesting a receptor site configuration whose components are adjacent. If multiple receptor sites are required for tight binding, these would be organized in close proximity to bind viruses side-by-side[100] without requiring significant lateral movement of the components with the plasma membrane.

A good correlation has been found between the grouping of viruses by their receptor specificity and their original classification into subgroups (Table 7).[101, 102] The three immunotypes of polioviruses compete for a common receptor on HeLa cells, for example, while the six group B coxsackieviruses compete for another common receptor. Further, these two types of receptors were found to be different from those that attached rhinoviruses, group A coxsackieviruses, and echoviruses, respectively. These studies were made possible by the finding that viruses in high titer could saturate the available receptor sites on a susceptible cell population and prevent the further attachment of related viruses that shared the same receptor specificity.[103] Evaluation of additional viruses for their receptor specificity could provide a further correlation between receptor families and the generic grouping of picornaviruses.

It was found, unexpectedly that coxsackievirus B3 shared a receptor with adenovirus type 2, and that coxsackievirus A21 shared a receptor with HRV 14.[104] So far these are the only dis-

TABLE 7. EVIDENCE FOR A CLOSE CORRELATION BETWEEN THE RECEPTORS FOR PICORNAVIRUS GENERA AND THEIR RESPECTIVE TAXONOMIC DESIGNATIONS, AS BASED ON INTERFERENCE OF VIRUS ATTACHMENT

Virus Types Used to Saturate Receptors	Interference of Attachment of Virus Immunotypes					
	Polio 1-3	Cox B1-6	Cox A21	HRV2	HRV14	E6
Polio 1-3	+	—	—	—	—	—
Cox B1-6	—	+	— [1]	—	—	—
Cox A21	—	—	+	—	+	NT
HRV 2 [2]	—	—	—	+	—	NT
HRV 14 [3]	—	—	+	—	+	NT

NOTES: HRV = Human rhinovirus; E6 = echovirus Type 6. HeLa cells were saturated with the interfering virus, and the cells were washed and tested for capacity to attach a second type of virus. + indicates interference of attachment of challenge virus; — indicates normal attachment of challenge virus; NT indicates not tested. 1) In addition, coxsackieviruses A13, A15, and A18 were not inhibited by coxsackievirus B5. 2) HRV-2 also inhibited attachment of HRV-1A and HRV-1B. 3) HRV-14 also inhibited attachment of HRV-3, 5, 15, 39, 41 and 51.

SOURCE: Reproduced, with permission, from R. L. Crowell and J-S. Siak, "Receptor for Group B Coxsackieviruses: Characterization and Extraction from HeLa Cell Membranes," in *Perspectives in Virology,* vol. 10, ed. M. Pollard (New York: Raven Press. 1978): 39-53.

parate findings for receptor grouping of related viruses; their significance remains unknown. It is tempting, however, to speculate that these unrelated viruses share a common tissue tropism that may influence the type of disease associated with each of these pairs of unrelated viruses. In this regard coxsackievirus A21 is well known as a cause of an upper respiratory tract infection similar to the common cold viruses in humans. By the same token it would be of interest to determine whether HRV 14 will grow in or attach to skeletal muscle tissue of the newborn mouse. In a similar relationship, both adenovirus type 2 and coxsackievirus B3 replicate in the nasopharynx and intestines of man, and both have been implicated in heart disease. To further test their receptor relationship it would be helpful to determine whether both viruses would attach to these tissues and to brain tissue of newborn mice. Thus the receptor grouping of viruses has provided another basis for determining relationships among viruses and provides an approach for examining the mechanism for specific viral tissue tropism.

We will now consider a number of studies that have focused on the distribution of picornavirus receptor sites on cells in culture derived from selected tissues. The results of these studies have revealed in a general way that there is correlation between the presence of receptors on those cells that are targets for virus replication, and an absence of receptors on cells that are insusceptible. Extrapolation of such data to the intact organs in vivo, however, is justified only under very limited circumstances. These circumstances include the use of cultured cells whose karyotypical, histological, and physiological characteristics are similar to their counterparts in vivo. Among the many reasons for this conservative point of view are the observations that picornavirus receptors are: 1) subject to change following removal of tissue from experimental animals and placing them in culture; [105-107] 2) inactivated by proteolytic enzymes that may subsequently be regenerated; [108, 109] and 3) formed on differentiated cells [110, 111] and on erythrocytes.[112]

On the other hand, many highly significant observations regarding virus-receptor interactions have been made using continuous cell lines. While these cells are usually transformed, they have provided readily available populations that are more uniform than primary cell cultures. Fortuitously, many of these continuous cell lines, HeLa and L-strain, for example, express a wide spectrum of virus receptors, although in some cases transformed cells may not express receptors characteristic of their normal counterparts.[113-115]

In an effort to approximate the cellular relationship found in vivo, as well as to satisfy the specific tropism of a virus with restricted host range, a number of investigators have used organ cultures to grow selected picornaviruses.[116] For example, some rhinoviruses have been cultured only in different organ cultures of trachea.[117] Similarly, coxsackieviruses A1 and A5 have been grown in differentiating primary fetal mouse muscle cultures,[118] whereas nondifferentiating mouse cultures have been refractory.[119] The technical problems inherent in organ cultures used for virological studies have been overcome in part by the development of continuous lines of cells that have retained the capacity to differentiate. Thus the L-8 rat muscle cell line has been shown to undergo changes in its pattern of susceptibility, as influenced by stages in differentiation.[120] In these studies, presumptive myoblasts were nonpermissive to Cox A2, whereas fusing myoblasts were susceptible. Cells at either the presumptive myoblast or fusing myoblast stage attached virus equally well. Thus the virus restriction imposed by the presumptive myoblasts occurred at the level of uncoating, or at a subsequent stage in the virus life cycle. The finding that cells undergoing differentiation exhibit differences in virus susceptibility is relevant to the in vivo condition. In this regard, rubella, CMV, and coxsackieviruses, among others, are known to cause congenital disease in humans, the incidence and severity of which depend on the stage of cellular differentiation at the time of infection. Unfortunately the role of receptors in determining viral tropism in these embryonic tissues has not been evaluated.

With respect to animal studies, attempts have been made to detect picornavirus receptor specificity on intact organ minces or homogenates. The objectives of these experiments were to determine whether the presence or absence of receptors correlated with known in vivo tropisms, and to develop suitable animal models for studies of this phenomenon. In this regard several investigators have found poliovirus receptors on human and monkey CNS tissue and intestine [121-124] in accord with the known in vivo targets of this virus (Table 2). The absence of detectable poliovirus receptors on human lung, heart, skeletal muscle,[125] or skin [126] reinforces the positive findings noted earlier, as does the absence of poliovirus receptors on mouse CNS tissue.[127] In the latter case, poliovirus does not usually replicate in mice unless specially adapted. Similar correlates were made in studies of porcine enteroviruses in pig tissues.[128]

Among the more important problems requiring resolution are conflicting observations regarding the presence of poliovirus receptors on human liver and on monkey heart and skeletal muscle.[129, 130] It may be that such diverse results are a reflection of variation between genetically dissimilar animals of the same species. Another issue deserving further experimental inquiry is the observation that, while human CNS tissue contained receptors to group B coxsackieviruses, receptors to the group A coxsackieviruses were absent. Additional instances observed by Holland, where the absence of virus receptors on tissues is at variance with their known predilection for these tissues in vivo, include the absence of group B coxsackievirus receptors on human intestinal tissue and on mouse heart tissue; [131] group A coxsackievirus and echovirus-8 receptors were found on the former tissue. The absence of group B coxsackievirus receptors on mouse heart tissue may have been due to the virus variant employed,[132] and/or the strain of mice studied,[133] since these receptors have been detected by others.[134]

In this regard it is significant to note that failure to detect receptor activity on cells by no means excludes their presence unless extensive studies have been conducted.[135] The finding that the receptors for the human enteroviruses exist on tissues not involved in their pathogenesis, as well as in species other than their natural host, has been extended to other picornaviruses such as bovine enterovirus (BE) [136] and FMDV.[137] For example, both human and bovine intestines have been found to contain receptors to BE. Conversely, CNS tissue and liver did not bind BE virus, although bovine CNS and liver tissues were found to contain receptors to this virus. The foregoing illustrate the ubiquitous occurrence of picornavirus receptors in nature. The fact that such receptors are not always involved directly in the pathogenesis of disease is underscored by their presence on red blood cells.[138] While the relationship of these sites to receptors on susceptible cells has not been determined,[139] it is possible that the receptors on erythrocytes may be analogous to heterophile antigens appearing on unrelated tissues or antigens.

With regard to pathogenesis, it is reasonable to suggest that the picornavirus receptors on red blood cells may serve to translocate certain of the hemagglutinating picornaviruses during the viremic phase in a semiprotected state: that is, these erythrocyte-bound viruses might be unavailable to inactivation by low levels of neutralizing antibody and assorted inhibitors. Nevertheless, the presence of so many nonhemagglutinating picornaviruses in nature minimizes the importance of erythrocyte receptors in picornavirus pathogenesis. In brief, we must conclude that our knowledge of the distribution of receptors on different cell types and tissues in the susceptible host is limited.

PERSPECTIVES AND SUMMARY

Results of many studies on human picornavirus receptor specificity have revealed that the several virus immunotypes arrange themselves into groups. These virus groups are markedly similar in composition to those constructed originally on the basis of the capacity of the respective viruses to cause specific disease syndromes and histopathology. It is relevant to note here that the current picornavirus classification scheme no longer considers viral pathogenesis a useful marker. Given the foregoing data it might be helpful if classification of new picornavirus immunotypes encompassed receptor and antigenic groupings in addition to immunotyping.

The role of receptors in picornavirus tropism has been well documented, although the function by which these different receptors serve cells is unknown. Since a strategem of viral chemotherapy might involve selective inactivation of viral receptors[140,141] it would be helpful to elucidate the relationship that functional receptors have to the well-being or life processes of the cell. It has been found that blockade of receptors on HeLa cells by anticellular serum or virus-antibody complexes (at the cell surface) fail to inhibit cell growth,[142] suggesting that these receptors are not involved in transport of essential nutrients.[143] Furthermore, inactivation of receptors by proteolytic enzymes did not result in cell death.[144] Continued efforts to identify cellular functions for these receptors is needed. Since HeLa cells are tumor cells, extrapolation of the results to diploid cell systems is not necessarily valid. The effects of receptor-blocking agents on differentiating cells possessing normal growth-regulatory mechanisms are not known and also should be tested.

It has been shown that viral receptors are controlled by genetic information located on one or more chromosomes. It would be helpful to identify which chromosomes contain the genes for each of the individual picornavirus receptors, as has been done for the poliovirus receptor gene, that is, chromosome 19.[145] Once the receptor genes are identified on given chromosomes it will be easier to search for another function, since only those genes that are linked need be evaluated. Conversely, genetic information located on chromosomes different from those with receptor genes, for example, HL-A antigens,[146] would be excluded.

Persons of genetically different lineage probably possess picornavirus receptors with subtle structural differences, that is, multicomponent receptor hypothesis (Figure 4). These differences may account for the selection of variant viruses from the large populations of viruses known to be produced during infection. For a given individual the variant virus may prove to be especially virulent, whereas for another person it will be avirulent. Unless a diagnostic test were developed, variant viruses may escape detection. It is proposed that genetically specified receptors may account for diverse expressions of disease among individuals.

Cultures of cells that more accurately reflect the in vivo condition are needed to analyze the factors that control expression of functionally active viral receptors. It is anticipated that such cultures would reveal a relationship between receptor expression and differing stages in differentiation.[147,148] Since primary cultures are heterogeneous, the development of cell-sorting techniques to obtain more homogeneous populations would be useful.[149] Application of the receptor-solubilizing method [150] to carefully dissected tissues, followed by analysis with affinity chromatography,[151] may be helpful in following the appearance of receptors during embryogenesis and development.

In summary, the importance of the hypothesis that receptors are determinants of picornavirus tropism in the pathogenesis of human and animal diseases has been advocated. Studies have been presented that support the contention that a correlation exists between the grouping of picornaviruses by their capacity to compete for specific receptors on HeLa cells and their original classification, based on disease production and tissue tropism. The specificity of virus-receptor interactions is too intriguing to dismiss as a mere accident of nature or laboratory artifact. We have, we hope, provided a thoughtful overview of the complexities of relating the results of in vitro studies to in vivo correlates without discouraging others from defining them further. While the available correlates between the presence or absence of receptors and the known tropism of the virus are not absolute, there is sufficient data to warrant further study.

NOTES

1. J. F. E. Newman, D. J. Rowlands and F. Brown, "A Physico-Chemical Subgrouping of the Mammalian Picornaviruses," *Journal of General Virology* 18 (1973): 171-80.

2. R. R. Rueckert, "On the Structure and Morphogenesis of Picornaviruses," in *Comprehensive Virology*, ed. H. Fraenkel-Conrat and R. R. Wagner (New York: Plenum Press, 1976): 131-213.

3. P. D. Cooper, "Genetics of Picornaviruses," in *Regulation and Genetics: Genetics of Animal Viruses. Comprehensive Virology*, ed. H. Fraenkel-Conrat and R. R. Wagner (New York: Plenum Press, 1977): 133-207.

4. D. Bodian, "Emerging Concept of Poliomyelitis Infection," *Science* 122 (1955): 105–08.

5. D. Bodian and D. M. Horstmann, "Polioviruses," in *Viral and Rickettsial Diseases of Man*, ed. F. L. Horsfall and I. Tamm (Philadelphia: J. B. Lippincott Co., 1965): 430-73.

6. F. Fenner, "The Pathogenesis and Ecology of Viral Infections," in *The Biology of Animal Viruses*, vol. II (New York: Academic Press, 1968): 475-535.

7. F. H. Top, K. M. Johnson, and P. F. Wehrle, "Enteroviruses: Poliomyelitis," in *Communicable and Infectious Diseases*, ed. F. H. Top and P. F. Wehrle (St. Louis: C. V. Mosby Co., 1976): 260-78.

8. D. M. Horstmann, "Acute Poliomyelitis: Relation of Physical Activity at the Time of Onset to the Course of Disease," *Journal of the American Medical Association* 142 (1950): 236-41.

9. R. M. Friedman, "Antiviral Activity of Interferons," *Bacteriological Reviews* 41 (1977): 543-67.

10. H. J. Eggers and A. B. Sabin, "Factors Determining Pathogenicity of Variants of Echo 9 Virus for Newborn Mice," *Journal of Experimental Medicine* 110 (1959): 951-67.

11. C. M. Kunin and N. E. Halmagyi, "The Relative Abundance of Viral Receptors. An Explanation of the Differential Susceptibility of Suckling and Adult Mice to Coxsackie B1 Infection," *Journal of Clinical Investigation* 40 (1961): 1055-56.

12. C. M. Kunin, "Virus-Tissue Union and the Pathogenesis of Enterovirus Infections," *Journal of Immunology* 88 (1962): 556-69.

13. Rueckert, "On The Structure" (See note 2).

14. D. Cavanaugh et al., "Immunogenic and Cell Attachment Sites of FMDV: Further Evidence for Their Location in a Single Capsid Polypeptide," *Journal of General Virology* 35 (1977): 149-58.

15. M. Breindl, "VP4, the D-Reactive Part of Poliovirus," *Virology* 46 (1971): 962–64.

16. K. Lonberg-Holm and L. Philipson, "Molecular Aspects of Virus Receptors and Cell Surfaces," in *Cell Membranes and Viral Envelopes*, ed. H. A. Blough and J. M. Tiffany (New York: Academic Press, 1979): forthcoming.

17. T. F. Wild, J. N. Burroughs, and F. Brown, "Surface Structure of Foot-and-Mouth Disease Virus," *Journal of General Virology* 4 (1969): 313-20.

18. H. L. Bachrach et al., "An Experimental Protein Vaccine for Foot-and-Mouth Disease," in *Perspectives in Virology*, vol. 10, ed. M. Pollard (New York: Raven Press, 1978): 147-58.

19. S. T. Beatrice and R. L. Crowell, "Induction of Type Specific Neutralizing and Precipitating Antibodies to Coxsackievirus B3 by the Viral Polypeptide, VP2," *Abstracts of the Annual Meeting of the American Society for Microbiology* (1976): 211.

20. Breindl, "VP4 D-Reactive Part" (See note 15).

21. A. Hamann et al., "Isoelectric Points of Polypeptides of Standard Poliovirus Particles of Different Serological Types and of Empty Capsids and Dense Particles of Poliovirus Type 1," *Journal of General Virology* 38 (1978): 567-70.

22. D. M. Moore and K. M. Cowan, "Effect of Trypsin and Chymotrypsin on the Polypeptides of Large and Small Plaque Variants of Foot-and-Mouth Disease Virus: Relationship to Specific Antigenicity and Infectivity," *Journal of General Virology* 41 (1978): 549-62.

23. L. C. McLaren, J. J. Holland, and J. T. Syverton, "The Mammalian Cell-Virus Relationship. I. Attachment of Poliovirus to Cultivated Cells of Primate and Non-Primate Origin," *Journal of Experimental Medicine* 109 (1959): 475-85.

24. J. J. Holland and L. C. McLaren, "The Mammalian Cell-Virus Relationship. II. Adsorption, Reception, and Eclipse of Poliovirus by HeLa Cells," *Journal of Experimental Medicine* 109 (1959): 487-504.

25. J. J. Holland, L. C. McLaren and J. T. Syverton, "The Mammalian Cell-Virus Relationship. IV. Infection of Naturally Insusceptible Cells with Enterovirus Ribonucleic Acid," *Journal of Experimental Medicine* 110 (1959): 65–80.

26. K. Lonberg-Holm and L. Philipson, "Early Interaction between Animal Viruses and Cells," *Monographs in Virology* 9 (1974): 1-148.

27. ———, "Molecular Aspects" (See note 16).

28. I. Zajac and R. L. Crowell, "Location and Regeneration of Enterovirus Receptors of HeLa Cells," *Journal of Bacteriology* 89 (1965): 1097-1100.

29. R. L. Crowell and J-S. Siak, "Receptor for Group B Coxsackieviruses: Characterization and Extraction from HeLa Cell Membranes," in *Perspectives in Virology,* vol. 10, ed. M. Pollard (New York: Raven Press, 1978): 39-53.

30. I. Zajac and R. L. Crowell, "Effect of Enzymes on the Interaction of Enteroviruses with Living HeLa Cells," *Journal of Bacteriology* 89 (1965): 574-82.

31. T. G. Roesing, P. A. Toselli, and R. L. Crowell, "Elution and Uncoating of Coxsackievirus B3 by Isolated HeLa Cell Plasma Membranes," *Journal of Virology* 15 (1975): 654-67.

32. B. Mandel, "Neutralization of Poliovirus: A Hypothesis to Explain the Mechanism and the One-Hit Character of the Neutralization Reaction," *Virology* 69 (1976): 500-10.

33. P. Boulanger and K. Lonberg-Holm, "Components of Nonenveloped Viruses Which Recognize Receptors," in *Virus Receptors,* ed. L. Philipson, L. Randall, and K. Lonberg-Holm (London: Chapman and Hall, Ltd., forthcoming).

34. Roesing, Toselli, and Crowell, "Elution and Uncoating" (See note 31).

35. J. De Sena and B. Mandel, "Studies on the In Vitro Uncoating of Poliovirus. II. Characteristics of the Membrane-Modified Particle," *Virology* 78 (1977): 554-66.

36. N. Guttman and D. Baltimore, "A Plasma Membrane Component Able to Bind and Alter Virions of Poliovirus Type 1: Studies on Cell-Free Alteration Using a Simplified Assay," *Virology* 82 (1977): 25-36.

37. M. L. McGeady and R. L. Crowell, "Stabilization of "A" Particles of Coxsackievirus B3 by a HeLa Cell Plasma Membrane Extract," *Abstracts of the Annual Meeting of the American Society for Microbiology* (1978) : 217; *Journal of Virology,* in press.

38. H. J. Eggers, B. Bode, and D. T. Brown, "Cytoplasmic Localization of the Uncoating of Picornaviruses," *Virology* 92 (1979): 211-18.

39. M. L. Fenwick and M. J. Wall, "Factors Determining the Site of Synthesis of Poliovirus Proteins: The Early Attachment of Virus Particles to Endoplasmic Membranes," *Journal of Cell Science* 13 (1973): 403-13.

40. K. Lonberg-Holm, L. B. Gosser, and E. J. Shimshick, "Interaction of Liposomes with Subviral Particles of Poliovirus Type 2 and Rhinovirus Type 2," *Journal of Virology* 19 (1976): 746-49.

41. B. Mandel, "The Fate of the Inoculum in HeLa Cells Infected with Poliovirus," *Virology,* 25 (1965): 152-54.

42. R. L. Crowell, "Comparative Generic Characteristics of Picornavirus-Receptor Interactions," in *Cell Membrane Receptors for Viruses, Antigens, and*

Antibodies, Polypeptide Hormones and Small Molecules, ed. R. F. Beers, Jr. and E. G. Bassett (New York: Raven Press, 1976): 179-202.

43. B. A. Phillips and R. Fennell, "Unusual Attachment Behavior Exhibited by a Urea-Resistant Mutant of Poliovirus Type 1," *Virology* 83 (1977): 295-304.

44. J. J. Holland and B. H. Hoyer, "Early Stages of Enterovirus Infection," *Cold Spring Harbor Symposia on Quantitative Biology* 27 (1962): 101-11.

45. C. E. Schwerdt and J. Fogh, "The Ratio of Physical Particles per Infectious Unit Observed for Poliomyelitis Virus," *Virology* 4 (1957): 41–52.

46. Mandel, "Fate of the Inoculum" (See note 41) .

47. W. K. Joklik and J. E. Darnell, "The Adsorption and Early Fate of Purified Poliovirus in HeLa Cells," *Virology* 13 (1961): 439-47.

48. M. L. Fenwick and P. D. Cooper, "Early Interactions between Poliovirus and ERK Cells. Some Observations on the Nature and Significance of the Rejected Particles," *Virology* 18 (1962): 212-23.

49. L. Philipson and S. Bengtsson, "Interaction of Enteroviruses with Receptors from Erythrocytes and Host Cells," *Virology* 18 (1962): 457-69.

50. R. L. Crowell and L. Philipson, "Specific Alterations of Coxsackievirus B3 Eluted from HeLa Cells," *Journal of Virology* 8 (1971): 509-15.

51. Ibid.

52. J. Enders, A. Holloway, and E. A. Grogan, "Replication of Poliovirus 1 in Chick Embryo and Hamster Cells Exposed to Sendai Virus," *Proceedings of the National Academy of Sciences* 57 (1967): 637-44.

53. T. Wilson, D. Papahadjopoulos, and R. Taber, "Biological Properties of Poliovirus Encapsulated in Lipid Vesicles: Antibody Resistance and Infectivity in Virus-Resistant Cells," *Proceedings of the National Academy of Sciences* 74 (1977): 3471-75.

54. Cited in Boulanger and Lonberg-Holm, "Nonenveloped Viruses" (See note 33).

55. B. A. Phillips and S. Wiemert, "In Vitro Assembly of Poliovirus. V. Evidence That the Self-Assembly Activity of 14S Particles Is Independent of Extract Assembly Factors(s) and Host Proteins," *Virology* 88 (1978): 92-104.

56. D. A. Miller et al., "Human Chromosome 19 Carries a Poliovirus Receptor Gene," *Cell* 1 (1974): 167-73.

57. I. Zajac and R. L. Crowell, "Differential Inhibition of Attachment and Eclipse Activities of HeLa Cells for Enteroviruses," *Journal of Virology* 3 (1969): 422-28.

58. B. Mandel, "The Use of Sodium Dodecyl Sulfate in Studies on the Interaction of Poliovirus and HeLa Cells," *Virology* 17 (1962): 288-94.

59. A. Levanon, A. Kohn, and M. Inbar, "Increase in Lipid Fluidity of Cellular Membranes Induced by Adsorption of RNA and DNA Virions," *Journal of Virology* 22 (1977):353-60.

60. M. F. Perutz, "Electrostatic Effects in Proteins," *Science* 201 (1978): 1187-91.

61. Holland and McLaren, "Adsorption, Reception, and Eclipse," (See note 24).

62. M. Fiala and G. E. Kenney, "Effect of Magnesium on Replication of Rhinovirus HGP," *Journal of Virology* 1 (1967): 489-93.

63. L. C. McLaren, J. J. Holland, and J. T. Syverton, "The Mammalian Cell-Virus Relationship. V. Susceptibility and Resistance of Cells In Vitro to Infection by Coxsackie A9 Virus," *Journal of Experimental Medicine* 112 (1960): 581-94.

64. F. Brown, B. Cartwright, and D. L. Stewart, "Further Studies on the Infection of Pig Kidney Cells by Foot-and-Mouth Disease Virus," *Biochimica et Biophysica Acta* 55 (1962): 768-74.

65. Crowell, "Comparative Generic Characteristics" (See note 42).

66. Crowell and Siak, "Receptor for Coxsackieviruses" (See note 29).

67. L. Philipson and P. W. Choppin, "Inactivation of Enteroviruses by 2,3-Dimercaptopropanol (BAL)," *Virology* 16 (1962): 405-13.

68. Zajac and Crowell, "Effect of Enzymes" (See note 30.

69. P. Kraemer, "Regeneration of Sialic Acid on the Surface of Chinese Hamster Cells in Culture. I. General Characteristics of the Replacement Process," *Journal of Cellular and Comparative Physiology* 68 (1966): 85-90.

70. Zajac and Crowell, "Effect of Enzymes" (See note 30).

71. D. A. Axler and R. L. Crowell, "Effect of Anticellular Serum on the Attachment of Enteroviruses to HeLa Cells," *Journal of Virology* 2 (1968): 813-21.

72. M. C. Timbury, "The Growth of Echovirus 1 in HEp-2 Cells Treated with Antibody to the Host Cells," *Journal of General Virology* 4 (1969): 531-39.

73. R. L. Crowell, "Determination of the Relative Receptor Concentration of HeLa Cells for Different Enteroviruses by Attachment Kinetics. *Bacteriological Proceedings* (1967): 140.

74. Crowell, "Comparative Generic Characteristics" (See note 42).

75. Ibid.

76. ————, "Specific Cell-Surface Alteration by Enteroviruses as Reflected by Viral-Attachment Interference," *Journal of Bacteriology* 91 (1966): 198-204.

77. Ibid.

78. L. Quersin-Thiry and E. Nihoul, "Interaction between Cellular Extracts and Animal Viruses. II. Evidence for the Presence of Different Inactivators Corresponding to Different Viruses," *Acta Virologica* (Prague) 5 (1961): 283-93.

79. K. Lonberg-Holm, R. L. Crowell, and L. Philipson, "Unrelated Animal Viruses Share Receptors," *Nature* 259 (1976): 679-81.

80. P. Talbot and F. Brown, "A Model for Foot-and-Mouth Disease Virus," *Journal of General Virology* 15 (1972): 163-70.

81. L. Hall and R. Rueckert, "Infection of Mouse Fibroblasts by Cardioviruses. Premature Uncoating and Its Prevention by Elevated pH and Magnesium Chloride," *Virology* 43 (1971): 152-65.

82. J. J. Holland, "Receptor Affinities as Major Determinants of Enterovirus Tissue Tropisms in Humans," *Virology* 15 (1961): 312-26.

83. ————, "Enterovirus Entrance into Specific Host Cells and Subsequent Alterations of Cell Protein and Nucleic Acid Synthesis," *Bacteriological Reviews* 28 (1964): 3-13.

84. C. M. Kunin and W. S. Jordan, "In Vitro Adsorption of Poliovirus by Noncultured Tissues. Effect of Species, Age and Malignancy," *American Journal of Hygiene* 73 (1961): 245-57.

85. C. M. Kunin, "Cellular Susceptibility to Enteroviruses," *Bacteriological Reviews* 28 (1964): 382-90.

86. McLaren, Holland, and Syverton, "Attachment of Poliovirus" (See note 23).

87. Holland, McLaren, and Syverton, "Infection of Insusceptible Cells" (See note 25).

88. G. D. Hsiung, "Laboratory Diagnosis of Viral Infections: General Principles and Recent Developments," *Mt. Sinai Journal of Medicine* 44 (1977): 1-26.

89. C. P. Li, M. Schaeffer, and D. B. Nelson, "Experimentally Produced Variants of Poliomyelitis Virus Combining In Vivo and In Vitro Techniques," *Annals of the New York Academy of Sciences* 61 (1955): 902-10.

90. P. W. Choppin and H. J. Eggers, "Heterogenicity of Coxsackie B4 Virus: Two Kinds of Particles Which Differ in Antibody Sensitivity, Growth Rate, and Plaque Size," *Virology* 18 (1962): 470-76.

91. A. D. Vizoso and F. K. Sanders, "Alteration of the Pathogenicity of Some Group B Coxsackie Viruses under Different Conditions of Passage. I. Virus Type 4," *Acta Virologica* (Prague) 8 (1964): 38-45.

92. J. E. Craighead, "Pathogenicity of the M and E Variants of the Encephalomyocarditis (EMC) Virus. I. Myocardiotropic and Neurotropic Properties," *American Journal of Pathology* 48 (1966): 333-45.

93. J. W. Yoon, T. Onodera, and A. Notkins, "Virus-Induced Diabetes Mellitus. XV. Beta Cell Damage and Insulin-Dependent Hyperglycemia in Mice Infected with Coxsackie Virus B4," *Journal of Experimental Medicine* 148 (1978): 1068-80.

94. V. D. Podoplekin, "Effect of Tissue Cultures on the Hemagglutinating Properties of ECHO Viruses and Mechanism of the Interaction of These Viruses with Erythrocytes," *Acta Virologica* (Prague) 8 (1964): 254-62.

95. R. L. Crowell and B. Goldberg, "Propagation and Assay of Group A Coxsackieviruses in RD Cells," *Abstracts of the Annual Meeting of the American Society for Microbiology* (1974): 208.

96. N. J. Schmidt, H. H. Ho, and E. H. Lennette, "Propagation and Isolation of Group A Coxsackieviruses in RD Cells," *Journal of Clinical Microbiology* 2 (1975): 183-85.

97. R. L. Crowell and B. J. Landau, "Picornaviridae: Enterovirus-Coxsackievirus" in *CRC Handbook in Clinical Laboratory Science,* vol. 1, ed. G. D. Hsiung and R. Green (West Palm Beach, Florida: CRC Press, 1978): 131-55.

98. Crowell, "Comparative Generic Characteristics" (See note 42).

99. Lonberg-Holm, Crowell, and Philipson, "Unrelated Animal Viruses" (See note 79).

100. Roesing, Toselli, and Crowell, "Elution and Uncoating" (See note 31).

101. Crowell, "Comparative Generic Characteristics" (See note 42).

102. Quersin-Thiry and Nihoul, "Interaction between Cellular Extracts" (See note 78).

103. Crowell, "Specific Cell-Surface Alteration" (See note 76).

104. Lonberg-Holm, Crowell, and Philipson, "Unrelated Animal Viruses" (See note 79).

105. Holland, "Receptor Affinities as Major Determinants" (See note 82).

106. L. Quersin-Thiry, "Action of Anticellular Sera on Virus Infections. I. Influence on Homologous Tissue Cultures Infected with Various Viruses," *Journal of Immunology* 81 (1958): 253-60.

107. B. C. Casto and W. M. Hammon, "Correlation between Propagation of a Bovine Enterovirus and Development of Receptor-Like Material in Tissue Mince," *Proceedings of the Society for Experimental Biology and Medicine* 132 (1969): 154-60.

108. Zajac and Crowell, "Effect of Enzymes" (See note 30).

109. N. H. Levitt and R. L. Crowell, "Comparative Studies of the Regeneration of HeLa Cell Receptors for Poliovirus T1 and Coxsackievirus B3," *Journal of Virology* 1 (1967): 693-700.

110. R. J. Goldberg and R. L. Crowell, "Susceptibility of Differentiating Muscle Cells of the Fetal Mouse in Culture to Coxsackievirus A13," *Journal of Virology* 7 (1971): 759-69.

111. R. Chairez, J. W. Yoon, and A. L. Notkins, "Virus-Induced Diabetes Mellitus. X. Attachment of Encephalomyocarditis Virus and Permissiveness of Cultured Pancreatic Beta Cells to Infection," *Virology* 85 (1978): 606-11.

112. Philipson and Bengtsson, "Interaction of Enteroviruses" (See note 49).

113. McLaren, Holland, and Syverton, "Susceptibility and Resistance of Cells" (See note 63).

114. H. Diderholm and Z. Dinter, "A Line of Polyoma-Transformed BHK21 Cells Unsusceptible to Foot-and-Mouth Disease Virus," *Virology* 26 (1965): 369-72.

115. E. Hahn and J. Fogh, "Increased Resistance of SV_{40} Transformed Human Amnion Cells to Poliovirus Infection," *Archiv für die Gesamte Virusforschung* 29 (1970): 343-60.

116. N. R. Blacklow, F. B. Rose, and R. A. Whalen, "Organ Culture of Human Aorta: Prolonged Survival with Support of Viral Replication," *Journal of Infectious Diseases* 131 (1975): 575-78.

117. B. Hoorn and D. A. J. Tyrrell, "A New Virus Cultivated Only in Organ Cultures of Human Ciliated Epithelium," *Archiv für die Gesamte Virusforschung* 18 (1966): 210-25.

118. B. J. Landau et al., "The Permissiveness of Differentiating Mouse Muscle-Cell Cultures to Infection by Group A Coxsackievirus Types 1 and 5," *Proceedings of the Society for Experimental Biology and Medicine* 141 (1972): 753-58.

119. P. E. Came and R. L. Crowell, "Studies of Resistance of Fetal Mouse Tissue in Culture to Coxsackie Group A Viruses," *Virology* 23 (1964): 542-52.

120. M. S. Cohen and R. L. Crowell, "Effect of Myogenic Differentiation of the L8 Rat Cell Line on Susceptibility to Coxsackievirus A2," *Abstracts of the Annual Meeting of the American Society for Microbiology* (1978): 233; *Journal of Virology*, in press.

121. Holland, "Receptor Affinities" (See note 82).

122. Kunin and Jordan, "In Vitro Adsorption" (See note 84).

123. D. H. Harter and P. W. Choppin. "Adsorption of Attenuated and Neurovirulent Poliovirus Strains to Central Nervous System Tissues of Primates," *Journal of Immunology* 95 (1965): 730-36.

124. M. LaPlaca, "Adsorption In Vitro of Bovine Enteroviruses by Non-Cultured Bovine and Human Tissues," *Nature* 199 (1963): 1211-12.

125. Holland, "Receptor Affinities" (See note 82).

126. Kunin and Jordan, "In Vitro Adsorption" (See note 84).

127. T. Francis and L. W. Chu, "The Interaction In Vitro Between Poliomyelitis Virus and Nervous Tissue Constituents," *International Congress of Microbiology* 2 (1953): 426.

128. J. B. Derbyshire and D. M. Jessett, "Receptor Activity for Porcine Enteroviruses in Pig Tissues," *Journal of Medical Microbiology* 2 (1969): 489-93.

129. Holland, "Receptor Affinities" (See note 82).

130. Kunin and Jordan, "In Vitro Adsorption" (See note 84).

131. Holland, "Receptor Affinities" (See note 82).

132. C. J. Gauntt et al., "Properties of Coxsackievirus B3 Variants Which Are Amyocarditic or Myocarditic for Mice," *Journal of Medical Virology*, in press.

133. B. J. Landau, "Replication of Coxsackievirus B3 in Primary Mouse Cell Cultures of Cardiac Origin," *Abstracts of the Annual Meeting of the American Society for Microbiology* (1978): 233.

134. Kunin and Jordan, "In Vitro Adsorption" (See note 84).

135. Crowell, "Comparative Generic Characteristics" (See note 42).

136. LaPlaca, "Adsorption In Vitro" (See note 124).

137. C. H. Campbell, "Relationship of Donor Age to In Vitro Production of Foot-and-Mouth Disease Virus by Mouse Kidney Cells," *Journal of Experimental Medicine* 121 (1965): 69-83.

138. C. Howe and L. T. Lee, "Virus-Erythrocyte Interactions," *Advances in Virus Research* 17 (1970): 1-50.

139. L. Philipson et al., "Purification and Chemical Analysis of the Erythrocyte Receptor for Hemagglutinating Enteroviruses," *Virology* 22 (1964): 580-90.

140. H. J. Eggers, "Selective Inhibition of Uncoating of Echovirus 12 by Rhodanine. A Study on Early Virus-Cell Interactions," *Virology* 78 (1977): 241-52.

141. K. Lonberg-Holm, L. B. Gosser, and J. C. Kauer, "Early Alteration of Poliovirus Virions in Infected Cells and Its Specific Inhibition," *Journal of General Virology* 27 (1975): 329-42.

142. Axler and Crowell, "Effect of Anticellular Serum" (See note 71).

143. C. Bradbeer, M. L. Woodrow, and L. I. Khalifah, "Transport of Vitamin B_{12} in *Escherichia coli*: Common Receptor System for Vitamin B_{12} and Bacteriophage BF 23 on the Outer Membrane of the Cell Envelope," *Journal of Bacteriology* 125 (1976): 1032-39.

144. Zajac and Crowell, "Effect of Enzymes" (See note 30).

145. Miller et al., "Human Chromosome 19" (See note 56).

146. P. Couillin and M. Fellous, "Récherche d'une correlation entré le systeme antigenique HL-A et les recepteurs viraux de trois enterovirus a l'aide de l'hybridization cellulaire," *Comptes rendus des seances de la Societé de Biologie* 168 (1974): 180-86.

147. Goldberg and Crowell, "Susceptibility of Differentiating Muscle" (See note 110).

148. Chairez, Yoon, and Notkins, "Virus-Induced Diabetes Mellitus" (See note 111).

149. G. M. Edelman, "Nonenzymatic Dissociations. B. Specific Cell Fractionation on Chemically Derivatized Surface," *Tissue Culture Methods and Applications* ed. P. F. Kruse, Jr. and M. K. Patterson, Jr. (Academic Press: New York, 1973): 29-36.

150. Crowell and Siak, "Receptor for Group B Coxsackieviruses" (See note 29).

151. E. Borriss and G. Koch, "Isolation of Poliovirus-Neutralizing Antibodies by Affinity Chromatography," *Virology* 67 (1975): 356-64.

MOLECULAR ASSOCIATION BETWEEN TRANSPLANTATION ANTIGENS AND AN EARLY ADENOVIRUS PROTEIN

Lennart Philipson, Håkan Persson, Sune Kvist, Lars Östberg, and Per A. Peterson *

INTRODUCTION

The importance of cellular immunity against virus infections was first realized from clinical observations of viral infections in immunodeficient patients, and subsequently from experimental findings with virus infections of thymectomized or antilymphocyte-serum-treated animals.[1] In recent years it has become clear that the cellular immune defense against virus infections displays genetic restriction. Cytolytic T-lymphocytes generated during a viral infection can only lyse infected target cells that share major histocompatibility antigens with the cells that originally stimulated the development of the cytolytic T-cells.[2]

This genetic restriction, which has been mapped to the loci controlling the expression of the transplantation antigens, suggests a mechanism by which cytolytic T-lymphocytes recognize the target cells. Either the virus product and the histocompatibility antigens are separate entities on the surface of the target cells, and the T-cells express independent receptors for the two types of molecules (*dual recognition*), or the virus and the transplantation antigens form a molecular complex recognized by one

* Dr. Persson, like Dr. Philipson, is a member of the Department of Microbiology; Drs. Kvist, Östberg, and Peterson are members of the Department of Cell Research, Biomedical Center, University of Uppsala. Only Dr. Philipson attended the conference.

receptor on the T-lymphocytes (*altered self*).[3] Evidence for both hypotheses have accumulated.

Cocapping experiments,[4] antiserum-induced inhibition of T-cell-dependent cytolysis,[5,6] and selective incorporation of H-2 antigens into enveloped viruses[7] may argue for the altered-self concept. The development of precursor T-cells to mature T-lymphocytes with cytolytic activity, however, requires recognition of the major histocompatibility antigens and occurs independently of recognition of nonself antigens,[8] a result that is compatible with the dual recognition concept.

To obtain unambiguous information about the molecular nature of the target recognition sites for the cytotoxic T-lymphocytes, detailed biochemical studies are needed. We have approached this problem by examining the adenovirus system as a model system, and have found that adenovirus transformed cell lines may provide tools for this analysis.

The adenovirus growth cycle in human cells has been divided into an early and late phase separated by the onset of viral DNA replication. During the early phase a limited portion corresponding to 30 percent of the viral genome is expressed in messenger RNA.[9] As many as thirteen virus-induced early polypeptides have been detected, but about ten of them have so far been shown to be encoded within the viral genome.[10] Only one of the virus-coded early polypeptides appears to be a glycoprotein located at the outside of the plasma membrane of the infected cells. Labeling of cells with ^{3}H-glucosamine early after adenovirus type 2 (Ad2) infection identified this early polypeptide, which has an apparent molecular weight of 19,000 (E19K).[11] The gene for the E19K polypeptide is probably coded at the right-hand end of the Ad2 genome within the *Eco*RIC restriction enzyme fragment: map positions 89.7-100.[12] Some Ad2-transformed cell lines, such as a hamster cell line (Ad2 HE4) and a hooded Lister rat cell line (A2T2C4), contain this region of the viral genome in an integrated state, and consequently also express the glycoprotein E19K at the external surface of the plasma membrane. Other cell lines, although transformed by adenovirus, fail to express E19K.[13] These findings are of importance in evaluating whether this protein constitutes the tumor-specific transplantation antigen (TSTA) of adeno-transformed cells.

An experimental system involving mouse adenovirus and inbred animals has recently established the importance of cellular immunity in adenovirus infection.[14] Three different approaches were used to study cell-mediated immunity to mouse adenovirus-infected cells: 1) inhibition of capsid protein synthesis in infected cells after incubation with immune spleen cells; 2) direct and indirect inhibition of migration of immune mouse peritoneal exudate cells, that is, macrophage migration inhibition tests; and, finally, 3) specific ^{51}Cr release from infected cells after incubation with immune spleen cells.

These studies reveal that immune lymphocytes do react with adenovirus-infected cells, and the reaction is obviously independent of synthesis of viral DNA or capsid antigens, suggesting that an early protein might be involved in the recognition mechanism. A surface antigen was probably responsible for the cell-mediated immune reaction. The reactivity of the immune spleen cells was detectable within a week after sublethal infection of the animals, reaching its peak level around ten days postinfection, whereupon it faded away in a few weeks. It was also demonstrated that the reacting cells among the host lymphocytes are T-cells, and that the antigen responsible for the recognition was specific for mouse adenovirus. Furthermore the cell-mediated lysis of mouse adenovirus-infected cells appears to require H-2 compatibility between the synthesizing target cell and the effector cell. Similar findings have been reported concerning cell-mediated immunity against cells infected with lymphocytic choriomeningitis,[15] vaccinia,[16] ectromelia,[17] and Sendai viruses.[18] Obviously the adenoviruses might serve as a convenient model to identify the polypeptides associated with the histocompatibility antigens in the cytolytic T-cell reaction with virus-infected target cells.

This report summarizes the findings, indicating that the adenovirus-encoded product E19K may elicit syngeneic cytolytic T-lymphocytes. It appears also that the E19K forms a ternary complex with the major histocompatibility antigen subunits.[19]

PURIFICATION AND CHARACTERIZATION OF THE HUMAN ADENOVIRUS TYPE 2 GLYCOPROTEIN

Our purification scheme for the Ad2 glycoprotein from HeLa cells is shown in Figure 1. A cycloheximide block (25 μg./ml.) was introduced from two to five hours after virus infection to enhance the relative abundance of viral mRNA. The block was reversed and the cells labeled with ^{35}S-methionine from five to

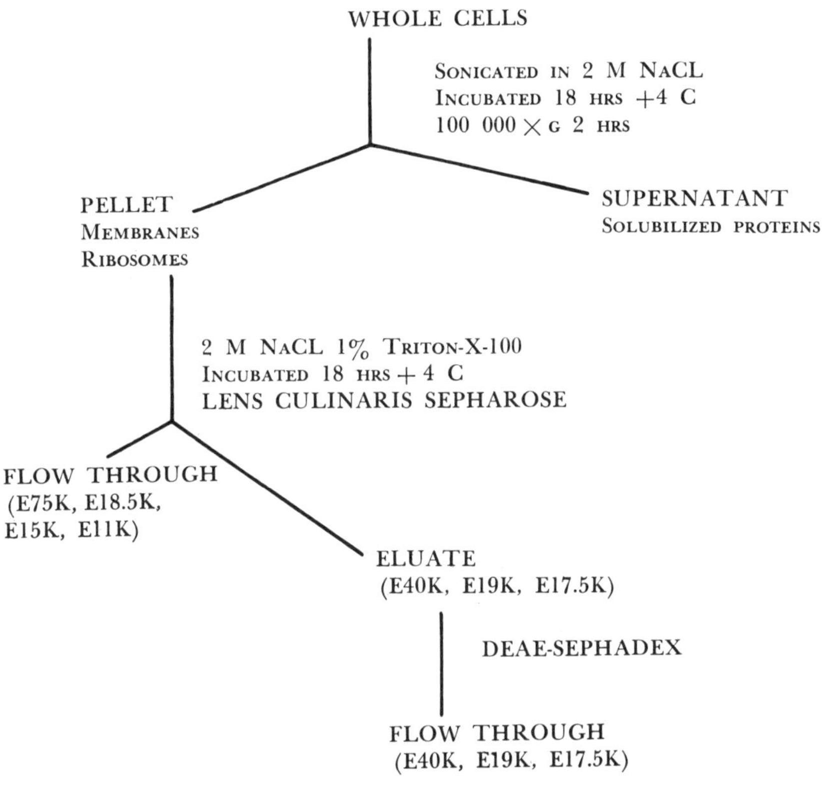

Figure 1. Scheme for purification of an Ad2 early glycoprotein. Ad2-infected HeLa cells were treated with cycloheximide (25 μg./ml.) between two to five hours postinfection. The cells were washed and labeled with ^{35}S-methionine in the presence of cytosine arabinoside (25 μg./ml.). At nine hours postinfection the cells were harvested, washed once in phosphate-buffered saline (PBS), and subjected to the outlined purification scheme.

nine hours postinfection in the presence of cytosine arabinoside (25 μg./ml.) to prevent viral DNA synthesis. Mock infected cells were treated in the same way.

Cell extracts were prepared and analyzed by sodiumdodecylsulphate polyacrylamide gel electrophoresis (SDS-PAGE) to identify Ad2 early-induced polypeptides. Several Ad2 early polypeptides that were absent or less conspicuous in mock infected cells were observed, as shown in Figure 2. The cell extracts were incubated in a high salt buffer, and subsequently centrifuged at high speed to separate the soluble from the insoluble protein. Four Ad2 early polypeptides, E19K, E18.5K, E17.5K, E11K, were highly enriched in the pellet fraction. Quantitative analysis showed that around 80 percent of the E19K polypeptide was recovered in the pellet. The E17.5K was also preferentially associated with the pellet. Figure 2 shows that no polypeptide corresponding to E19K was found in the mock infected pellet fraction. The mock infected material was therefore not analyzed further. The pellet fraction was solubilized overnight in a high salt buffer containing 1 percent of the neutral detergent Triton X-100. Solubilized material was fractionated by lectin (*Lens culinaris*) affinity chromatography, where around 40 percent of the radioactivity bound to the column. SDS-PAGE revealed that the E19K and the E17.5K had a high affinity for the lectin column. About 60 percent of the bound material was recovered in the E19K polypeptide, which was enriched 20-fold after this purification step.

In the final purification step, DEAE-Sephadex chromatography, E19K did not bind to the column. The flow-through material was therefore collected and analyzed by SDS-PAGE. Three polypeptides, E40K, E19K, and E17.5K, were detected, with 90 percent of the radioactivity in the E19K polypeptide. Tryptic peptide analysis of these three proteins showed they are related to each other. The E40K is probably a dimer of the E19K, and the E17.5K is probably a precursor of the mature glycoprotein. This purification scheme yielded a 200-fold enrichment of the E19K polypeptide, with an estimated recovery of 12 percent.

Figure 3 shows a stained polyacrylamide gel of the purified protein. In order to study the kinetics of synthesis of the E19K,

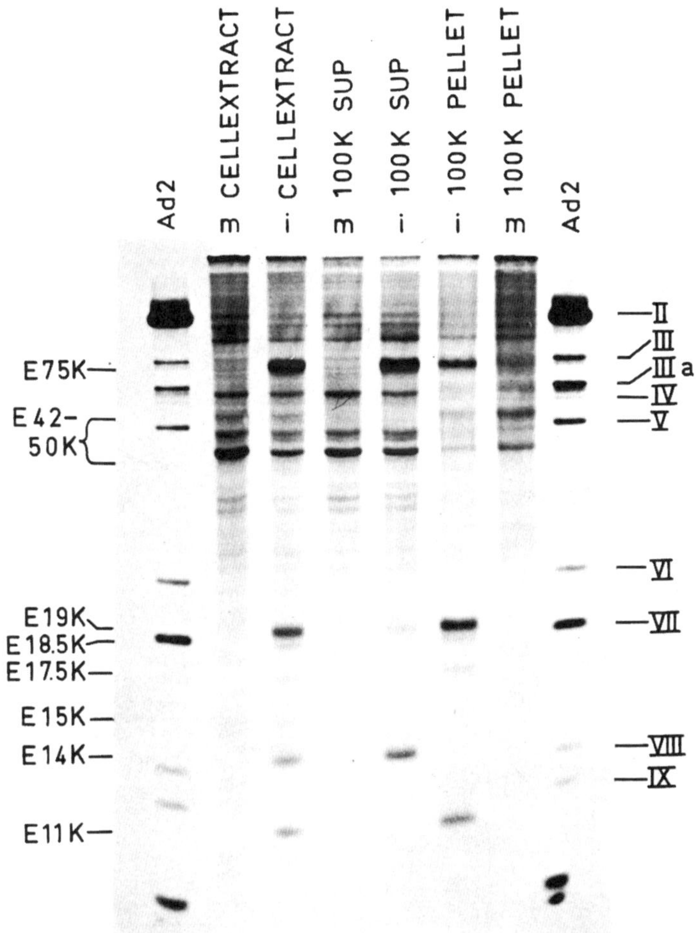

Figure 2. SDS-PAGE of material from different purification steps. Ad2-infected or mock-infected cells were treated with cycloheximide and labeled with ^{35}S-methionine. Cell extracts and the 100K supernatant and pellet fractions were prepared. The 100K pellet fraction was solubilized in a high salt buffer containing 1% Triton X-100. A 100 μl. portion of the samples was precipitated with 10% TCA, washed, and dissolved in gel sample buffer. An equal amount of cell equivalents from each sample was analyzed in a 13% SDS-polyacrylamide gel. Electrophoresis was carried out at 100 V for ten hours, and the gel was analyzed by fluorography. (m) = material from mock infected cells; (i) = that from Ad2-infected cells. 100K sup denotes material from the supernatant; 100K pellet denotes material that solubilized from the pellet. An adenovirus type 2 ^{35}S-methionine-labeled virus marker was also included (Ad2).

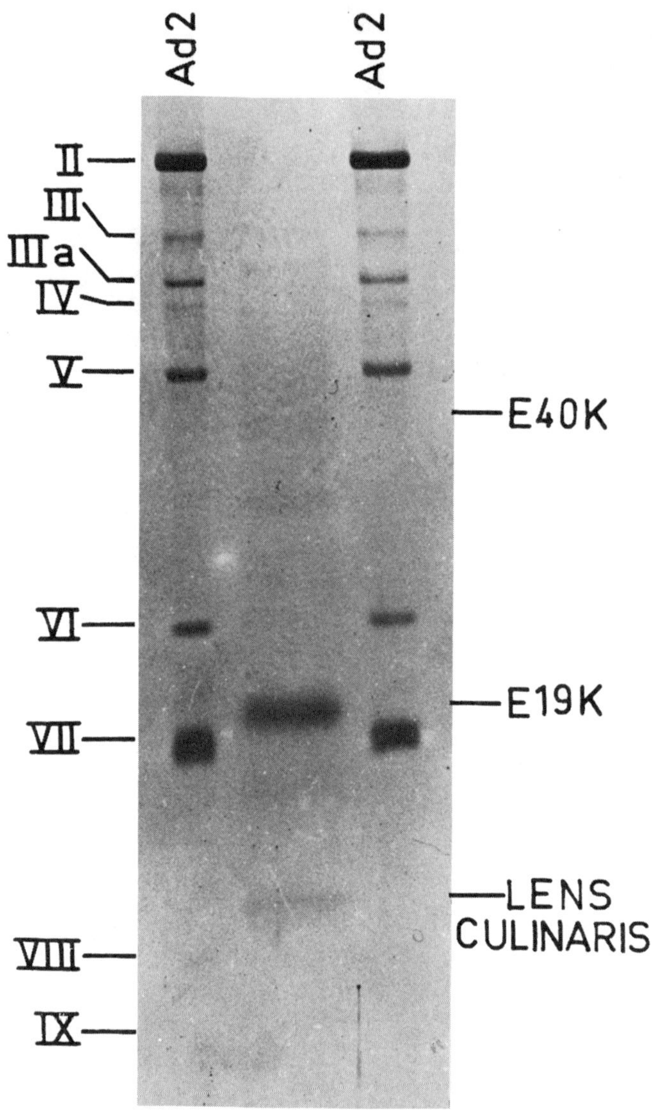

Figure 3. SDS-PAGE analysis of the purified E19K protein. Ad2 refers to a labeled Ad2 virus marker. *Lens culinaris* denotes the size of the lectin used for affinity chromatography in one of the purification steps.

cell extracts were prepared from cells that had been pulse-labeled for thirty minutes with ^{35}S-methionine at different times after Ad2 infection. An antiserum against the highly purified E19K was used to immunoprecipitate the protein. The E19K polypeptide was detected at two hours, and the rate of synthesis reached a maximum at four hours postinfection. The E17.5K showed the same kinetics as the E19K polypeptide. Three hamster embryo cell lines transformed with UV-inactivated adenovirus type 2 ($Ad2HE_1$, $Ad2HE_4$, $Ad2HE_5$) and two Ad2-transformed rat cell lines (A2F19 and A2T2C4) were analyzed for the presence of the glycoprotein. Whole cell extracts were immunoprecipitated with the antiserum, and only one hamster cell line ($Ad2HE_4$) and one rat cell line (A2T2C4) expressed this polypeptide. Immunoprecipitation with a serum prepared in syngeneic rats by injection of A2T2C4 cells revealed E19K polypeptide in the latter two cell lines.

INHIBITION OF T-LYMPHOCYTE-DEPENDENT LYSIS OF A2T2C4 CELLS

Spleen cells obtained from hooded Lister rats, primed in vivo with the A2T2C4 cells, could lyse the cells provided the primed spleen cells had been restimulated in vitro with the target cells prior to cytolysis. Lysis was observed as an effector to target cell ratio as low as 3:1. No cytolytic activity of the spleen cells was recorded when the target cells were primary fibroblasts from the hooded Lister rats or when the adenovirus-transformed hamster cells, Ad2 HE_4, were also carrying the E19K polypeptide. Cell-mediated cytolysis of the A2T2C4 cells was enhanced when B-cells were removed, either by nylon-wool columns or by anti-rat IgG serum. Together these results suggest that T-lymphocytes were responsible for the cytolytic activity.

To examine whether the T-lymphocyte-dependent lysis of the A2T2C4 cells recognized virus-genome-coded products, in addition to the recognition of the syngeneic AgB antigens, the cytolytic action of T-lymphocytes was investigated in the presence of various concentrations of antiserum. Figure 4 shows that a rabbit anti-AgB antiserum efficiently blocked lysis; normal rabbit

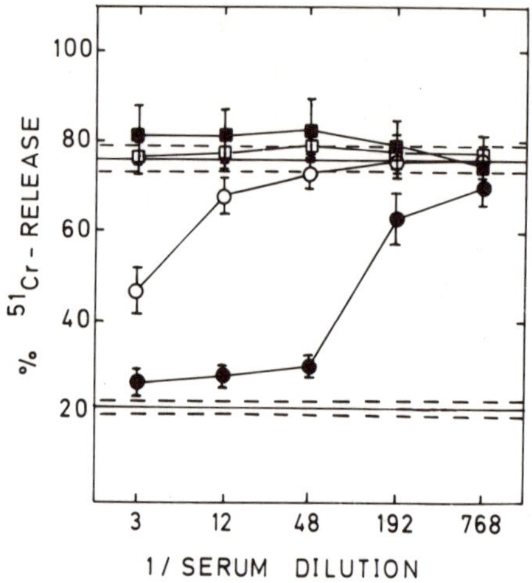

Figure 4. Inhibition of T-cell-dependent cytolysis of A2T2C4 cells. The indicated dilutions of a rabbit antiserum against AgB antigens (●), syngeneic rat antiserum against the A2T2C4 cells (○), normal rabbit serum (□), and normal rat serum (■) were incubated with 1×10^8 ^{51}Cr-labeled A2T2C4 cells. After thirty minutes at +4°C., educated rat spleen T-lymphocytes were added at an effector to target cell ratio of 20:1. The incubations proceeded for six hours at 37°, and the ^{51}Cr released was measured after centrifugation. A release of 100 percent represents the amount of radioactivity recovered in the supernatant after Triton X-100 treatment of the target cells. The solid and the broken horizontal lines denote the mean and the standard deviation of the ^{51}Cr release in the presence (upper) and in the absence (lower) of the cytolytic T-lymphocytes alone. The experimental points and the vertical bars are the mean ± the standard deviation of analyses performed in quadruplicate.

serum did not affect lysis. A syngeneic antiserum against A2T2C4 cells also inhibited lytic activity of the T-cells, although this antiserum was only effective at high concentrations. As expected, normal rat serum did not display inhibitory capacity.

IMMUNOPRECIPITATION OF GLYCOPROTEINS

Cellular glycoproteins were solubilized by neutral detergents followed by lectin affinity chromatography. When the glycoproteins from A2T2C4 cells were analyzed, two polypeptide chains with molecular weights of 45,000 and 19,000 were immunoprecipitated, with the syngeneic antiserum raised against the transformed A2T2C4 cells as shown in Figure 5. Similar results were

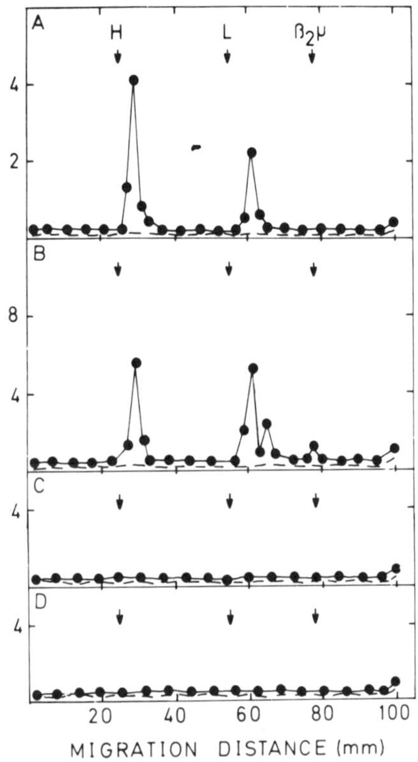

Figure 5. SDS-PAGE of A2T2C4 cell proteins reactive with a syngeneic antiserum raised against the same cells. Galactose oxidase and ^{3}H-sodiumborohydride-treated A2T2C4 cell surface molecules (A), ^{3}H-tyrosine-labeled molecules from the A2T2C4 cells (B), primary rat fibroblasts (C), and rat spleen cells (D) were isolated by lectin-affinity chromatography and precipitated with the antiserum. The solid lines depict the radioactive profiles of molecules precipitated by the antiserum; the broken lines show molecules nonspecifically precipitated with a normal rat serum. The arrows denote the migration positions of marker IgG heavy (H) and light (L) chains and β_2-microglobulin (β_2m).

obtained irrespective of whether the glycoproteins were labeled with ^{125}iodine or with ^{3}H-sodiumborohydride. When immunoprecipitations were performed with ^{3}H-tyrosine glycoproteins, two additional components with molecular weights of 17,500 and 12,000 were visualized.

The syngeneic A2T2C4 serum obviously only recognizes molecules from adenovirus-transformed cells, since ^{3}H-tyrosine-labeled glycoproteins from spleen cells or from primary fibroblasts were not precipitated by this antiserum. A rabbit antiserum against highly purified papain-solubilized AgB antigens of spleen cells reacted only with AgB antigen subunits in spleen and fibroblast cells. This antiserum precipitated two additional polypeptides from the A2T2C4 cells, 19,000 and 17,500 in molecular weight, however, corresponding to the size of the E19K and the E17.5K. Precipitation of the E19K component by the rabbit anti-AgB antiserum seemed to be unique for the transformed cells.

EVIDENCE FOR A TERNARY COMPLEX BETWEEN THE TRANSPLANTATION ANTIGENS AND THE E19K

It was of importance to investigate whether the E19K was coprecipitated also with antisera of individual AgB antigen subunits. Figure 6 shows that an AgB alloantiserum precipitated the

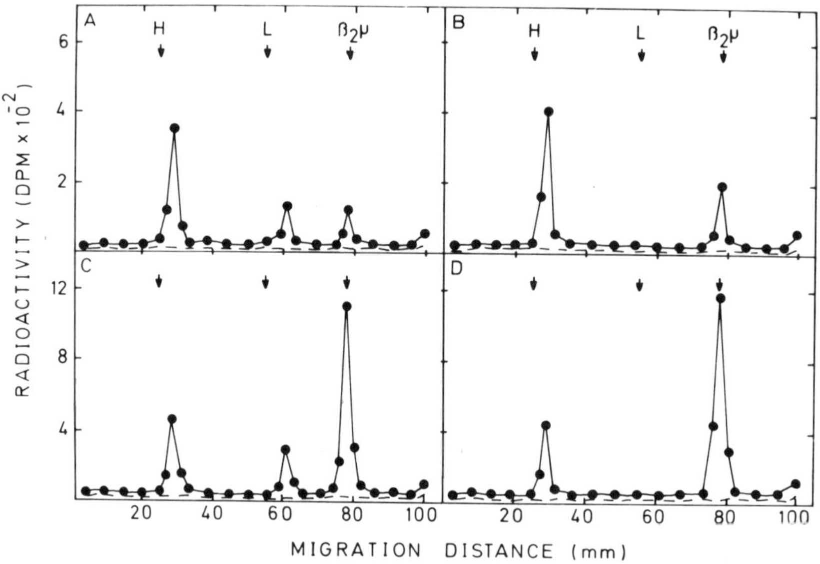

Figure 6. SDS-polyacrylamide gel electrophoreses of ^{3}H-tyrosine-labeled A2T2C4 cell glycoproteins (A and C), and ^{3}H-tyrosine-labeled hooded Lister primary rat fibroblast glycoproteins (B and D), isolated by indirect immunoprecipitations with an alloantiserum directed against hooded Lister rat AgB antigens (A and B), and with a rabbit antiserum against rat β_2m (C and D). Solid lines represent molecules specifically precipitated by the antisera; broken lines denote radioactivity recovered in indirect immunoprecipitates obtained with normal rat serum (A and B) and normal rabbit serum (C and D). The arrows show the migration positions for the marker IgG heavy (H) and light (L) chains and for β_2m.

E19K, together with the AgB antigen subunits from the ^3H-tyrosine-labeled glycoproteins obtained from A2T2C4 cells, but only the AgB antigen chains were visualized from the labeled molecules derived from primary rat fibroblasts. The same results were obtained with the antiserum against β_2-microglobulin. These results are consistent with the idea that the two AgB antigen subunits formed a ternary complex with the E19K.

DISCUSSION

In the adenovirus-transformed rat cell line A2T2C4, the AgB antigens appear to bind to a 19,000 viral glycoprotein. This polypeptide is encoded in the right-hand end of the viral genome, and it is expressed in some but not all virus-transformed cell lines. After detergent solubilization of the membrane glycoproteins a substantial amount of the AgB antigens was coprecipitated with an antiserum recognizing the E19K. Likewise, antisera directed against the AgB antigens and their subunits coprecipitated the E19K protein.

Quantitative aspects of this complex formation are difficult to assess. It is possible that the E19K replaces the β_2-microglobulin, which obviously is present in equimolar amounts with the AgB antigen heavy chain on the cell surface, since less β_2-microglobulin was precipitated with the anti-E19K serum than with the anti-β_2-microglobulin serum. Antibodies may, however, compete for binding to the interacting sites of any complex, which would interfere with quantitative assessments of the interactions between the AgB antigens and the E19K.

The present analysis indicates that the syngeneic rat antiserum reacts only with one cell-surface protein and its precursor on the A2T2C4 cells. It is therefore tempting to suggest that this protein may be involved in tumor rejection of adenovirus-transformed cells.[20] Since several transformed cell lines do not express the E19K, however, and since only a limited portion of the left-hand end of the adenovirus genome is required for induction and maintenance of transformation,[21] another TSTA of adenoviruses may reside in the protein(s) encoded in the left end of the viral genome.

This problem can only be attacked experimentally by determining whether the E19K is engaged in the rejection of the A2T2C4 cells and other adenovirus-transformed cell lines. Thus immunization with the E19K should impede the tumor growth if this protein is the adenovirus TSTA. The findings that the AgB antigens and the E19K, after solubilization of the membrane glycoproteins with neutral detergents, also form a complex provide suggestive evidence that the T-lymphocytes might recognize altered self rather than possess a dual recognition mechanism for the target cells. Furthermore, the identification of a close association between the E19K and the transplantation antigens explains the mechanism of cell-mediated immunity to virus-infected cells in the adenovirus system.

Analysis of kinetics of polypeptide synthesis in Ad2-infected cells suggests that the E19K is one of the first proteins expressed on the genome during productive infection. It is tempting to speculate that this protein might have a specific function in membranes generating a favorable environment for adenovirus gene expression. An analysis of cell-mediated immunity in mouse cells infected with mouse adenovirus may also establish that this protein is involved in the T-cell cytotoxicity against virus-infected cells. The present study opens the field for a closer inspection of the role of virus-coded membrane proteins in the productive infection of the virus, as well as of the host defense mechanism involving recognition of tumor-specific transplantation antigens and induction of cell-mediated immunity.

NOTES

1. E. F. Wheelock and S. T. Toy, "Participation of Lymphocytes in Viral Infections," *Advances in Immuology* 16 (1973): 123-84.

2. P. C. Doherty, R. V. Blanden, and R. M. Zinkernagal, "Specificity of Virus-Immune Effector T-Cells for H-2K or H-2D Compatible Interactions: Implications for H-Antigen Diversity," *Transplantation Reviews* 29 (1976): 98-123.

3. R. M. Zinkernagel and P. C. Doherty, "The Concept That Surveillance of Self Is Mediated via the Same Set of Genes That Determines Recognition of Allogenic Cells," *Cold Spring Harbor Symposia on Quantitative Biology* 41 (1977): 505-10.

4. R. Henning, J. W. Schrader, and G. M. Edelman, "Antiviral Antibodies Inhibit the Lysis of Tumour Cells by Anti H2-Sera," *Nature* 263 (1976): 689-91.

5. J. W. Schrader and G. M. Edelman, "Participation of the H2-Antigens of Tumor Cells in Their Lysis by Syngeneic T-Cells," *Journal of Experimental Medicine* 143 (1976): 601-14.

6. A. H. Hale et al., "Vesicular Stomatitis Virus Glycoprotein Is Necessary for H-2 Restricted Lysis of Infected Cells by Cytotoxic T-Lymphocytes," *Proceedings of the National Academy of Sciences* 75 (1978): 960-74.

7. J. E. Bubbers, S. Chen, and F. Lilly, "Non-Random Inclusion of H-2K and H-2D Antigens in Friend Virus Particles from Mice of Various Strains," *Journal of Experimental Medicine* 147 (1978): 340-51.

8. R. M. Zinkernagel et al., "On the Thymus in Differentiation of 'H-2 Self Recognition' by T Cells: Evidence for Dual Recognition?" *Journal of Experimental Medicine* 147 (1978): 882-96.

9. J. Flint, "The Topography and Transcription of the Adenovirus Genome," *Cell* 10 (1977): 153-66.

10. M. L. Harter and J. B. Lewis, "Adenovirus Type 2 Early Proteins Synthesized In Vitro and In Vivo: Identification of the 38,000 to 50,000 Molecular Weight Protein Encoded by the Left End of the Adenovirus Type 2 Genome," *Journal of Virology* 26 (1978): 736-49.

11. M. Ishibashi and J. V. Maizel, Jr., "The Polypeptides of Adenoviruses. VI. Early and Late Glycopeptides," *Virology* 58 (1974): 345-61.

12. Harter and Lewis, "Adenovirus Type 2" (See note 10).

13. H. Persson, C. Signäs, and L. Philipson, "Purification and Characterization of an Early Glycoprotein from Adenovirus Type 2 Infected Cells," *Journal of Virology*, in press.

14. H. Uetake and T. Inada, "Mechanism of Induction of Cellular Immunity to Virus Infections: Experiments with Adenoviruses," *Annals of Microbiology* (Institute Pasteur) 128B (1977): 517-30.

15. R. M. Zinkernagel and P. C. Doherty, "Restriction of In Vitro T-Cell-Mediated Cytotoxicity in Lymphocytic Choriomeningitis within a Syngeneic or Semiallogeneic System," *Nature* 248 (1974): 701-02.

16. U. Koszinowski and R. Thomsen, "Target Cell-Dependent T-Cell-Mediated Lysis of Vaccinia Virus-Infected Cells," *European Journal of Immunology* 5 (1975): 245-51.

17. I. D. Gardner, N. A. Bowern, and R. V. Blanden, "Cell-Mediated Cytotoxicity against Ectromelia-Infected Target Cells. III. Role of the H-2 Gene Complex," *European Journal of Immunology* 5 (1975): 122-27.

18. P. C. Doherty and R. M. Zinkernagel, "Specific Immune Lysis of Paramyxo Virus-Infected Cells by H-2-Compatible Thymus-Derived Lymphocytes," *Immunology* 31 (1976): 27-32.

19. S. Kvist et al., "Molecular Association between Transplantation Antigens and a Cell-Surface Antigen in an Adenovirus Transformed Cell Line," *Proceedings of the National Academy of Sciences,* in press.

20. H. O. Sjögren, J. Minowada, and J. Ankerst, "Specific Transplantation Antigens of Mouse Sarcomas Induced by Adenovirus Type 12," *Journal of Experimental Medicine* 125 (1967): 689-701.

21. P. H. Gallimore, P. A. Sharp, and J. Sambrook, "Viral DNA in Transformed Cells. II. A Study of the Sequences of Adenovirus 2 DNA in Nine Lines of Transformed Rat Cells Using Specific Fragments of the Viral Genome," *Journal of Molecular Biology* 89 (1974): 49-72.

THE ROLE OF HOST PROTEASES IN THE INTERACTIONS BETWEEN THE MEMBRANES OF VIRUSES AND CELLS *

Andreas Scheid

Several groups of viruses, subsumed under the term enveloped viruses, possess a membrane that is composed of a lipid bilayer and virus-specified protein. In myxoviruses and paramyxoviruses, the lipid bilayer is derived from the host cell plasma membrane during virus maturation, and three proteins are associated with the membrane: [1] a nonglycosylated protein (M protein) lines the inner surface of the membrane, and two glycoproteins (HA and NA in myxoviruses, HN and F in paramyxoviruses), form spike-like projections on the outer surface. Such virus membranes are regarded as advantageous models for the study of the structure and biogenesis of biological membranes. Several features make this system attractive for analysis of such general questions as the biosynthesis of membrane proteins or protein lipid interactions; [2] viral membranes contain a small number of protein species with great abundance of the individual proteins, as opposed to the very complex protein composition, such as the plasma membrane, encountered in cellular membranes; and protein, lipid, and carbohydrate can be varied independently by the selection of the appropriate cell type, virus, and growth medium.

In addition to serving as a model for the study of general

* Research described in this paper was supported by research grants PCM78-09091 from the National Science Foundation; CA-18213 from the National Cancer Institute; and AI-05600 from the National Institute of Allergy and Infectious Diseases.

phenomena in membrane biology, the viral membrane is a target for the legitimate curiosity of the virologist because of its important functions in the life cycle of the virus. Thus the membranes of myxoviruses and paramyxoviruses carry the activity for attachment of the virus to cells (hemagglutinating activity); and the enzyme neuraminidase, which can destroy the receptor for virus attachment by hydrolysing the ketosidically linked neuraminic acid on glycoproteins or glycolipids, and which is important for virus release. The viral membrane also carries activity required for the entry of the virus genome into the target cell, that is, for virus penetration; and, in the case of paramyxoviruses, two other biological phenomena, virus-induced hemolysis and cell-cell fusion, are functions of the virus membrane.

The activity of the paramyxovirus membrane responsible for penetration and for hemolysis and cell-cell fusion requires proteolytic processing by cellular protease to become active, and the presence or absence of a suitable protease may decide whether a tissue or a host is susceptible to infection by a given virus. Since this observation was made in our laboratory [3] and in that of M. Homma,[4] we have continued an active interest in the proteolysis dependence of viral infectivity, cell-cell fusion, and hemolysis because of the biological consequences for host range and tissue tropism of these viruses, and because this finding has opened new approaches for studying the molecular mechanism of virus penetration and virus-induced membrane fusion. This report will summarize the more recent results, which are described in detail elsewhere.

THE PARAMYXOVIRUS PROTEIN RESPONSIBLE FOR VIRUS ATTACHMENT

To obtain individual proteins of the paramyxovirus membrane for characterization of their biological activity, for immunogens, or for studies of protein structure, procedures have been developed that allow the isolation of the proteins without detectable denaturation.[5-7] In the presence of nonionic detergents, such as Triton X-100, the membrane is disrupted and the glycoproteins are selectively solubilized. This procedure can be modified to also allow the isolation of the nonglycosylated membrane protein, M, which is solubilized with the detergent extraction if salt is present at this stage, and which then can be selectively made to precipitate by lowering the ionic strength through dialysis. Separation of the two glycoproteins from each other was accomplished initially by velocity sedimentation on sucrose gradients, and subsequently by affinity chromatography. These procedures give individual glycoproteins in good yield, and biological activity can be demonstrated under suitable conditions, as will be indicated later.

One of the two isolated glycoproteins, the HN protein, could be identified as the active receptor-binding (hemagglutinin) protein, and the same protein possesses the neuraminidase activity.[8, 9] The neuraminidase is fully active even in the presence of the detergent, and hemagglutinating activity, which requires the multivalency of the hemagglutinating agent, becomes demonstrable when the HN protein is allowed to aggregate by removal of the detergent. Both activities involve specificity for ketosidically linked N-acetylneuraminic acid, and the available evidence supports the view that, in paramyxoviruses, receptor-binding is a function of the active site of the enzyme neuraminidase.[10] This feature of paramyxoviruses, the association of both functions with one protein, is different from the situation found in influenza viruses, in which the hemagglutinating activity and neuraminidase are functions of two distinct spike glycoproteins.[11] In the natural environment of the virus there are a great variety of potential substrates for attachment and neuraminidase action; the biological significance of the neuraminidase is that it enables the virus to inactivate targets for virus attachment, which, in the absence of the enzyme, would bind the virus irreversibly, foiling its chances to bind to and penetrate cells. Such biological dead-end situations could result from virus attachment to extracellular glycoproteins, other virions, or the surface of the cell from which the virus has matured. The neuraminidase can also destroy receptors on the target cell surface, causing the attached virus to elute; during the phase of reversible attachment, however, other interactions occurring between the virus and the cell membrane may render the binding irreversible. The evidence discussed later suggests that in paramyxoviruses this event is fusion between the viral and cell membranes.

Like other spike glycoproteins of enveloped viruses, the HN protein is anchored in the virus envelope, and this is thought to result from the interaction of a hydrophobic portion of this protein, located at the base of the spike, with the lipid of the viral membrane. This has been inferred not only from the arrangement of the spike on the virion, but from the solubility properties of the spike glycoproteins.[12] The spike glycoproteins are soluble in the presence of nonionic detergents through micelle formation between the hydrophobic base and the detergent, and removal of detergents causes the spikes to aggregate by their bases into rosette-like structures. In the case of the HN protein one can demonstrate the hydrophobicity of the base directly, because this protein can be conveniently dissected with protease into a small hydrophobic and a large water-soluble segment.[13] Treatment of HN of the paramyxovirus SV5 (M.W. ~ 64,000) with chymotrypsin removes a polypeptide segment of ~ 5,000, and with the loss of this small segment the larger product of cleavage (M.W. ~ 59,000) has become soluble in the absence of detergent. The large fragment still contains the active site of the neuraminidase, but it cannot hemagglutinate because of the inability to aggregate into multivalent structures. This suggests that the small segment represents the portion of the molecule that is responsible for the aggregation of the intact HN, as well as for insertion of the spike into the lipid bilayer of the virus envelope. The extent of the hydrophobic interaction between spike and lipid is not known; the size of the hydrophobic peptide, however, is sufficient to span the entire lipid bilayer.

THE PARAMYXOVIRUS PROTEIN RESPONSIBLE FOR VIRUS PENETRATION AND MEMBRANE FUSION

When the paramyxovirus glycoproteins were first available in purified form, biological activity could be demonstrated only with the HN protein, not with the other glycoprotein, which since then has been dubbed F. At that time we hypothesized that the

F protein may be involved in the other activities of the virus envelope,[14] that is, virus-induced hemolysis and cell-cell fusion. Since then, evidence has proven this hypothesis to be correct. The initial failure to demonstrate activity with the isolated F protein is due to the fact that it must be inserted into lipid to express activity, and it has recently been possible to obtain hemolysis-reactive particles by reconstitution from phospholipid and purified F protein.[15]

The involvement of the F protein in hemolysis, in cell-cell fusion, and in the initiation of infection first became apparent in studies with a defective system in which lack of these activities in the virus progeny could be correlated with lack of processing of the F protein.[16, 17] Thus when Sendai virus is grown in the usual laboratory host, that is, in the allantoic sac of the chick embryo, the virus possesses all the activities usually associated with paramyxoviruses; when the virus was grown in bovine kidney (MDBK) cells, however, the virus progeny is noninfectious and lacks hemolyzing and cell-fusing activities. Comparison of the proteins in the viruses from the two hosts revealed that the MDBK-grown virus contained a glycoprotein (F_0), which is larger than the corresponding protein in the egg-grown virus (F_1), and that treatment of F_0-containing virus with trypsin cleaves this protein to yield F_1. Most important, cleavage of F_0 rendered the MDBK-grown virions infectious and active in hemolysis and cell-cell fusion. Thus activation of these processes is dependent on cleavage of the F protein by a cellular enzyme, which is present in the chick embryo but not in MDBK cells, and the cleaved F protein must be involved in infectivity, virus-induced hemolysis, and cell-cell fusion.[18]

An important clue with regard to the mechanism of virus penetration can be derived from the fact that cleavage of F is a requirement common to infectiousness and hemolyzing as well as to the fusing activity of the virus. The step in infectivity that involves the F protein must be beyond virus attachment, which is not affected by the cleavage of F, and before transcription of the RNA; therefore this step must be penetration. The only conceivable process that could jointly account for the expression of all three activities is fusion between viral and target membranes in all three instances (Figure 1). The participation of the F protein in these activities provides strong support for the previously advanced hypothesis that penetration involves fusion between viral and cell membranes.[19]

PROTEIN CLEAVAGE AS A DETERMINANT OF HOST RANGE AND VIRULENCE

The requirement for F protein cleavage for virus infectivity has important implications for the host range of paramyxo-

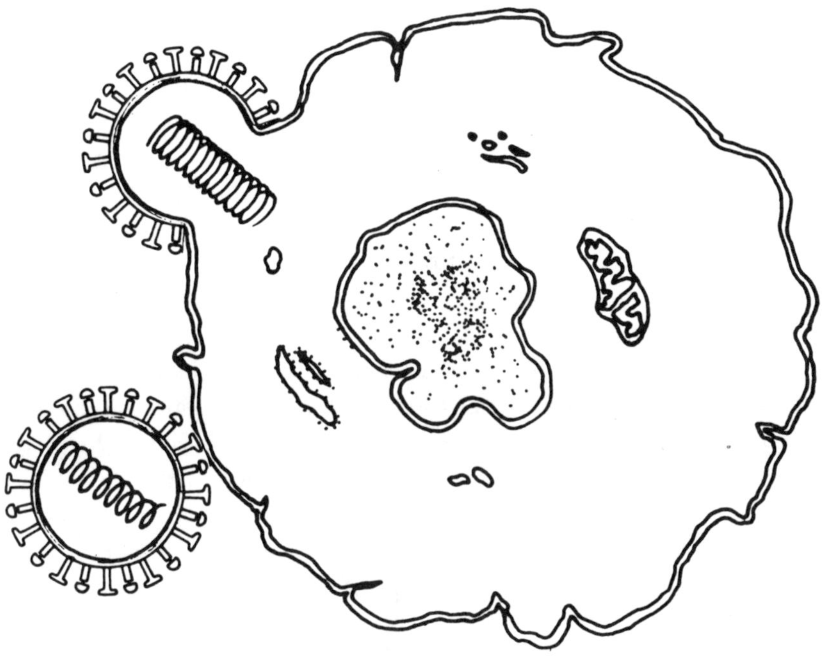

Figure 1. Diagram of the role of the F protein in the biological activities of paramyxoviruses. The F protein spikes cause the membrane of attached virus (*lower left*) to fuse with the plasma membrane of the cell (*upper left*). This introduces the helical nucleocapsid, which contains the genome RNA, into the cytoplasm (virus penetration). Also, this mechanism results in the insertion of viral membrane into the plasma membrane, which can lead to leakiness, for hemoglobin (hemolysis), for example, or to fusion with the plasma membrane of another cell (cell-cell fusion). Fusion only occurs with virions containing F that has been cleaved by host protease. If the F protein has not been cleaved, fusion between the virus and the cell membrane does not take place, even though virus attachment, mediated by the another spike glycoprotein, HN, is possible.

viruses.[20, 21] The two host systems described earlier may illustrate the two extreme situations. In the permissive system, that is, the chick embryo, the virus produced is fully active and can penetrate other cells, and virus spread and multiple cycle replication are possible. In contrast, the lack of suitable protease renders a host, such as MDBK cells, nonpermissive. The cells can be infected if active virus is introduced from an outside source. Because the progeny virus contains F protein that is not cleaved, however, it is not able to infect other cells; infection of the tissue stops after a single cycle of virus replication. Thus permissiveness of a host for virus replication is dependent on a match between the protease available in a host and the susceptibility to cleavage of the F protein.

The match between host protease and F protein susceptibility could be an important determinant for host range if the specificities are limited in both cases, that is, if the range of protease specificities in the host is limited, and if the susceptibility of the F protein is restricted to a limited range of protease specificities. Such specificity does indeed exist, as was first demonstrated with wild type (wt) Sendai virus and mutants with altered susceptibilities of proteases.[22, 23] As discussed earlier, trypsin cleaves the F_0 protein of Sendai virus, but other proteases such as chymotrypsin and elastase do not cleave F, nor do they activate the virus. It has been possible to isolate mutants of Sendai virus cleaved by proteases that do not activate the wt virus. To obtain such mutants, virus was mutagenized with nitrous acid and passed in MDBK cells in the presence of specific proteases, such as chymotrypsin or elastase. Virus was selected that could replicate through multiple cycles in the presence of those enzymes. Such mutants, designated protease activation (pa) mutants, were isolated and characterized with regard to the susceptibility of their F protein to cleavage and activation by various proteases. One mutant, pa-cl, isolated in the presence of chymotrypsin, is activated by chymotrypsin or by elastase; this mutant has lost the susceptibility of wt virus to trypsin. With elastase present during mutant selection, two classes of mutants were obtained, one activated by elastase only, the other by either elastase or by trypsin. Other pa mutants obtained are activated by plasmin or thermolysin.

Several important biological implications arise from the isolation and the properties of these mutants. They establish the validity of the conclusions derived from the results with wt virus, that is, that host-dependent cleavage of F_0 is required for the activation of infectivity, cell fusion, and hemolysis, and that the host range and tissue tropism of a given virus is determined by the availability of the appropriate protease in the host. In addition to documenting the restriction in F protein susceptibility to protease, the pa mutants have provided tools to assay the range of protease specificities available in an intact host.[24] Thus, as indicated, wt Sendai virus grows well in the allantoic sac of the chicken embryo; mutuant pa-cl cannot undergo multiple cycle replication in the embryo, however, because the appropriate enzyme is not available. Only if activating enzyme is supplied by injecting chymotrypsin or elastase into the allantoic sac at the time of infection does this mutant grow to high titers. The allantoic membrane apparently contains a protease with tryptic specificity, whereas elastolytic or chymotryptic specificities are lacking. This approach should allow a detailed mapping of protease specificities in other tissues. These experiments with pa mutants in the chicken embryo demonstrate that the range of protease specificities in an intact host may be limited, and that mutation resulting in a change in F protein susceptibility can therefore result in a change in the host range of the virus.

The demonstration that permissiveness of a host for virus replication depends on the match between protease and F protein susceptibility has raised several interesting and potentially important possibilities. The infection by a paramyxovirus of a tissue or organ not usually affected by that virus could conceivably be due to either the emergence of a virus mutant with susceptibility to a protease present in that tissue, or to the occurrence of a protease not normally elaborated by that tissue. Another implication of practical importance concerns virus isolation and the possible effect of the inherent change in hosts on the isolate. Thus when virus is isolated from a human or from an infected animal, the laboratory host may select for a mutant or minor component of the population, rather then yield the virus that was predominant in the initial isolate. This might be one factor in the rapid attenuation for man that occurs upon isolation and passage of virus in cell culture.

Based on the observations with Sendai virus and the protease activation mutants, we postulated that virus virulence in the natural host could depend on the susceptibility of the virus to cleavage,[25] and suggested as examples the virulent and avirulent strains of Newcastle disease virus (NDV), which had been isolated over the years and which up to that time had resisted efforts to identify the basis for the differences in virulence for the chicken. Dependence of virulence on susceptibility to cleavage was subsequently clearly shown by Y. Nagai and coworkers,[26] who examined both virulent and avirulent strains of NDV and found that in all instances virulence correlated with the ability of chick embryo fibroblasts to cleave the F protein of the virus. These investigators also found that in two of the avirulent strains, not only the F protein but a HN precursor failed to be cleaved; such a HN precursor has not, however, been found with other paramyxoviruses. Thus virus virulence in the adult animal is dependent on proteolytic activation of the virus, as predicted in previous studies in cell culture and chick embryos.

The implications of the host-dependent proteolytic activation of infectivity extend beyond host range, tropism, and virulence. The question arises as to whether, in some persistent infections, the failure to produce infectious virus may be due to the lack of the protease required for virus activation. Persistent virus infections in which little or no infectious virus is produced are common in cultured cells; they also occur in animals, and at least in one instance in man, that is, in subacute sclerosing panecephalitis due to persistent infection with measles virus. Rescue of virus from persistent infections has on occasion been accomplished by cocultivation of the infected tissue with other cells, and one function that could be supplied by the "permissive cell" is the protease capable of activating the virus.

THE CLEAVAGE SITE OF THE F PROTEIN

Because of the biological importance of the cleavage we have investigated the structure of this protein further, with emphasis on the region of the molecule around the cleavage site.[27,28] It could be shown with three different paramyxoviruses, Sendai, SV5, and NDV, that cleavage of the precursor F_0 (M.W. ~ 65,000) yields two polypeptides, F_1 (M.W. ~ 50,000) and F_2 (M.W. ~ 15,000), which are linked by disulfide bonds. Both polypeptides have carbohydrate side chains. The orientation of the molecule with regard to the viral membrane is shown in Figure 2. The hydrophobic base by which it is attached to the membrane contains the end of F_1; the amino terminus, which is blocked in Sendai virus, is found on F_2.

To further investigate the structure at the cleavage site, the amino acid sequence has been determined, beginning at the N-terminus, which is generated on F_1 by cleavage. The F_1 polypeptide of three different viruses has been investigated, and the first sixteen residues are as follows:

SV5 Phe-*Ala*-Gly-*Val*-Val-Ile-Gly-*Leu*-*Ala*-Ala-Leu-Gly-Val-Ala-Thr-Ala

Sendai Phe-*Phe*-Gly-Ala-Val-Ile-Gly-*Thr*-*Ile* -Ala-Leu-Gly-Val-Ala-Thr-Ala

NDV Phe-*Ile* -Gly-Ala-*Ile* -Ile-Gly-*Gly* -*Val*-Ala-Leu-Gly-Val-Ala-Thr-Ala

The N-terminal sequences at the cleavage site are strikingly similar. Variations occur only at five positions, and most of the differences can be explained by point mutations. Thus the N-terminal sequences of F_1 are homologous, and proteolytic cleavage has occurred at identical sites in the three viruses. The high degree of conservation in these sequences suggests that specific sequence in this region is necessary for biological activity; additional evidence in support of this has been obtained with the pa mutants. The altered protease susceptibility of these mutants could be due to a change in tertiary structure, making a different cleavage site available, or amino acid substitution could have led to an altered

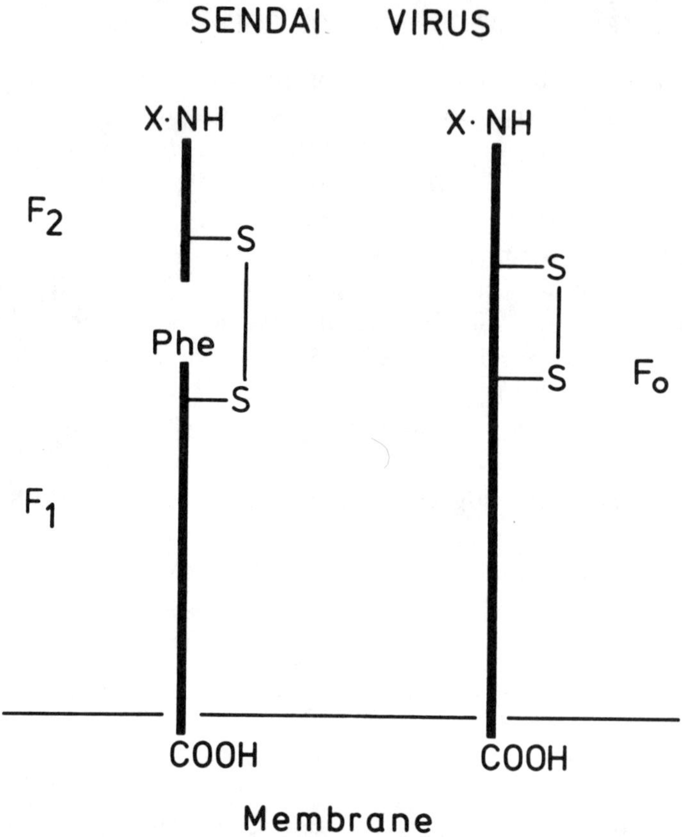

Figure 2. Diagram of the structure of the F protein, as discussed in the text. Proteolytic cleavage, which activates F, involves a single cut in the polypeptide chain of the precursor F_0, exposing phenylalanine as a new N-terminus on the F_1 polypeptides in Sendai virus as well as in the two other paramyxoviruses.

susceptibility at the same position where cleavage occurs in the parent virus. Studies with pa-cl showed a N-terminal sequence of F_1 identical to that of wt virus, supporting the conclusions derived from the comparison of the three different paramyxoviruses, that is, the apparent requirement for cleavage at a sharply defined site and the need for the specific amino acid sequence for the expression of biological activity.

Participation of this region of the F molecule in the reaction with target membranes is suggested, not only by the conservation of this sequence in the different viruses but by the hydrophobic nature of this polypeptide stretch. There is no charged amino acid in this region, and the amino acid residues provide for a hydrophobicity not usually found in sequences of such length in other proteins. The hydrophobicity of the newly exposed N-terminus now raises the possibility that the mode of F protein action on membranes involves a direct interaction of this region with the target membrane lipid bilayer.

This view is further supported by another line of evidence. We noted that oligopeptide inhibitors that prevent cell fusion and virus penetration by measles virus, SV5, and Sendai virus [29-31] resemble in their amino acid sequence the N-terminal region of F_1, and reasoned that their inhibitory activity might be due to this similiarity. To investigate this possibility, oligopeptides analogous to the N-termini of F_1 have been synthesized, and the structural requirements with regard to amino acid sequence, chain length, and N-terminal terminal groups on the synthetic oligopeptides are being investigated.[32] Such studies so far have confirmed the necessity for specific sequences in the oligopeptide inhibitor, and promise to shed light on the mechanism of paramyxovirus-induced membrane fusion, as well as on the mechanism of membrane fusion in general.

The significance of the cleavage region for F protein activity is supported by another line of evidence derived from extensive work done with influenza viruses. In influenza viruses one of the spike glycoproteins, the hemagglutinin (HA), is cleaved by cellular proteases or by the serum protease plasmin into two disulfide-linked subunits (HA_1 and HA_2).[33,34] Even though this cleavage does not affect the hemagglutinating activity of HA, it could be shown that, with regard to penetration, the situation is analogous to the need for F protein cleavage in paramyxoviruses, that is, that cleavage of the influenza virus HA is necessary for infectivity.[35,36] Comparison of the N-termini at the cleavage site of F and HA [37] shows a significant homology in the first eight residues, indicating a functional relatedness of the proteins of these different groups of viruses, and further emphasizing the biological importance of the cleavage region of these molecules for the expression of biological activity.

SUMMARY

Interactions of paramyxoviruses with cell membranes are mediated by two virus-coded glycoproteins of the virus membrane, HN and F. The HN protein is responsible for the attachment of virions to the membrane of the target cell. The F protein mediates fusion between the virus membrane and the cell membrane, an essential step for the infection process, as well as for the expression of other biological activities of paramyxoviruses, that is, cell-fusion and hemolysis. The activity of F is dependent on the proteolytic processing of a precursor (F_0) to yield two disulfide-linked subunits (F_1 and F_2); unless this cleavage occurs the virions are inactive and not infectious. Viruses differ in their susceptibility to specific proteases, as demonstrated with Sendai virus and its protease activation (pa) mutants. Findings with these viruses have indicated that the host range, tissue tropism, multiple-cycle replication in a host, and ability of paramyxoviruses to spread within a host and to cause disease are dependent on the availability of an appropriate protease capable of cleaving the viral glycoprotein. The importance of such cleavage in the virulence of virus in animals has been demonstrated with Newcastle disease virus in chickens.

Studies on the structure of the paramyxovirus glyproteins indicate they are inserted with a hydrophobic region of the molecule in the lipid bilayer of the virus membrane. Analysis of the primary structure of the F protein at the cleavage site reveals striking sequence homology among three different paramyxoviruses, indicating a requirement for cleavage at a sharply defined site and the importance of a specific primary sequence for biological activity. This region is highly hydrophobic, suggesting that it may interact with the lipid bilayer of the target cell membrane during virus penetration cell fusion or hemolysis. Oligopeptides that resemble the N-terminal sequence at the cleavage site can act as inhibitors of the activities. Further studies of this system should provide knowledge regarding the mechanisms involved in viral penetration and in membrane fusion in general.

NOTES

1. P. W. Choppin and R. W. Compans, "Reproduction of Paramyxoviruses," in *Comprehensive Virology,* vol. 4, ed. H. Fraenkel-Conrat and R. R. Wagner (New York: Plenum Press, 1975): 95-178.

2. P. W. Choppin et al., "Structure and Assembly of Viral Membranes," in *Membrane Research,* ed. C. F. Fox (New York: Academic Press, 1972): 163-79.

3. A. Scheid and P. W. Choppin, "Identification of Biological Activities of Paramyxovirus Glycoproteins. Activation of Cell Fusion, Hemolysis and Infectivity by Proteolytic Cleavage of an Inactive Precursor Protein of Sendai Virus," *Virology* 57 (1974): 475-90.

4. M. Homma and M. Ohuchi. "Trypsin Action on the Growth of Sendai Virus in Tissue Culture Cells. III. Structural Difference of Sendai Viruses Grown in Eggs and in Tissue Culture Cells," *Journal of Virology* 12 (1973): 1457-65.

5. A. Scheid et al., "Isolation of Paramyxovirus Glycoproteins. Association of Both Hemagglutinating and Neuraminidase Activities with the Larger SV5 Glycoprotein," *Virology* 50 (1972): 640-52.

6. A. Scheid and P. W. Choppin, "Isolation and Purification of the Envelope Proteins of Newcastle Disease Virus," *Journal of Virology* 11 (1973): 263-71.

7. ———, "The Hemagglutinating and Neuraminidase Protein of a Paramyxovirus: Interaction with Neuraminic Acid in Affinity Chromatography," *Virology* 62 (1974): 125-33.

8. Scheid et al., "Paramyxovirus Glycoproteins" (See note 5).

9. Scheid and Choppin, "Purification of Envelope Proteins" (See note 6) .

10. ———, "Protein of a Paramyxovirus" (See note 7).

11. D. Bucher and P. Palese, "The Biologically Active Proteins of Influenza Virus: Neuraminidase," in *The Influenza Viruses and Influenza,* ed. E. D. Kilbourne (New York: Academic Press, 1975): 83-123.

12. Choppin and Compans, "Reproduction of Paramyxoviruses" (See note 1).

13. Scheid et al., "Studies on the Structure and Function of Paramyxovirus Glycoproteins," in *Negative Strand Viruses and the Host Cell,* ed. B. W. J. Mahy and R. D. Barry (London: Academic Press, 1978): 181-93.

14. A. Scheid et al., "Isolation of Paramyxovirus Glycoproteins" (See note 5).

15. M.-C. Hsu, A. Scheid, and P. W. Choppin, "Reconstitution of Membranes with Individual Paramyxovirus Glycoproteins and Phospholipid in Cholate Solution," *Virology* 95 (1979): 476-91.

16. Scheid and Choppin, "Identification of Biological Activities" (See note 3).

17. Homma and Ohuchi, "Trypsin Action" (See note 4).

18. Scheid and Choppin, "Identification of Biological Activities" (See note 3).

19. C. Morgan and C. Howe, "Structure and Development of Viruses as Observed in the Electron Microscope. IX. Entry of Parainfluenza Virus Type I (Sendai) Virus," *Journal of Virology* 2 (1968): 1122-32.

20. Scheid and Choppin, "Identification of Biological Activities" (See note 3).

21. S. M. Silver, A. Scheid, and P. W. Choppin, "Loss on Serial Passage of

Rhesus Monkey Kidney Cells of Proteolytic Activity Required for Sendai Virus Activation," *Infection and Immunity* 20 (1978): 235-41.

22. A. Scheid and P. W. Choppin, "Protease Activation Mutants of Sendai Virus: Activation of Biological Properties by Specific Proteases," *Virology* 69 (1976): 265-77.

23. ———, "Activation of Cell Fusion and Infectivity by Proteolytic Cleavage of a Sendai Virus Glycoprotein," in *Proteases and Biological Control,* ed. E. Reich, D. B. Rifkin, and E. Shaw (Cold Spring Harbor, New York: Cold Spring Harbor Laboratory, 1975): 645-59.

24. ———, "Protease Activation Mutants" (See note 22).

25. ———, "Activation of Cell Fusion" (See note 23).

26. Y. Nagai, H.-D., Klenk, and R. Rott, "Proteolytic Cleavage of the Viral Glycoproteins and Its Significance for the Virulence of Newcastle Disease Virus," *Virology* 72 (1976): 494-508.

27. Scheid et al., "Paramyxovirus Glycoproteins" (See note 13).

28. A. Scheid and P. W. Choppin, "Two Disulfide-Linked Polypeptide Chains Constitute the Active F Protein of Paramyxoviruses," *Virology* 80 (1977): 54-66.

29. F. Nicolaides et al., "Potential Antiviral Agents. Carbobenzoxy Di- and Tri-Peptides Active Against Measles and Herpes Virus," *Journal of Medical Chemistry* 11 (1968): 74-79.

30. E. Norrby, "The Effect of a Carbobenzoxy Tripeptide on the Biological Activities of Measles Virus," *Virology* 44 (1971): 599-608.

31. M. C. Graves, S. S. Silver, and P. W. Choppin, "Measles Virus Polypeptide Synthesis in Infected Cells," *Virology* 86 (1978): 254-63.

32. C. D. Richardson, A. Scheid, and P. W. Choppin, "The Inhibition of Paramyxovirus Replication by Oligopeptides," *Abstracts of the American Society of Microbiology,* in press.

33. S. G. Lazarowitz, R. W. Compans, and P. W. Choppin, "Influenza Virus Structural and Non-Structural Proteins in Infected Cells and Their Plasma Membranes," *Virology* 46 (1971): 830-43.

34. S. G. Lazarowitz, A. R. Goldberg, and P. W. Choppin, "Proteolytic Cleavage by Plasmin of the HA Polypeptide of Influenza Virus," *Virology* 56 (1973): 172-80.

35. S. G. Lazarowitz and P. W. Choppin, "Enhancement of the Infectivity of Influenza A and B Viruses by Proteolytic Cleavage of the Hemagglutinin Polypeptide," *Virology* 68 (1975): 440-54.

36. H.-D. Klenk et al., "Activation of Influenza A Viruses by Trypsin Treatment," *Virology* 68 (1975): 426-39.

37. J. J. Skehel and M. D. Waterfield, "Studies on the Primary Structure of the Influenza Virus Hemagglutinin," *Proceedings of the National Academy of Sciences* 72 (1975): 93-97.

MEASLES VIRUS PROTEINS AND RECEPTORS *

Calderon Howe

Persistent infection with rubeola (measles) virus is now recognized as a major factor in the pathogenesis of subacute sclerosing panencephalitis (SSPE). An analogous relationship to infection with measles virus, and perhaps other viral agents, is strongly suspected in multiple sclerosis. In the latter condition, pathogenic mechanisms are evidently further complicated by genetic considerations centering on the major histocompatibility complex.

These important developments have heightened interest in the molecular biology of measles virus.[1] The virus closely resembles paramyxoviruses in its capacity to agglutinate and to lyse erythrocytes and to cause endogenous, infection-associated cell fusion (endofusion), as well as exogenous, rapid cell fusion (exofusion). Endofusion is a primary characteristic of the cytopathological effect caused by measles virus in cultured cells. Exofusion can be effected by exposure of cells to virus in high concentration. Several important characteristics set measles virus apart from the paramyxoviruses. The first is the absence of neuraminidase from the measles virion, in contradistinction to the paramyxoviruses, in which the major envelope glycoprotein (HN) contains neuraminidase as well as hemagglutinin. The presence of the enzyme accounts for the absence of N-acetylneuraminic acid (NANA) from the mature paramyxoviral particle.

* The work reported herein was supported by grant AI 10945 from the National Institutes of Health and by the Edward G. Schlieder Educational Foundation. The electron micrographs were taken by J. E. Coward. The author is indebted to T. Bächi, T. W. Fenger, and L. T. Lee for making available their unpublished data.

A second major difference between measles and paramyxoviruses is that, lacking neuraminidase, the virion contains NANA residues,[2] which, like other carbohydrates, are of cellular origin. Accordingly, treatment of measles virus with soluble neuraminidase abolishes its capacity to agglutinate erythrocytes and greatly reduces infectivity for Vero cells. While their location on the glycoprotein is not certain, NANA residues increase the efficiency of viral attachment to cells.

A third feature that sets measles virus apart from the paramyxoviruses has to do with corresponding cell receptors. Paramyxoviruses attach to cell surface components that contain terminally linked NANA and react with erythrocytes of avian as well as most mammalian species through similar receptor groupings. Measles virus, while it does not require NANA in cell receptors, agglutinates only primate and not human or other mammalian or avian erythrocytes. Neuraminidase treatment of rhesus monkey erythrocytes greatly increases their susceptibility to agglutination by measles virus, and desialylation of soluble erythrocyte glycoprotein does not alter its capacity to inhibit measles viral hemagglutination. Together, these findings suggest that the receptors for measles virus on primate erythrocytes differ qualitatively from those for paramyxoviruses.

We will consider here the results of interrelated investigations of the hemagglutination-hemolysis-fusion activities of measles virus from three standpoints: 1) analysis of protein constituents on or in the viral envelope, which mediate viral interactions with cells; 2) attempts to identify cell receptors for virus, particularly those on erythrocytes; and 3) examination, by immune electron microscopy, of the interactions of measles virus with primate erythrocytes. Since erythrocytes do not support viral replication, interactions are restricted to viral attachment and penetration through the membrane, and are therefore analogous to the first steps in the initiation of productive viral infection in susceptible nucleated cells.

It has been the general experience that rigorous purification of measles virus is extremely difficult, due to the fragility and pleomorphism of the virus. Moreover, the physical methods most often used to dislodge virions from infected cells invariably yield membrane fragments containing viral envelope antigens. These biologically active subunits may to varying extent contaminate even gradient "purified" virus. We have therefore attempted to identify, by immune precipitation, the virus-coded envelope proteins as they occur on the surface of infected cells.[3]

Rabbit antiserum was prepared to the LEC strain of measles virus. Before use, the antiserum was extensively absorbed with lyophilized Vero cells, as well as with monolayers of uninfected Vero cells. For analysis of viral antigens, monolayers of Vero cells were infected with measles virus at a multiplicity of ten. After incubation for thirty to fifty-four hours the cells were washed, resuspended in PBS, and surface-labeled with ^{125}I by the lactoperoxidase method.[4] They were then solubilized in 0.5 percent NP-40 for fifteen minutes at room temperature. Nuclei were removed by low-speed centrifugation, and the low-speed supernatants were centrifuged at 85,000 $\times$ g. The supernatants containing most of the solubilized membrane material were mixed with antiserum and stored overnight at 4°C. Resulting immune precipitates were pelleted by centrifugation of the primary mixture at 85,000 $\times$ g for sixty minutes through a 20 percent sucrose barrier. Precipitates were solubilized in SDS and 2-mercaptoethanol, or in SDS alone, and analyzed by polyacrylamide gel electrophoresis (PAGE). Autoradiography of gels was carried out by standard procedures.

Table I summarizes our findings, along with those of other workers who have analyzed glycoproteins in virions by SDS-PAGE. It can be seen that the various direct analyses of virions, including our own, are in general agreement. Besides the nucleoprotein and M protein, reducing (R) gels reveal a major envelope glycopolypeptide of M.W. 76-80k; the hemagglutinin (H), an unglycosylated polypeptide (F_1, M.W. 42-43k) distinct from cell actin; and a smaller glycopeptide (F_2, approximately 15-20k). In nonreducing (NR) gels the latter two migrate as a single band in the 58-62k region.[5,6]

Our analyses of virus-coded antigens on the surface of infected cells are in substantial agreement with the results of

PAGE analysis of purified virus. The major glycoprotein, which because of uncertainty as to its identity we have designated VGP1, is probably the viral H glycoprotein of the same molecular weight as that found on isolated virus. In a time-sequence analysis from four to fifty-four hours postinfection, VGP1 and 2 were expressed together as early as twenty-four hours. VGP2, however, was not seen before twenty-four hours. In cells metabolically labeled with ^{14}C glucosamine or ^{3}H fucose twenty-four to forty-eight hours postinfection, both VGP1 and 2 were labeled, indicating that both were glycosylated. VGP1 and 2 were also expressed on the surface of HEp-2 cells infected with the Edmondston strain of measles virus (Figure 1), and in the membranes of Vero and HeLa cells persistently infected with the LEC strain of measles virus (not shown).

Human immunoglobulin, which had a higher titer of measles HI antibody, precipitated the same two antigens from solubilized membranes of Vero cells productively infected with LEC strain virus. The same preparation of human immunoglobulin was conjugated with ferritin by J. E. Deas.

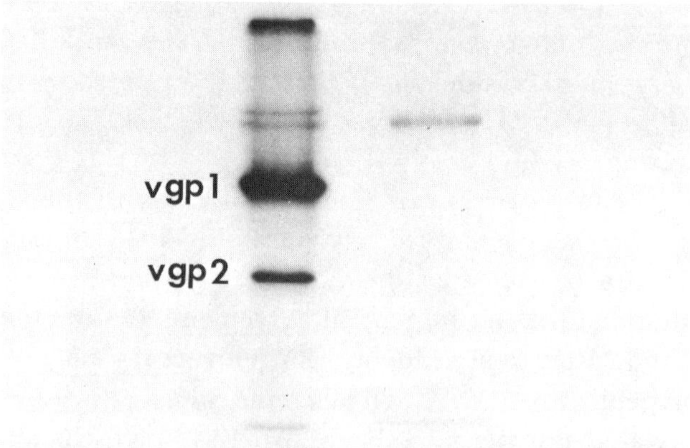

Figure 1. Measles viral envelope antigens in membranes of HEp-2 cells, surface labeled with ^{125}I forty-two hours after infection with virus (Edmonston strain). Labeled membranes were solubilized with NP40 and precipitated with rabbit antibody to virus. Specific precipitates were analyzed by SDS-PAGE autoradiography. *Left,* specific precipitate of infected cells showing VGP 1 (H) and VGP 2 (?F_1); *right,* normal uninfected cells.

Figure 2 shows a persistently infected Vero cell, on the surfact of which viral antigen is identified by direct reaction with the ferritin-labeled antibody. These morphological observations underscore the specificity of the precipitates obtained by reaction of human or rabbit antibody with virus-coded antigens on the surface of three different cell types infected with LEC or Edmondston strains of virus.

Analyses of surface components of cells following treatment with either trypsin or chymotrypsin suggest that VGP2 consists of at least two polypeptides, one of which may be derived from VGP1. Figure 3 shows autoradiographed reducing (R) and nonreducing (NR) PAGE patterns of specific precipitates derived from Vero cells labeled with ^{125}I and trypsinized forty-two hours

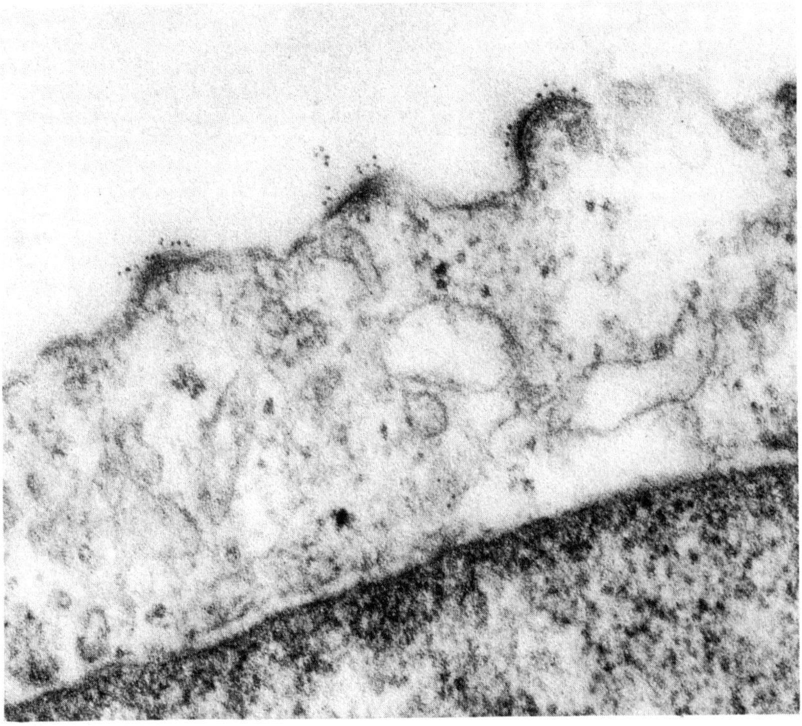

Figure 2. Vero cells persistently infected with measles virus. Viral antigen is identified at discretely thickened sites on the membrane by reaction with ferritin-labeled human antibody to virus. Magnification, x 95,500.

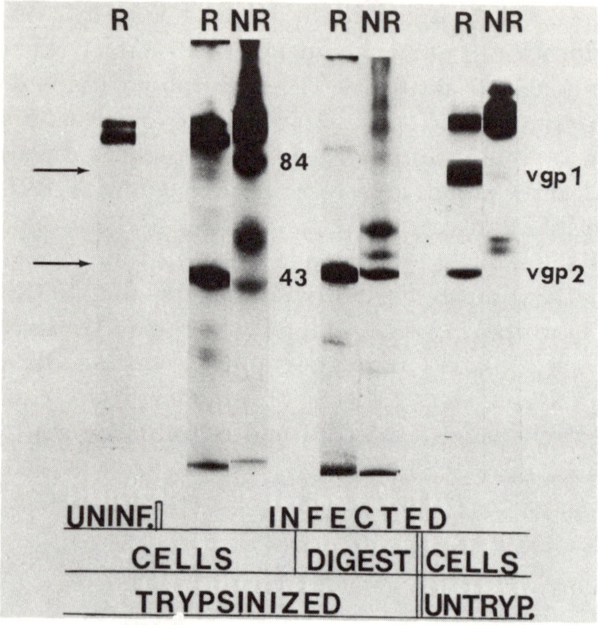

Figure 3. Normal uninfected Vero cells and cells thirty hours after infection with measles virus were labeled with ^{125}I and treated with trypsin (20 μg./ml.) in phosphate-buffered saline at pH 7.2. After removal of trypsin, specific precipitates of solubilized membranes and the tryptic digests were analysed by PAGE and radioautography, along with membranes of infected untrypsinized cells. R = reducing; NR = nonreducing conditions.

postinfection (LEC strain of virus). Immune precipitates of tryptic digests were similarly analyzed. As noted in Table 1, trypsinization removed precipitable VGP1 from the cells, and, as shown in Figure 3, VGP2 increased in intensity in R gels. Other workers also found that trypsinization of virus (20 μg. trypsin/ml.) yielded a 40k component and nothing in the H region.[7] At higher trypsin concentrations, however, only much smaller molecular weight components were recovered.

Comparison of the NR gel with the R gel of trypsinized cells reveals that, on reduction, the 84k protein yielded a fragment that comigrated with VGP2 (?F_1 protein, 43k). In the NR gel of the digest from infected cells, the 60k component may represent F protein, which on reduction also yields a 43k component (F_1)

and a smaller polypeptide (?F_2). This finding further substantiates the supposition that VGP2 consists of two components, only one of which, derived from VGP1, may be glycosylated. This would explain the discrepancy between our findings and those of other workers who have shown that only the smaller F_2 protein is glycosylated (Table I).

The foregoing cleavage patterns are subject to several possible interpretations, depending on the existence and distribution of interchain or intrachain S-S bonds within the complex(es) reactive with viral antibody. H may be the 200k component, which on reduction yields the 78k component. Despite tryptic cleavage of the molecule at or near the intrachain S-S junction, the molecule remains intact. On reduction in gels, the intrachain disulfide bonds are broken, yielding 43k and 35k fragments. An analogous interpretation has been invoked in structural studies of orthomyxoviral hemagglutinin.[8]

Alternatively, interchain disulfide bonds holding the macromolecule together may be proximal to the site of tryptic cleavage. This leaves an antigenic component of 84k attached to the cells. Distal portions, bearing antigenic determinants, are seen in R gels of precipitates from the tryptic digest. The 43k component seen in the R gel of the digest consists of the 43k component and two others of larger molecular weight that appear in NR gels of the same material. While the origin of these several 43k components is still not certain, they may all be derived from VGP1 or 2. In preliminary experiments we have obtained evidence that VGP1, after treatment with chymotrypsin, yields VGP 2.*

It appears, then, that, except for the presence of NANA residues in the envelope glycoprotein(s), measles virus resembles paramyxoviruses in having a major glycoprotein with H activity and a second glycoprotein (F), some or all of which may be involved in fusion reactions.

The receptor for measles virus has not been previously defined, although, as already indicated, NANA is not required as

* T. W. Fenger and C. Howe, 1978: unpublished observations.

it is in receptors for some of the paramyxoviruses. Taking into account the sharp species specificity of measles virus in agglutinating only primate erythrocytes, we have compared rhesus monkey erythrocyte ghost proteins with those of two other species (human and sheep) whose erythrocytes are not agglutinated. PAGE analysis of whole membranes from each of the three species showed consistent differences (Figure 4, Gh). Rhesus monkey erythrocyte membranes contain five bands that stain with PAS and with the resorcinol reaction for NANA. We have numbered the bands consecutively according to size. These designations do not correspond with the bands found in human preparations, in which PAS-1 is glycophorin A. Only three PAS bands were found in the sheep membrane preparations.

Water-soluble glycoproteins were isolated from all three species of erythrocytes by treatment of hemoglobin-free membranes with lithium diiodosalicylate-phenol (LIS-ph).[9] Each species of glycoprotein, after solubilization with LIS-ph, aggregated into a distinctive and reproducible pattern of larger mo-

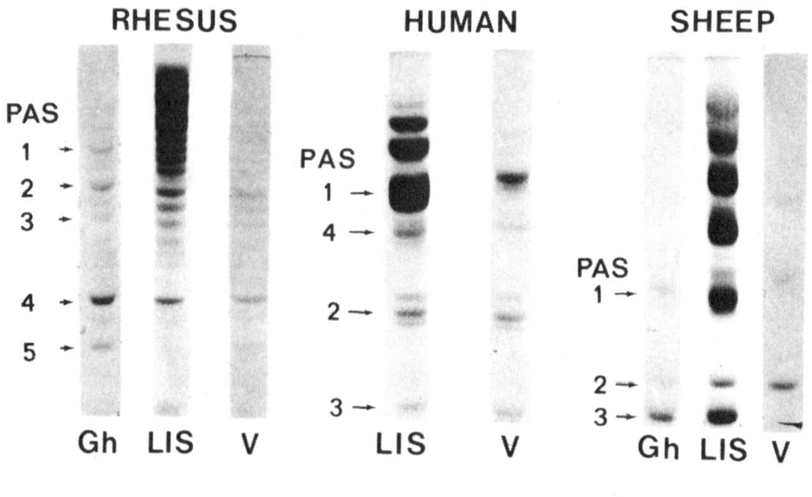

Figure 4. PAGE analysis (periodic acid Schiff [PAS] stain) of hemoglobin-free membranes (Gh) of simian, human, and sheep erythrocytes, and LIS-ph glycoproteins derived from them (LIS). Glycoproteins of each species were mixed with virus, which was then examined for adsorption of individual erythrocyte glycoprotein components (V).

lecular weight components (Figure 4, LIS). Of the three species of soluble LIS-ph glycoprotein, only that from rhesus monkey erythrocytes contained receptor activity, as measured by inhibition of viral agglutination of rhesus monkey erythrocytes (Table 2). We therefore isolated the individual glycoprotein bands by elution after primary separation in preparative PAGE.[10] Residual SDS was removed from eluted fractions by dialysis against isopropanol in TRIS buffer. Each of the eluted bands inhibited HA (Table 2), PAS 1, 2, and 4 being slightly more active than PAS 3 or 5. In PAGE analyses it was determined that the LIS-ph glycoproteins were composed of two "monomeric" forms designated PAS 4, which reaggregates to form PAS 1 and 2, and PAS 5, which reaggregates to form PAS 1 and 3.

To determine the distribution of reactivity with virus among the several components, LIS-ph glycoproteins were mixed with whole virus, layered over a 25 percent sucrose barrier, and centrifuged at 45,000 rpm for forty-five minutes. The resulting pellets, after being washed twice with PBS, were analyzed in SDS reducing gels. Proteins alone did not yield any pellet under the sucrose barrier, and the pellet of virus alone gave no PAS stain. Pellets recovered from the mixtures of virus with glyco-

TABLE 2. HI ACTIVITY OF ERYTHROCYTE GLYCOPROTEINS

Fraction	μG Inhibiting 8 HA Units Measles Virus
LIS-ph Glycoproteins	
Human	> 108
Rh monkey	12
Sheep	> 110
PAGE Eluates	
Rh monkey	
PAS 1	11
PAS 2	10
PAS 3	19
PAS 4	8
PAS 5	13
Human glycophorin A	> 50

proteins, however, contained PAS 2 and 4 adsorbed to virus (Figure 4, V). This reaction was inhibited by specific viral antibody. In further confirmation of this specific adsorption reaction, PAGE profiles of LIS-ph glycoproteins after absorption with virus (not shown) demonstrated a corresponding deficit of PAS 4 and a reduction of higher molecular weight components. This deficit was even more pronounced after absorption of the glycoproteins with infected cells. It thus appears that PAS 2 and 4 carry most of the receptor activity, since they bind preferentially to virus as well as to viral antigens in the membranes of infected cells.

Similar experiments were done with both human and sheep LIS-ph glycoproteins, with somewhat discordant results. Despite the fact that neither preparation initially inhibited viral agglutination of rhesus monkey erythrocytes, individual components in both species of glycoprotein adsorbed directly to virus. Only one glycoprotein component (PAS 2) was adsorbed from the sheep preparation (Figure 4, sheep, V). From the human LIS-ph glycoproteins, however, the upper portion of PAS 1 and some each of PAS 2, 3, and 4 adsorbed to virus (Figure 4, human, V). Analyses of human erythrocyte glycoproteins has revealed that PAS 2 and 3 are subunits of PAS 1,[11] the principal integral protein (glycophorin A). The viral adsorption pattern suggests that only the larger components in PAS 1 react with virus, and may therefore represent aggregated forms of PAS 2 and 3.

From the foregoing data it is evident that certain of the glycoproteins isolated in soluble form from human or sheep erythrocytes bind to virus, but do so in a manner that does not interfere with the ability of H glycoprotein on the virus to attach to rhesus monkey erythrocytes.

With the paramyxoviruses, exemplified by Sendai virus, attachment to cells through NANA-containing glycoprotein receptors is a necessary prelude to subsequent hemolysis/fusion or penetration. We used a monolayer system to analyze the reaction of measles virus with rhesus monkey erythrocytes,* and, for

* T. Bächi and C. Howe, 1978: unpublished observations.

comparison, the reaction of Sendai virus with human erythrocytes. Washed cells of either species were attached to plastic dishes precoated with poly-L-lysine. Viruses were adsorbed to monolayers in the cold for sixty to ninety minutes and then washed free of unadsorbed virus. On raising the temperature to 37°C, individual cells were lysed by measles virus and patches of contiguous cells underwent fusion to form stable polymembranes that retained some hemoglobin (Figure 5,b). Polymembranes increased in size and were stable after overnight incubation at 37°C (Figure 6). These reactions were inhibited by specific antibody to

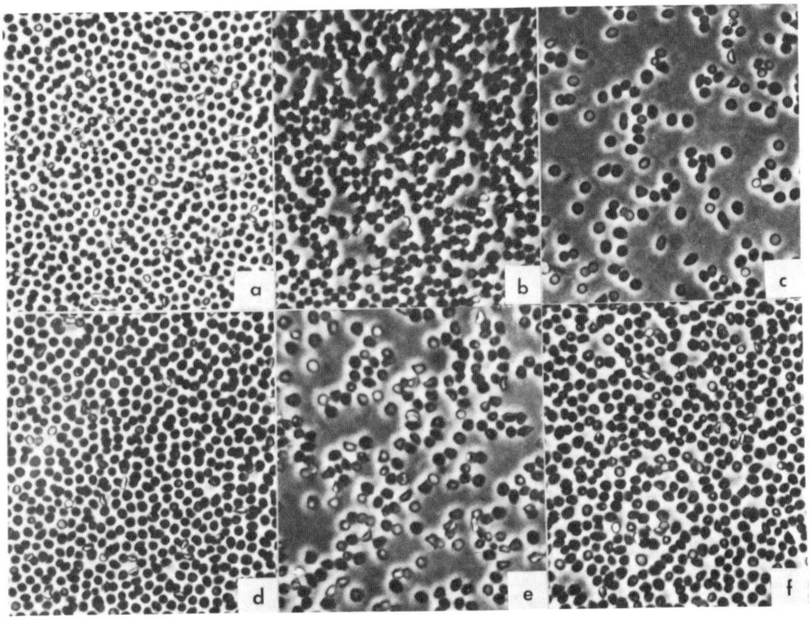

Figure 5. Simian (a, b, c) and human (d, e, f) erythrocytes attached in monolayers to plastic Petri dishes pretreated with poly-L-lysine (T. Bachi and C. Howe, 1979: unpublished observations). a) Normal simian erythrocytes; b) hemolysis/fusion of simian cells by measles virus; c) enhancement of measles virus hemolysis/fusion by pretreatment of cells with *Clostridium perfringens* neuraminidase; d) normal human erythrocytes; e) hemolysis/fusion of human cells by Sendai virus; and f) inhibition of Sendai virus hemolysis/fusion by pretreatment of cells with *Clostridium perfringens* neuraminidase.

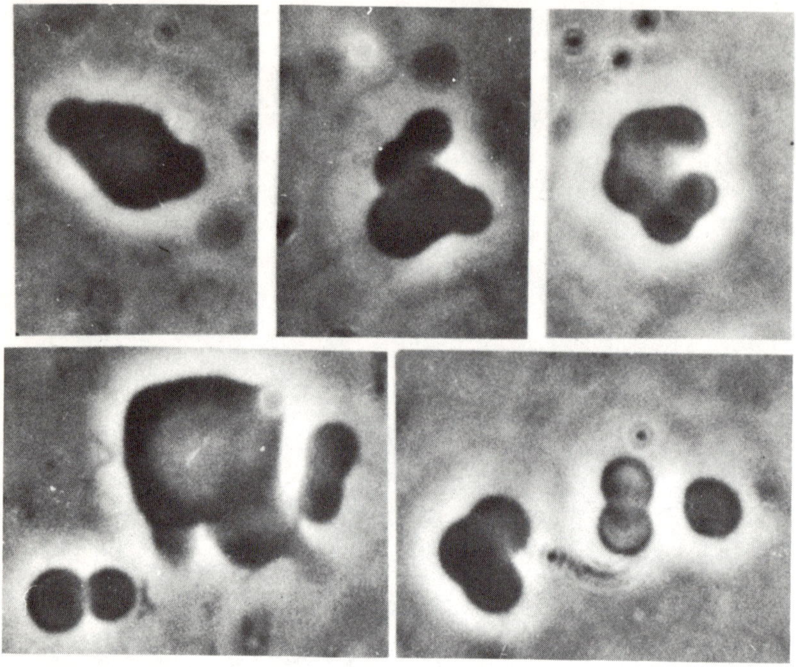

Figure 6. Polymembranes formed by fusion of rhesus monkey erythrocytes with concentrated measles virus at 37°C.

measles virus and by LIS-ph glycoprotein from rhesus monkey erythrocyte membranes (not shown).

Human erythrocytes to which Sendai virus had been adsorbed were similarly lysed and fused, as has been previously reported by others[12] (Figure 5,e). Hemolysis/fusion by Sendai virus, however, was inhibited by pretreatment of the human erythrocyte monolayers with neuraminidase (Figure 5,f). Neuraminidase treatment of the rhesus monkey erythrocyte monolayers to which measles was subsequently adsorbed did not inhibit the reaction, and indeed appeared to enhance it, as with hemagglutination (Figure 5,c). This result was in conformity with the divergent character of the receptors for measles and Sendai viruses, respectively.

The reaction of gradient purified measles virus (LEC strain) with simian erythrocytes in suspension has been analyzed by immune electron microscopy and freeze-etching.* Following agglutination of cells, hemolysis occurred to varying extent, accompanied by fusion of viral envelope to cell membrane, centrifugal intramembranal spread of viral envelope antigens, and discharge of nucleocapsid into the cytoplasm. These reactions resulted also in virus-induced cell-to-cell fusion and the formation of stable polymembranes that retained hemoglobin.

Figure 7, a shows the virions attached to cell membranes in the cold after direct staining with ferritin-labeled human antibody to virus. Only the viral particles are tagged with antibody. The envelope of the larger of the two particles has been damaged, allowing the penetration of antibody to the interstices of the virion. The other particle appears to be intact, the specific reaction with antibody being limited to the envelope. The apparent interposition of antibody between virion and cell surface can be ascribed to the plane of section. The curvature of the cell membrane, however, which conforms with the viral contour, suggests that both particles were attached.

Figure 7, b shows the reaction at a later stage of incubation at 37 °C. Antibody is excluded from one attachment site by close contact between virion and cell. In Figure 7, c, pieces of viral nucleocapsid are seen to be merging with the erythrocytoplasm. Some hemoglobin has escaped from one erythrocyte, as shown by the contrast in cytoplasmic density between the cells. At an even later stage no morphologically recognizable viral remnants are seen, yet antigen persists in patches along the apposed membranes, which are interrupted at several points where cell-to-cell fusion has occurred (Figure 7, d).

This sequence of reactions with erythrocytes shows that measles virus behaves in much the same manner as Sendai virus,[13] and reflects the general functional similarity of the corresponding viral envelope proteins. The results described with measles virus provide substantial evidence that the virus gains entry to

* T. Bächi and C. Howe, 1978: unpublished observations.

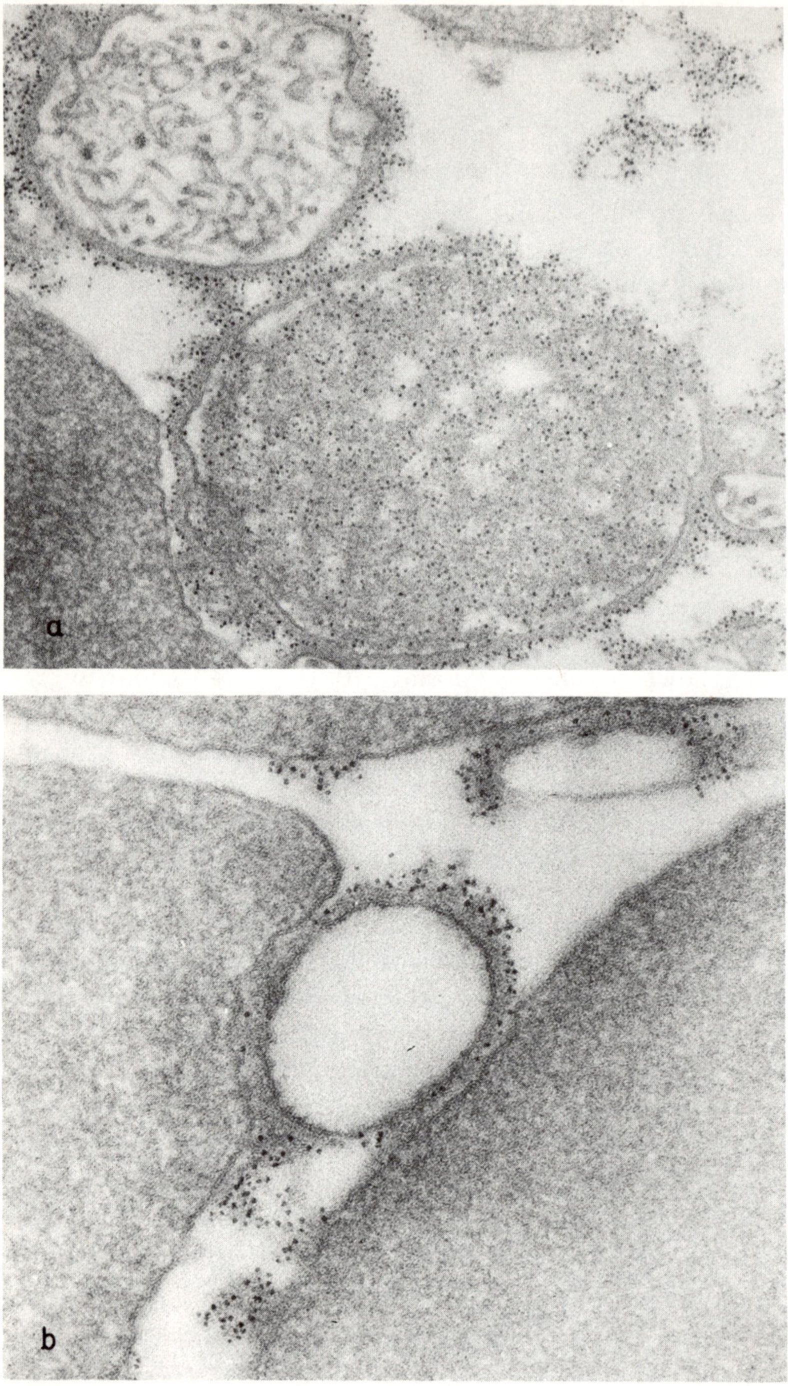
a
b

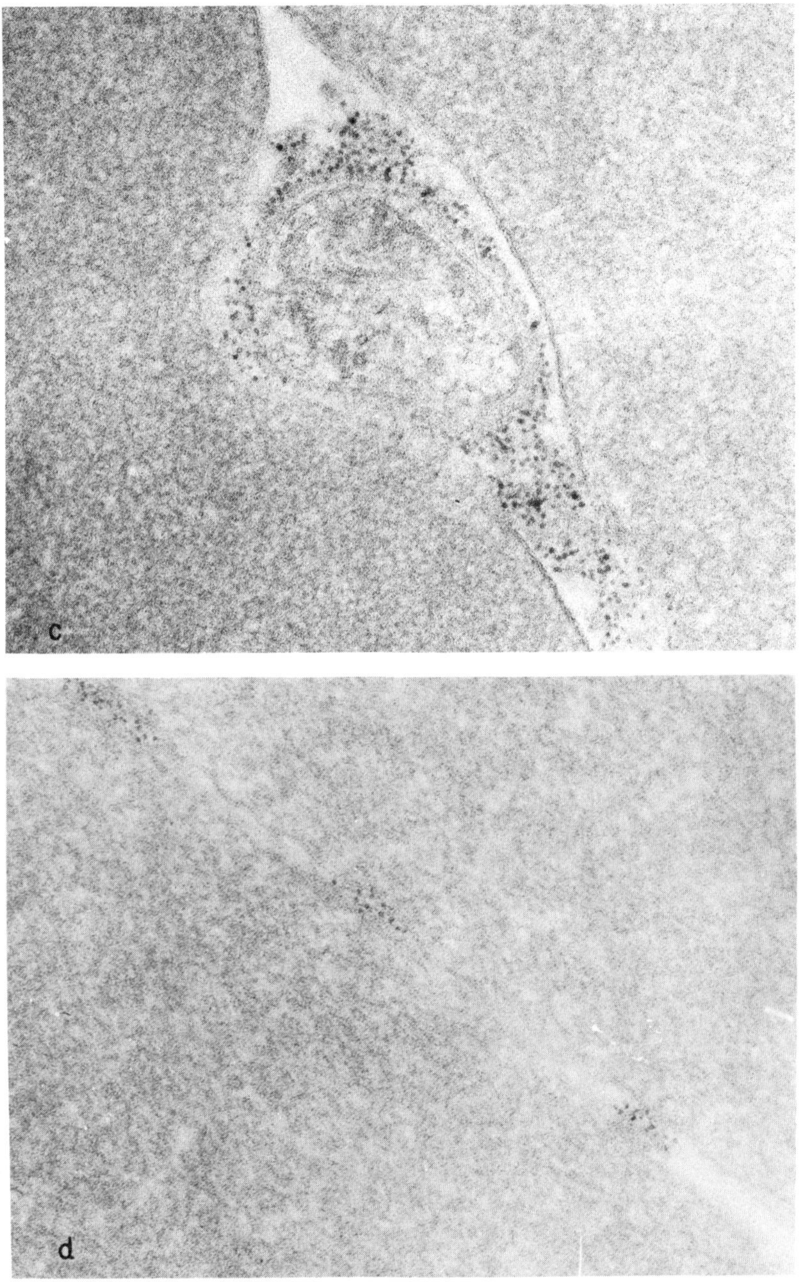

Figure 7. Reaction of concentrated measles virus with rhesus monkey erythrocytes. At successive stages following addition of virus, cells were stained with unlabeled rabbit antibody to virus, followed by ferritin-labeled goat antibody to rabbit globulin (7a, b, d,), or with ferritin-labeled human antibody to virus (7c). For description of individual micrographs, see text. Magnification, x 119,000.

susceptible cells by a similar fusion mechanism, as has already been suggested by other workers.[14]

NOTES

1. L. K. B. Fraser and S. J. Martin, "Measles Virus and Its Biology," in *Experimental Virology*, ed. T. W. Tinsley and F. Brown (London: Academic Press, 1978).

2. P. Dore-Duffy and C. Howe, "N-acetylneuraminic Acid (NANA) in Measles Virus," *Proceedings of the Society for Experimental Biology and Medicine* 157 (1978): 622-25.

3. T. W. Fenger, J. W. Smith, and C. Howe, "Analysis of Immunoprecipitated Surface Glycoproteins in Measles Virions and in Membranes of Infected Cells," *Journal of Virology* 28 (1978): 292-99.

4. M. Morrison, "The Determination of the Exposed Proteins on Membranes by the Use of Lactoperoxidase," *Methods in Enzymology* 32B (1974): 103-09.

5. M. C. Graves, S. M. Silver, and P. W. Choppin, "Measles Virus Polypeptide Synthesis in Infected Cells," *Virology* 86 (1978): 254-63.

6. J. M. Hardwick and R. H. Bussell, "Glycoproteins of Measles Virus Under Reducing and Nonreducing Conditions, *Journal of Virology* 25 (1978): 687-92.

7. D. A. J. Tyrell and E. Norrby, "Structural Polypeptides of Measles Virus," *Journal of general Virology* 39 (1978): 219-29.

8. R. Rott and H.-D. Klenk, "Structure and Assembly of Viral Enevlopes," in *Cell Surface Reviews*, vol. 2. *Virus Infection and the Cell Surface*, ed. G. Poste and G. L. Nicolson (Amsterdam: North-Holland Publishing Company, 1977).

9. V. T. Marchesi, H. Furthmayr, and M. Tomita, "Molecular Features of a Sialoglycoprotein of the Human Erythrocyte Membrane Which Carries Influenza Virus Receptors," in *Cell Membrane Receptors for Viruses, Antigens, and Antibodies, Polypeptide Hormones and Small Molecules*, ed. R. F. Beers, Jr. and E. G. Bassett (New York: Raven Press, 1976).

10. T. W. Fenger and C. Howe, "Characterization of Measles Virus Receptors from Rhesus Money Erythrocytes," *Proceedings of the Society for Experimental Biology and Medicine* (Accepted for publication).

11. L. T. Lee et al., "Comparative Immunochemical Study of Human Erythrocyte Glycoproteins," *Proceedings of the Society for Experimental Biology and Medicine* 158 (1978): 530-36.

12. T. Bächi, G. Eichenberger, and H. P. Hauri, "Sendai Virus Hemolysis: Influence of Lectins and Analysis by Immune Fluorescence," *Virology* 85 (1978): 518-30.

13. T. Bächi, M. Aguet, and C. Howe, "Fusion of Erythrocytes by Sendai Virus Studied by Immunofreeze-Etching," *Journal of Virology* 11 (1973): 1004-12.

14. M. Dubois-Dalcq and L. T. Barbosa, "Immunoperoxidase Stain of Measles Antigen in Tissue Culture," *Journal of Virology* 12 (1973): 909-18.

15. W. E. Mountcastle and P. W. Choppin, "A Comparison of the Polypeptides of Four Measles Virus Strains," *Virology* 78 (1977): 463-74.

16. W. W. Hall and S. J. Martin, "The Biochemical and Biological Characteristics of the Surface Components of Measles Virus," *Journal of general Virology* 22 (1974): 363-74.

17. ———, "The Structural Proteins of Measles Virus," in *Negative Strand Viruses,* ed. B. W. J. Mahy and R. D. Barry (London: Academic Press, 1975): 89-103.

18. B. K. Rima, E. Gould, and S. J. Martin, "Variations in Proteins Synthesized During the Establishment of Persistent Infection with Measles Virus," in *Negative Strand Viruses and the Host Cell,* ed. R. D. Barry and B. W. J. Mahy (London: Academic Press, 1977).

II. RECEPTORS AND IMMUNE SYSTEMS

USE OF RADIOCHEMICAL TECHNIQUES FOR PRIMARY STRUCTURAL ANALYSIS OF MOUSE AND RABBIT HISTOCOMPATIBILITY ANTIGENS

T. J. Kindt, J. E. Coligan,
E. S. Kimball, B. Ewenstein,
H. Uehara, J. Martinko,
and S. G. Nathenson *

INTRODUCTION

Although studies of membrane-related phenomena have dominated several areas of research in recent years, there is a striking paucity of information concerning the primary structure of molecules isolated from the surfaces of mammalian cells. Complete amino acid sequence data are available only for a few membrane proteins: glycophorin from red blood cell membranes,[1] cytochrome b_5 from microsomal membranes,[2] and rhodopsin from retinal cells.[3] An important group of membrane proteins for which primary structural data are needed are those encoded by the major histocompatibility complex (MHC). The primary reason for this information gap appears to be lack of sufficient amounts of material for detailed structural analysis. Several promising experimental approaches have the potential to remedy this deficiency. One of these, the use of proteins radiolabeled by short-term cell culture with radioactive amino acids, was used in the present study, which is directed toward the complete amino

* Drs. Coligan and Kimball, like Dr. Kindt, are associated with the Laboratory of Immunogenetics, National Institute of Allergy and Infectious Diseases, National Institutes of Health. Drs. Ewenstein, Uehara, Martinko, and Nathenson are with the Department of Microbiology and Immunology, Albert Einstein College of Medicine, Bronx, New York. Only Dr. Kindt participated in the conference.

acid sequence of a cell surface glycoprotein encoded by a MHC gene of the mouse. This method has also been applied in studies of a rabbit MHC antigen with some preliminary success.

Because much has been written about MHC genes that are present in all species thus far studied,[4,5] a general description is not warranted in this manuscript. The importance of this gene complex to health and disease has been underscored by numerous studies showing relationships between MHC types and various disease states. The mouse H-2 locus, the most widely studied animal histocompatibility gene complex, has been reported to encode no less than sixty different traits ranging from control of immune responsiveness to body weight.[6]

Figure 1 depicts a representation of the murine seventeenth chromosome. Underneath the various loci are listed the subunit structures and molecular weights of the products that have been thus far characterized. It is interesting that five of the products have identical subunit structures with a large chain of 44,000 molecular weight in association with a smaller chain with a molecular weight of 12,000. In the case of four of these molecules the small chain is beta 2-microglobulin (β_2m), a low molecular-weight protein first observed in the urine of patients suffering from chronic kidney disease.[7] β_2m is not encoded at the MHC, but appears to be a constituent of all histocompatibility antigens thus far examined.[8] The T/t antigens are associated with a low molecular-weight chain that is similar but not identical to β_2m.

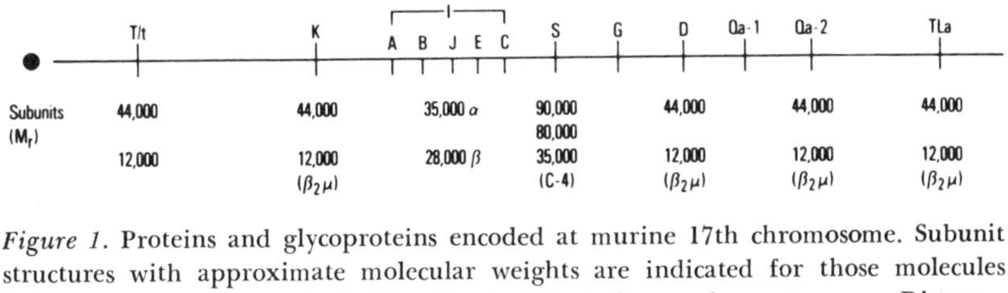

Figure 1. Proteins and glycoproteins encoded at murine 17th chromosome. Subunit structures with approximate molecular weights are indicated for those molecules that have been characterized. The closed circle indicates the centromere. Distances are not drawn to scale.

SOURCE: Adapted from E.S. Vitetta and J.D. Capra, "The Protein Products of the Murine 17th Chromosome: Genetics and Structure," *Advances in Immunology* 26 (1978): 147-93.

Of paramount interest to immunologists are the molecules encoded by the K and D genes, which serve as recognition units for allograft rejection and for cell-mediated cytolysis of cells infected with viruses. The molecule chosen for initial study was the glycoprotein encoded by the murine H-2 K^b gene. The reasons for choosing this molecule were partly of a technical nature, which include a good supply of specific antisera and the availability of a cell line bearing this molecule, which gives high incorporation rates of radiolabeled amino acids into the desired protein. A more important reason for choosing the K^b glycoprotein, however, is the fact that there are no less than nine mutant mouse strains of the b haplotype, which differ from the parent strain by mutations at the K locus.[9] It should be possible to correlate primary structural differences among the K^b molecules in the mutant strains to variations in their biological reactivities. These variations include differences in serologic reactivity, in histogenic activity, and in in vitro cellular responses.

METHODOLOGY EMPLOYED

The application of classical techniques in protein chemistry to amino acid sequence analysis of the K^b antigen, which is approximately 360 amino acids long, would require hundreds of milligrams of purified protein. The use of the latest technologies for peptide separation and sequence analysis, along with a bit of good luck, could lower this requirement possibly by a factor of ten. The difficulty in meeting this need was indicated in a recent study, where 1 kg. of mouse liver yielded 2-3 mg. of a mixture of K and D antigens.[10] The use of cell lines that express large amounts of the desired antigens has proven helpful in studies of human HLA antigens.[11] Such cell lines that hyperproduce histocompatibility antigens are not currently available for other species, however, thus limiting the general applicability of this approach. An additional complication involves facilities for growth of Kg amounts of cells. A more universal technique involves the use of intrinsically radiolabeled proteins obtained from cells cultured in the presence of radioactive amino acids. The cells may be lysed in the presence of nonionic detergents and specific antisera used to precipitate the desired radiolabeled antigen from the lysate. This approach was used in early studies by S. G. Nathenson and coworkers,[12] who compared tryptic peptide maps of H-2 molecules from mice of different MHC haplotypes. It was later shown that molecules obtained by this technique could be subjected to automated sequence analysis. Application of this technique in recent studies of the K^b glycoprotein has allowed assignments for 155 residues in 170 positions of this molecule.[13] All amino acids with the exception of aspartic acid have been successfully incorporated in isotopically labeled form into the molecule at levels sufficiently high to perform amino acid sequence analysis.

General Structural Features of Murine Transplantation Antigens

The major transplantation antigens encoded by the K and the D loci of the mouse are found on almost all cells and are in the highest concentration on lymphoid cells. These molecules exist as membrane-bound glycoproteins and, when isolated, are insoluble in aqueous media. Digestion of the cells with papain releases the antigens in a soluble form that has the full range of immunological activities, although the molecule is smaller by approximately 7,000 daltons. As mentioned earlier (Figure 1), the K and D molecules, along with several other proteins encoded by the MHC, are noncovalently associated with β_2m, a protein of 12,000 molecular weight. The interaction of β_2m with the histocompatibility antigens is sufficiently stable to allow precipitation of these molecules by antisera directed against β_2m. Such treatment does not yield a pure product, however, because both K and D antigens, as well as certain other cell-surface proteins encoded at or near the MHC, are associated with β_2m. These molecules include the TL and the Qa-2 antigens (Figure 1).

The molecular weight of the H-2 heavy (H) chain is approximately 44,000, of which about 5,000 may be attributed to carbohydrate groups. The H chain, derived from molecules that have been isolated by papain cleavage, has a molecular weight of approximately 37,000; whereas the size of β_2m from these molecules is not perceptibly altered.

Radiolabeling of H-2 K^b Molecules

A primary consideration in radiosequence analysis of a protein concerns the choice of radioactive amino acids to be added to cell culture in order to radiolabel the protein. Tritium-labeled amino acids were chosen in our study because of the high specific activity and low cost relative to ^{14}C amino acids. The question of how many amino acids are to be added to a single culture is not easily answered and requires experimental data for an intelligent solution. The extremes of adding only one or, alternatively, all twenty amino acids are usually not practical. With a single amino acid labeled in the protein, all peptides would have to be purified and their sequences determined a minimum of twenty times. This would represent a huge effort. With twenty amino acids radiolabeled, complete sequence determinations for each peptide could be done after one or two isolations. The derivatives of the radioactive amino acids from the automated sequencer, however, would require very precise analysis, and any significant inconsistencies among the specific activities of the radioactive amino acids would lead to errors in identification. A compromise is certainly indicated.

In original studies, groups of six or seven amino acids were chosen for incorporation.[14] The relative amount of each radioactive amino acid incorporated into cellular proteins was determined by sequence analysis of β_2m, a byproduct of an early purification step for the H-2 molecules. Murine β_2m has a known amino terminal sequence for approximately thirty-five residues, and these residues include all but two of the twenty amino acids normally found in mammalian proteins. Analysis of radiolabeled β_2m molecules indicated that certain amino acids (alanine, glutamic acid, glutamine) have relatively low incorporation rates in the presence of other radiolabeled amino acids. The PTH derivatives of glycine and serine are problematic, in that normal conversion procedures lead to loss of radioactivity in these derivatives. Because of these problems the amino acids mentioned above were incorporated in individual experiments (Table 1), and the molecules radiolabeled with these residues were analyzed separately.

The sequence analysis of β_2m, and subsequently that of the

TABLE 1. COMBINATION OF RADIOLABELED[1] AMINO ACIDS ANALYZED FOR H-2K^b

Group	
I	Phe, Leu, Lys, Cys, Pro,[2] Ala[2]
II	His, Trp, Thr, Ile, Val, Ser,[2] Gly[2]
III	Glu, Pro, Arg
IV	Gly, Ser
V	Gln, Glu, Pro
VI	Met, Tyr
VII	Ala, Cys

NOTES: 1) All amino acids, except Cys and Met, were radiolabeled with ^3H. 2) Incorporated at levels too low for reliable sequence analysis.

H-2 molecule, has shown that with the EL-4 cell line certain amino acids are biosynthetically interconverted during growth in culture. The most important of these interconversions are listed in Table 2. More recent data using rabbit cell lines show that the serine to glycine and glycine to serine interconversions also occur in these cultured cells.

It was further observed that higher percentages of certain radiolabeled amino acids were utilized for protein biosynthesis. The amino acid with the best incorporation was arginine, with as much as 80 percent of the added radioactivity observed in cell protein. At the bottom of the list would be aspartic acid, which has an incorporation rate of around 1 percent. This difficulty

TABLE 2. INTERCONVERSION OF ^3H AMINO ACIDS IN EL-4 CELL CULTURES

Added to Culture	Radioactive Amino Acids in β_2m[1]	Relative Activity
Glu	Glu	1.0
	Pro	0.64
	Arg	0.18
Gln	Gln	1.0
	Glu	0.22
	Pro	0.09
Ser	Ser	1.0
	Gly	0.1
Gly	Gly	1.0
	Ser	0.5

NOTE: 1) Detected by NH_2 terminal sequence analysis of the β_2m separated from H-2K^b by gel filtration.

with aspartic acid has made it the only amino acid for which data are not available in the K^b molecule. Asparagine has only recently been available in high specific activity, but there appears to be no difficulty with its incorporation. The average incorporation level is approximately 30 percent for the other amino acids.

Purification of the H-2 K^b Antigen

After the tumor cells have been cultured in the presence of the radioactive amino acid mixture, the cells are lysed and surface molecules solubilized by treatment with a nonionic detergent. Nuclei and debris are removed from the lysate by ultracentrifugation followed by isolation of a glycoprotein fraction by affinity chromatography using a column consisting of a lectin from *Lens culinaris* (LcH) coupled to Sepharose. To the glycoprotein fraction obtained by elution of the LcH column with alpha methyl mannoside is added mouse antiserum directed against the private specificity (H-2.33) of the K^b glycoprotein. Next is added a sufficient quantity of a xenoantiserum directed against mouse immunoglobulin to precipitate the anti-H-2.33 and its bound radiolabeled antigen. The EL-4 cell line has no immunoglobulin to interface with secondary immunoprecipitation. Consequently, no other radiolabeled proteins are present in the resulting precipitate. Therefore at this point a complete purification of the K^b molecule from the other radiolabeled components of the EL-4 cell line has been accomplished. The precipitate is next dissolved in dilute formic acid to dissociate complexes, and the β_2m is separated from the H chain by gel filtration. These steps are diagramed in Figure 2.

Isolation of Peptides from H-2K^b

In order to obtain peptides for amino acid sequence analysis of the K^b glycoprotein, the molecule was cleaved with cyanogen bromide. This treatment results in cleavage at the C-terminal side of most methionine residues in a protein. Chromatography of the CNBr digest of K^b on Sephadex G-100 in the presence of guanidine hydrochloride resulted in three major fractions. The largest of these, designated I, was subjected to reduction and

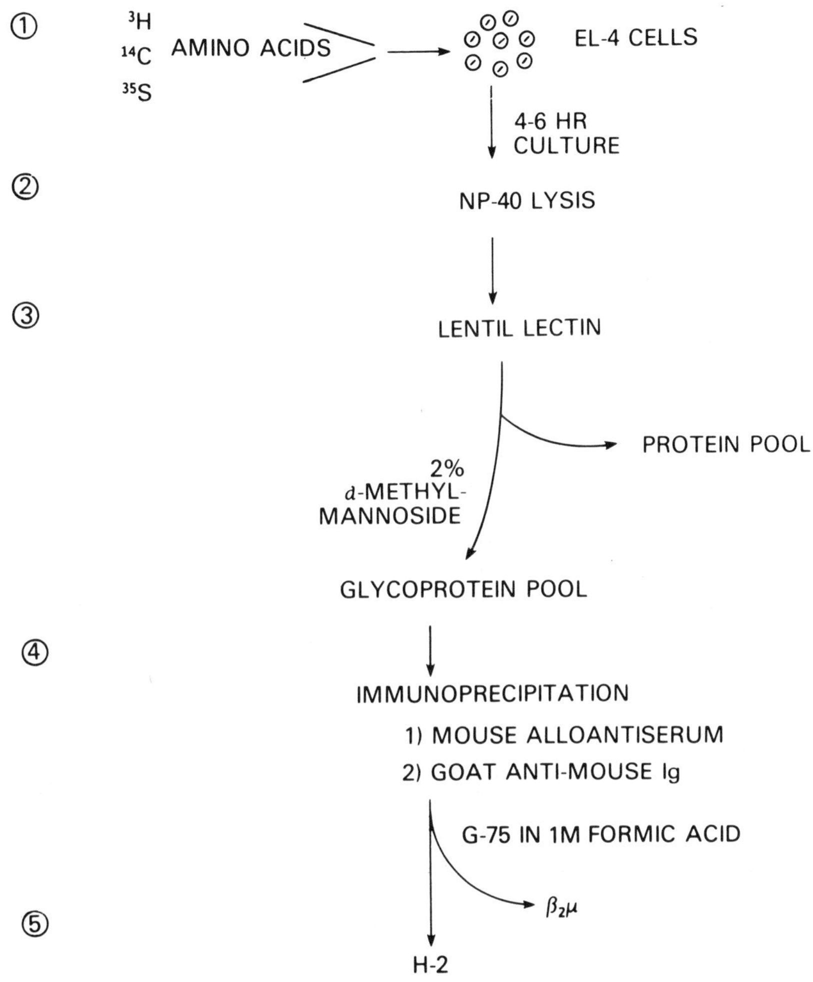

Figure 2. Procedure for preparation of H-2K^b used for radiochemical determination of primary structure.

alkylation in order to cleave any disulfide bonds among the large peptides in this mixture. Following this treatment, the pool I peptides were again subjected to chromatography on the same Sephadex G-100 column. This resulted in two fractions: one contained a pure peptide designated Ic; the other contained more than one peptide and was designated Iab. It was observed that the molecular weight of peptide Ic changed when the molecule was subjected to digestion with papain, indicating that this peptide included the C-terminus of the papain-solubilized molecule and the probable site at which the K^b molecule was bound to the cell membrane. The other fraction, Iab, was subjected to chromatography on DEAE Sephadex, which yielded two pure glycopeptides (Table 3).

The pool III from the original column was subjected to chromatography on G-50 Sephadex, and three fractions, designated IIIc, IIIa, and IIIn, each containing a pure peptide, were obtained. The peptide designated IIIc was not present in digests of papain-treated K^b molecules. Amino acid sequence analysis indicated that fraction IIIn contained the peptide present at the amino terminus of the K^b molecule. It was further shown by sequence analysis that IIIa contained the peptide immediately to the C-terminus of IIIn beginning at position 24 in the intact molecule.[15] Pool V from the original G-100 column contained at least three peptides that could be separated by gel filtration chromatography on G-25 Sephadex. These peptides were not present in papain-digested molecules and have not been studied in detail as yet.

Table 4 lists the information concerning the major CNBr fragments of H-$2K^b$. As indicated by the results described, fractions Ib, Ia, and Ic are linked to one another by disulfide bridges, and Ia and Ib are both glycopeptides, as indicated by experiments using labeled carbohydrates and by the fact that these peptides will bind to the lentil lectin column. The alignment of the fragments, as shown in Table 4, was deduced from a variety of data that included isolation and sequence analysis of tryptic overlap peptides radiolabeled in methionine.[16] The alignment is furthermore confirmed by homology of amino acid sequence to that of HLA molecules.

Amino Acid Sequence Analysis of CNBr Fragments of H-$2K^b$

The major CNBr fragments isolated, as described, which contained one of the combinations of amino acids listed in

TABLE 3. PEPTIDES ISOLATED FROM CYANOGEN BROMIDE DIGEST OF H-2K^b

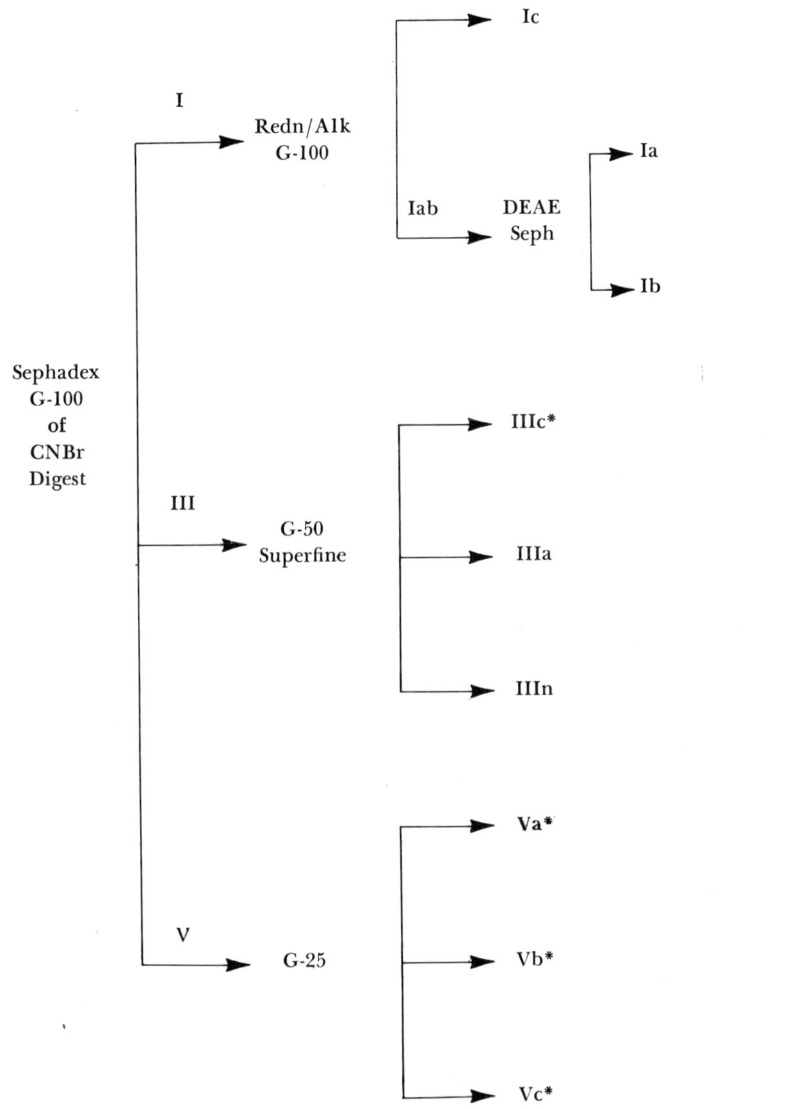

NOTE: Numerals with small letters indicate peaks detected by radioactivity in column effluent that were shown by gel electrophoresis and sequence analysis to contain single peptides. Asterisks indicate peptides present only for molecules not treated with papain. Peptide Ic is smaller when isolated from papain-treated molecules.

TABLE 4. CHARACTERIZATION OF MAJOR CNBR FRAGMENTS OF H-2K^b

Fragment	M_r	Spanning Positions	NH_2 Terminus	Carbohydrate
IIIn	2,500	1-23	Gly Pro His	—
IIIa	3,200	24-52	Glu Val Gly	—
Ib	11,500[1]	53-138	Glu Gln Glu	+
Ia	12,000[1]	139-228	Ala Ala Leu	+
Ic	5,800[1]	229-283	Glu Leu Val	—

NOTE: 1) Molecular weights were estimated for these fragments.

Table 1, were subjected to thirty to fifty cycles of automated Edman degradation. PTH amino acids were analyzed by high pressure liquid chromatography using methods developed in our laboratory that resolve the PTH derivatives of all twenty amino acids.[17] As indicated in Table 1, all amino acids with the exception of aspartic acid and asparagine have been determined for the amino terminal positions of the major CNBr fragments of K^b.

Results of the amino terminal sequence analyses of the CNBr fragments are shown in Table 5. The dashes indicate positions at which no assignment could be made from our data. It is assumed that many of these positions will be occupied by aspartic acid or asparagine residues because these amino acids have not yet been analyzed. The lengths of the CNBr fragments, with the exception of IIIn and IIIa, which were determined directly, were estimated by homology to HLA antigens when possible or by the molecular weight obtained for the fragment in question. The position of the carbohydrate units in peptide Ib has been assigned to position 86 (from the amino terminus). There is a second carbohydrate moiety present in peptide Ia, but this unit has not yet been localized.

HOMOLOGY OF H-2K^b TO OTHER HISTOCOMPATIBILITY ANTIGENS

Comparison of H-2K^b amino terminal sequence with the data available for other mouse histocompatibility antigens indicates a high degree of homology among the amino termini of these allelic products. For example, comparison with H-2K^k indicates only

TABLE 5. AMINO ACID SEQUENCE DATA FOR CNBR FRAGMENTS OF H-2K[b]

Fragment IIIn

5 10 15 20

GLY PRO HIS SER LEU ARG TYR PHE VAL THR ALA VAL SER ARG PRO GLY LEU GLY GLU PRO ARG TYR MET

Fragment IIIa

25 30 35 40 45 50

GLU VAL GLY TYR VAL --- --- THR GLU PHE VAL ARG PHE --- SER --- ALA GLU(ALA)PRO ARG TYR GLU PRO ARG(ALA)ARG TRP MET

Fragment Ib

5 10 15 20 25

GLU GLN GLU GLY PRO GLU TYR TRP GLU ARG GLU THR GLN LYS ALA LYS(GLY) --- GLU GLN SER PHE ARG VAL --- LEU ARG THR LEU

30 35 40 86

LEU(GLY)TYR TYR --- GLN SER LYS GLY --- HIS THR ILE(GLN)··········[MET]

CH_2O

Fragment Ia

5 10 15 20 25

ALA ALA LEU ILE THR LYS --- LYS TRP GLU GLN ALA GLY GLU ALA GLU ARG LEU ARG ALA TYR LEU GLU GLY THR CYS VAL GLU

30 35 89

TRP LEU ARG ARG --- LEU LYS···········[MET]

Fragment Ic

5 10 15 20 25

GLU LEU VAL GLU THR ARG PRO ALA GLY --- GLY THR PHE GLN LYS TRP ALA SER --- VAL VAL PRO LEU GLY LYS GLU --- TYR

30 35 40 53

TYR THR CYS(HIS) --- TYR GLN GLN (GLY)LEU PRO --- ···········[MET]

NOTE: Positions in fragments IIIn and IIIa are numbered from the amino terminus of the intact molecule; other fragments are numbered individually. Dashes indicate positions where no positive assignment has been made; parentheses indicate tentative assignments that will require confirmation.

three differences among the amino terminal twenty-three residues for a homology of approximately 85 percent. Comparison of our sequence of H-2^b, which was obtained from the EL-4 cell line with H-2K^b isolated from spleens of C57Bl/6 mice by R. G. Cook and coworkers,[18] reveals two differences in amino terminal positions. There is a lysine at position 19, where we have found glutamic acid, and isoleucine at 23, where methionine was observed by direct analysis, as well as by susceptibility to CNBr digestion. Data obtained by sequence analysis of K^b obtained from C57Bl/6 spleens labeled with methionine indicated the presence of this residue at position 23.* It is tempting to speculate that these differences have arisen from mutations that have been fixed in sublines of the C57Bl/6 mouse strain.

An extensive comparison of amino acid sequence data for human and mouse histocompatibility antigens is now possible because of the structural data recently reported by H. L. Ploegh and coworkers.[19] The regions of HLA-A2 or -B7, for which extensive continuous amino acid sequences have been reported, are compared to the amino acid sequences for corresponding segments of H-2K^b (Table 6). Comparison of K^b with A2 at positions 1 through 96 shows that the amino acids at 62 of the 89 positions that are comparable are identical, for a homology of 70 percent. Further comparison of the K^b molecule between positions 211 to 249 to similar positions for the B7 molecule indicate homology of 79 percent, with 27 of the 34 compared positions being identical. In these comparisons Asx (aspartic acid or asparagine) was substituted in all positions in H-2K^b where it was felt that if any other amino acid residue occurred it would have been detected by our sequence analysis. Of the 9 presumptive Asx residues that have been inserted in the K^b sequence, 7 are homologous to those present in the HLA. In all but three—positions 45, 238, and 244—the differences between the HLA and H-2K^b molecules can be accounted for by single base changes in the DNA sequences. One insertion, at position 229 in the B7 sequence, was required to maintain maximum homology.

* S. G. Nathenson: unpublished data.

TABLE 6. COMPARISON OF H-2K^b TO HUMAN HLA ANTIGENS

	10 20 30
H-2K^b	G P H S L R Y F V T A V S R P G L G E P R Y M E V G Y V *B* *B* T E F V R
HLA-A2	– S — M —— F – S ———— R ——— F I A ————————
	40 50 60 70
K^b	F *B* S *B* A E A P R Y*E P R A R W M E Q E G P E Y W E R E T Q K A K G *B*
A2	———— A S Z – M ——— P — I ——— X — X X – X X X — V — A H
	80 90
K^b	E Q S F R V *B* L R T L L G Y Y *B* Q S K G G S H T I Q
A2	A H T V ——— G — R ————— Z A ——— V —
	211 220 230 240
K^b	E L V E T R P A G *B* G T F Q K W A S *B* V V P L G K E X Y*Y T C H X Y*Q
HLA-B7	——————— R ————— A [ ]—— X – X X X R ——— V Q H
	249
K^b	Q G L P
B7	E ——

NOTE: For this comparison, positions in the K^b sequence, for which no assignment has been made, were assumed to be aspartic acid or asparagine (italicized *B*). Asterisks denote substitutions needing more than a single base change in DNA sequence; bracket indicates an insertion introduced to maximize sequence homology.

Present efforts are being concentrated on completing the amino acid sequence of H-2K^b by preparing smaller peptides, by methods such as tryptic digestion, from the CNBr fragments. These efforts are greatly facilitated by earlier work of Nathenson and coworkers on the tryptic digestion of the histocompatibility antigens.[20] In addition, we have begun to study the amino acid sequence of the H-2D^b molecule and, in collaboration with Rod Nairn, the H-2D^d molecule. The latter studies will be expedited by the availability of radiolabeled material from the K^b study.

Analysis of Histocompatibility Antigens of the Rabbit

A stringent test of the universal applicability of the radiochemical method described above for amino acid sequence determination of cell surface glycoproteins was carried out in attempts to determine the primary structure of histocompatibility antigens of the rabbit. This antigenic system is not nearly as well defined as that of the H-2 complex, and no specific sera are available for products of the subloci within the rabbit MHC. A cell line (RL-5) was available, however, which had been derived from rabbits homozygous at their MHC. These cells were radiolabeled and RLA molecules were isolated by a variation of the procedure used for isolation of the mouse H-2K^b. After the RL-5 cells were radiolabeled with amino acids they were lysed, and the supernatant was subjected to chromatography on a lentil lectin-sepharose column. The glycoprotein fraction obtained by this procedure was then passed over a column to which was coupled antirabbit β_2m. The antibody column was eluted with the dissociating agent, ammonium thiocyanate. Electrophoresis of the molecules obtained by this procedure indicated that a uniform peak with molecular weight of approximately 43,000, along with β_2m, were the only materials obtained by elution of the antibody column.

The β_2m was separated from the glycoprotein by gel filtration chromatography and the RLA molecule(s) was subjected to amino acid sequence analysis. Thus far, data are available for nine different amino acids that have been incorporated into the RLA molecule. Results to date, which allow twenty-six assignments for the amino terminal thirty-five positions, indicate about 90 percent homology between the rabbit RLA molecules and the mouse K^b and human HLA molecules (Table 7).[21] It is puzzling that a single amino acid sequence has been obtained by this procedure, which

TABLE 7. COMPARISON OF AMINO ACID SEQUENCE DATA FOR MOUSE, HUMAN, AND RABBIT HISTOCOMPATIBILITY ANTIGENS

H-2K^b	G P H S L R Y F V T A V S R P G L G E P R Y M E V G Y V * * T E F V R
HLA-A2	—S——M———F—S———— R———F I A———D D———
RLA-11	G S H S * R Y F Y * S V S R P G L G * P R F * * V G Y V * * * * F V R

NOTE: Asterisks indicate positions for which no positive assignment has been made.
SOURCE: H-2K^b is from J. E. Coligan et al., "Further Structural Analysis of the Murine H-2K^b Glycoprotein Using Radiochemical Methodology," *Molecular Immunology* (1979), in press; HLA-A2 is from H. L. Pleugh et al., "Structure of HLA Antigens," in *Biological Markers of Neoplasia: Basic and Applied Aspects,* ed. H. Ruddon (New York: Elsevier North Holland Press, 1978): 201-11; RLA-11 is from E. S. Kimball, J. E. Coligan, and T. J. Kindt, "Structural Characterization of Antigens Encoded by Rabbit RLA-11 Histocompatibility Genes," *Immunogenetics* 8 (1979): 201-11.

relies solely on the association of the histocompatibility antigens with β_2m. By analogy to information available for mouse MHC products (Figure 1), one would expect several β_2m-associated molecules to be isolated in this procedure, including analogues of the K and D specificities, as well as QA and TL antigens. The results of our analyses indicate that the latter molecules are not present in the fraction obtained, or that they are present in such low amounts they are not detected by the amino acid sequence analysis. It is further possible that the isolated molecules, at least the K and the D analogues, could be homologous at their amino terminals and thus be indistinguishable by sequence analysis.

Comparison of the data obtained for the rabbit histocompatibility antigen to that obtained for histocompatibility antigens from various other species [22] shows that there is no position where the rabbit protein has an amino acid that has not been demonstrated for an antigen from one of the other species. As indicated by E. S. Vitetta and J. D. Capra,[23] in certain positions (3, 7, 15, 27) the same amino acid has been reported for histocompatibility antigens of all species thus far studied. The high degree of homology of the rabbit sequence to the others strongly indicates that we have by our indirect procedure isolated a rabbit histocompatibility antigen, which, by the limited analyses thus far carried out, is a single molecular species.

CONCLUSIONS

This report has documented progress toward the determination of the complete amino acid sequence of the murine MHC H-2K^b molecule and preliminary structural analysis of rabbit RLA molecules, both of which are cell-surface proteins. Radiochemical methodology employed here has allowed assignment of 153 residues in 170 positions of the K^b molecule. The approach used has been very similar to that which would be used for a nonradiolabeled molecule and has involved preparation of CNBr peptides and tryptic peptides. Application of similar methodology to a study of rabbit histocompatibility antigens has allowed assignments at 26 of the amino terminal 35 positions, and has shown a high degree of homology between this sequence and that obtained for mouse and human histocompatibility antigens, a finding that suggests a common evolutionary origin for these proteins. It is our contention that the methods described here can be applied for primary structural analysis of any cell surface protein that can be radiolabeled in cell culture and for which a reasonable purification procedure can be devised. The success in the present studies may be attributed to the availability of specific antisera for use in purification procedures.

NOTES

1. M. Tomita, H. Furthmayer, and V. T. Marchesi, "Primary Structure of Human Glycophorin A. Isolation and Characterization of Peptides and Complete Amino Acid Sequence," *Biochemistry* 17 (1978): 4756-69.

2. J. Ozols and C. Gerard, "Primary Structure of the Membranous Segment of Cytrochrome b_5," *Proceedings of the National Academy of Sciences* 74 (1977): 3725-29.

3. C. Cardet, A. Tardieu, and V. Luzzati, "Shape and Size of Bovine Rhodopsin: A Small-Angle X-Ray Scattering Study of a Rhodopsin-Detergent Complex," *Journal of Molecular Biology* 105 (1976): 383-407.

4. J. Klein, *Biology of the Mouse Histocompatibility-2 Complex* (New York: Springer Verlag, 1975) .

5. D. Gotze, *The Major Histocompatibility System in Man and Animals* (New York: Springer Verlag, 1977) .

6. J. Klein, "H-2 Mutations: Their Genetics and Effect on Immune Functions," *Advances in Immunology* 26 (1978): 55-146.

7. I. Barggård and A. G. Bearn, "Isolation and Properties of a Low Molecular Weight β_2-Globulin Occurring in Human Biological Fluids," *Journal of Biological Chemistry* 243 (1968): 4095.

8. E. S. Vitetta and J. D. Capra, "The Protein Product of the Murine 17th Chromosome: Genetics and Structure," *Advances in Immunology* 26 (1978): 147-93.

9. Klein, "H-2 Mutations" (See note 6) .

10. O. Henriksen, E. A. Robinson, and E. Appella, "Structural Characterization of H-2 Antigens Purified from Mouse Liver," *Proceedings of the National Academy of Sciences* 75 (1978): 3322-26.

11. C. Terhorst et al., "Further Structural Studies of the Heavy Chain of HLA Antigens and Its Similarity to Immunoglobulins," *Proceedings of the National Academy of Sciences* 74 (1977): 4002-06.

12. J. L. Brown et al., "Notable Diversity in Peptide Compositions of Murine H-2K and H-20 Alloantigens," *Biochemistry* 13 (1974): 3174.

13. J. E. Coligan et al., "Further Structural Analysis of the Murine H-$2K^b$ Glycoprotein Using Radiochemical Methodology," *Molecular Immunology* 16 (1979): 3-8.

14. B. M. Ewenstein et al., "Primary Structure of Murine Major Histocompatibility Complex Alloantigens: Isolation, Biochemical Characterization, and Preliminary Alignment of CNBr Fragments from the H-$2K^b$ Glycoprotein," *Proceedings of the National Academy of Sciences* 75 (1978): 2909-13.

15. J. E. Coligan et al., "Primary Structure of Murine Major Histocompatibility Complex Alloantigens: Amino Acid Sequence Studies of the Cyanogen Bromide Fragments of the H-2K^{b} Glycoprotein, *Proceedings of the National Academy of Sciences* 75 (1978): 3390-94.

16. H. Uehara et al., manuscript submitted for publication.

17. F. T. Gates, III, J. E. Coligan, and T. J. Kindt, "The Complete Amino Acid Sequence of Rabbit Beta-2-Microglobulin," *Biochemistry* 18 (1979): 2267-72.

18. R. G. Cook et al., "Structural Studies on Protein Products of Murine Chromosome 17. III. Partial Amino Acid Sequence of an H-2K^{q} Molecule," *Journal of Immunology* 121 (1978): 1015-19.

19. H. L. Ploegh et al., "Structure of HLA Antigens," in *Biological Markers of Neoplasia: Basic and Applied Aspects,* ed. H. Ruddon (New York: Elsevier North Holland Press, 1978): 201-11.

20. Brown et al., "Peptide Compositions" (See note 12) .

21. E. S. Kimball, J E.. Coligan, and T. J. Kindt, "Structural Characterization of Antigens Encoded by Rabbit RLA-11 Histocompatibility Genes," *Immunogenetics* 8 (1979): 201-11.

22. Vitetta and Capra, "Murine 17th Chromosome" (See note 8).

23. Ibid.

EXPRESSION AND FUNCTION OF B-LYMPHOCYTE RECEPTORS *

Daniel Levitt and
Max D. Cooper †

The B-cell lineage, having as its principal responsibility the production of a diverse array of antibodies, is an excellent model for the study of eukaryotic cell differentiation. This cell lineage exists in all vertebrates and provides a highly versatile recognition system important to immune defense. B lymphocytes representing intermediate stages of development along this cellular pathway have been studied most extensively because they are easily obtained in suspension; exhibit clonal diversity; and can be induced to respond with motility, proliferation, and differentiation into antibody-secreting plasma cells.

The surface of the B lymphocyte has received special attention. Entire books, sections of books, and numerous manuscripts have dealt with the plethora of membrane molecules and their postulated roles in regulation of the antibody response. Despite this intense scrutiny, however, the functions of most of the B cell-surface components remain a mystery and are still keenly debated. Certain molecules on the cell membrane clearly receive signals from the external environment and initiate changes in B-cell metabolism and antibody expression. Although characterization and assignment of distinct duties to the diverse array of

* The studies reported in this paper were supported by grants CA 16673 and CA 13148 awarded by the National Cancer Institute; and 1 T32 AI 07051 awarded by the National Institute of Allergy and Infectious Diseases, Department of Health, Education, and Welfare.

† The authors wish to thank Mrs. Summer King for her untiring assistance with this manuscript, and Mrs. Susie Gray for her help in planning and drawing the illustrations.

B-cell receptors are still incomplete, we will attempt to describe some of these molecules with regard to their appearance during B-cell ontogeny and involvement in B-cell differentiation.

SURFACE OF A MATURE B CELL

Figure 1 illustrates the variety of determinants that have been recognized on the surface of B lymphocytes. The earliest cells of B lineage do not possess all these components, nor do the plasma cells, the most mature cells of the lineage. Thus presence or absence of these markers can be used to define stages in B-cell development. Since functional roles have been suggested for many of these surface components, it is evident that regulation of B-lymphocyte responses to external signals may be modulated in a variety of ways and by a great number of factors.

ONTOGENY OF B LYMPHOCYTE RECEPTORS

Surface Immunoglobulins (sIg)

The earliest recognizable cell in the mammalian B-cell pathway lacks sIg detectable by immunofluorescence [1,2] or by func-

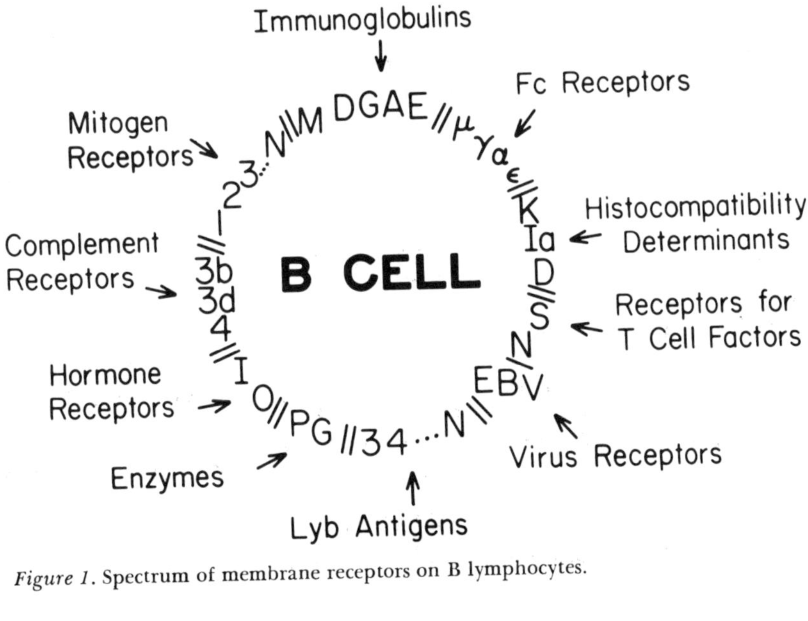

Figure 1. Spectrum of membrane receptors on B lymphocytes.

tional assays,[3-6] but it does possess stainable cytoplasmic IgM (Figure 2). This cell, designated the pre-B cell, may express sIgM with a very brief transit time.[7,8] Pre-B cells can be recognized in the murine fetal liver by eleven days of gestation,[9] and in humans by nine weeks of development.[10] Using a functional in vitro assay, precursors of B cells have been found in the mouse placenta as early as the ninth day of gestation.[11]

Although pre-B cells are nearly "bald" with regard to surface receptors characteristic of their B-lymphocyte progeny, they represent an interesting phase in the life history of this cell lineage. Cells entering this differentiation compartment undergo rapid division, during which clonal diversity may be generated. Pre-B cells can be shown to be clonally diverse with respect to their expression of immunoglobulin light-chain types, that is, κ or λ, light-chain alleles, and V-region idiotypes. For reasons that will become more evident later, we think that the paucity of functional surface receptors on pre-B cells may be an important protective device against premature clonal abortion [12] by multivalent antigens in the environment.

Expression of sIgM begins around the sixteenth day of mouse development; sIgM+ B cells first appear in the liver, and soon thereafter in spleen and bone marrow.[13-16] Comparable studies in human fetuses demonstrate sIgM+ cells in the liver during the ninth week of gestation,[17] and in guinea pigs by day fifty of a sixty-eight-day gestation.[18]

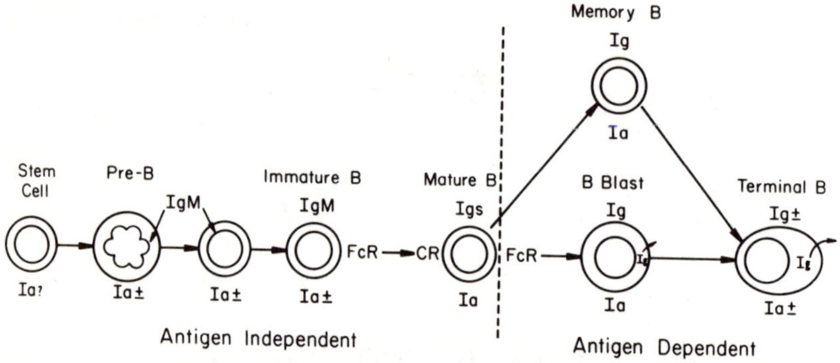

Figure 2. Pathway of B-cell development.

Generation of isotype diversity then follows in a pattern that is similar for both mice and humans (Figure 3). In mice, for example, all the earliest B lymphocytes bear sIgM only. These immature sIgM+ cells are pivotal in the generation of isotype diversity, since inhibition of their development by treatment with anti-IgM antibodies prevents the appearance of B cells of all isotypes. In mice, around the time of birth a few sIgM+ cells begin to express sIgG, and shortly thereafter sIgM+ .sIgA+ and sIgM+ .sIgD+ "doubles" appear. Few of these cells double label for IgG, IgA and IgD. By a week after birth the pattern has changed, so that almost all sIgG+ and sIgA+ B lymphocytes exhibit both sIgM and sIgD. With increased postnatal exposure to antigens, sIgA+ cells and sIgG+ cells appear that express only those single isotypes.

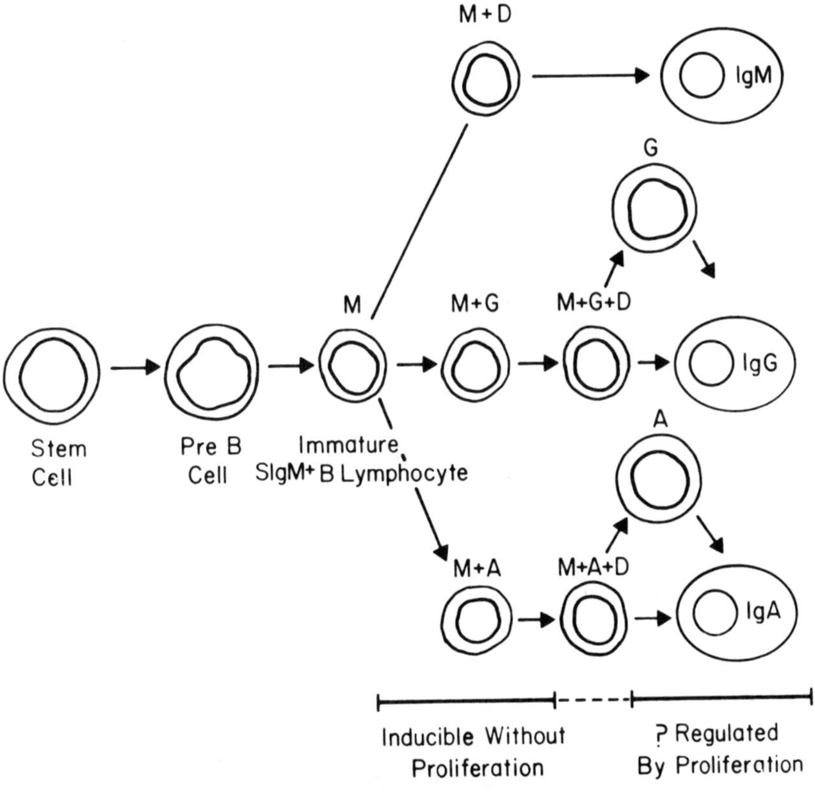

Figure 3. Sequential generation of surface isotypes on B lymphocytes.

The latter cells are usually larger lymphocytes that probably have been stimulated by environmental antigens or mitogens. Throughout life the predominant B-cell phenotype exhibits both sIgM and sIgD receptors.[19, 20]

It is important to stress the observation that generation of isotype diversity expressed on the surface of B lymphocytes does not require influences exerted by environmental antigens and T lymphocytes.[21] On the other hand, expression of different immunoglobulin isotypes can be modified in B lymphocytes and mature plasma cells by anti-immunoglobulins, antigenic stimulation, and T-cell factors.[22-25] It has also been demonstrated that B lymphocytes that can be triggered by the mitogen lipopolysaccharide (LPS) to differentiate to become IgM-, IgG-, and IgA-producing plasma cells all bear sIgM at the outset, but express the second isotype, IgG or IgA, en route to becoming plasma cells secreting antibodies of those classes.

The number of IgM receptors on resting mature B lymphocytes is estimated to be approximately 10^5 per cell; [26] immature B lymphocytes may have a relatively higher average density of sIgM receptors.[27]

Fc Receptor

Since the initial demonstration of an Fc receptor for IgG on B lymphocytes,[28, 29] numerous methods have been used to detect these receptors on B cells from various species.[30] Distinct lymphocyte subpopulations will bind Fc γ in a manner that is dependent on the subclass of IgG, the method of preparing IgG (rosette techniques versus heat aggregation versus antigen-antibody complexes), and exposure of cells to certain enzymes (trypsin, neuraminidase). As examples, Fc γ receptors are detected on virtually all the immature and mature B lymphocytes (but not pre-B cells) by large, heat-induced aggregates of human IgG,[31] whereas human IgG antibodies coated on Rh^+ erythrocytes fail to bind even to mature B lymphocytes,[32] while binding efficiently to monocyte/macrophages and "null" lymphocytes.

Initial observations using mouse bone-marrow cells suggested that the γFcR might be a specific marker for mature B lymphocytes [33, 34] which was then lost during terminal plasma cell differentiation. This idea is supported by the observation that the majority of immature B lymphocytes from neonatal mouse spleen and human peripheral blood lack Fc receptors detectable by rabbit immunoglobulin-alloantibody complexes conjugated to fluorescein. The percentage of such FcR^+ B cells increases rapidly, so that by adult life almost all B lymphocytes possess FcR.[35] Studies of the ontogeny of μ, α, or ϵ FcR on B lymphocytes have not been reported, although all have been found on B lymphocytes from mature individuals.

Complement Receptor

Demonstration of a receptor on B lymphocytes for breakdown products of the third component of complement (C3) [36, 37] soon led to an examination of their appearance during ontogeny in mice. Complement-receptor (CR)-bearing sIg^+ lymphocytes were virtually undetectable in the spleen before two or more weeks after birth.[38, 39] The time of expression of this receptor appears to be governed by a MCH-linked gene(s). This marker, coupled with the somewhat earlier but gradual emergence of γFcR, was postulated as important distinguishing criteria between "early" and "late" B cells. Complement receptors have also been utilized in attempts to define subpopulations of B lymphocytes with different functional capacities. This property will be discussed in later sections.

Histocompatibility Antigens

Histocompatibility antigens encoded by a group of linked genes on chromosome 6 in humans, and chromosome 17 in mice, have been shown to be intimately involved in regulation of B cell/T cell interactions in the immune response.[40,41] As defined in mice, the I subregion consists of multiple loci that code for I-associated (Ia) surface antigens, many of which appear to be of crucial importance for regulation of antibody responses. Although a few studies have suggested that appearance of Ia on mouse B lymphocytes closely parallels acquisition of sIg,[42] it has also been demonstrated that immature sIg^+ B cells do not express Ia determinants, but may be easily induced to do so.[43,44] This observation implies that an initial population of Ig^+Ia^- B lymphocytes may normally give rise to Ig^+Ia^+ cells. Recent data reveal that Ia is indeed lacking from splenic and liver B lymphocytes in fetal and neonatal mice and gradually appears on the cell surface during the first two weeks of life (Figure 4).[45] The H-2K antigen is present in detectable levels on a portion of B cells in fetal and neonatal tissue, but does not reach adult values until several days after birth. As would be expected from these results, murine pre-B cells lack Ia markers. Thus Ia antigens, along with sIg, FcR, and CR, can be used to delineate subpopulations of mouse B lymphocytes.

In humans, the counterpart to Ia antigen associated with the HLA-D locus is found early in gestation on cells of the B lineage,* and probably is expressed even on hemopoietic stem cells. Pre-B cells have surface molecules easily detected by fluoresceinated antibodies to an Ia-like antigen (anti-p23,30). It thus seems apparent that human B lymphoid cells may differ from their murine counterparts with regard to the expression of certain

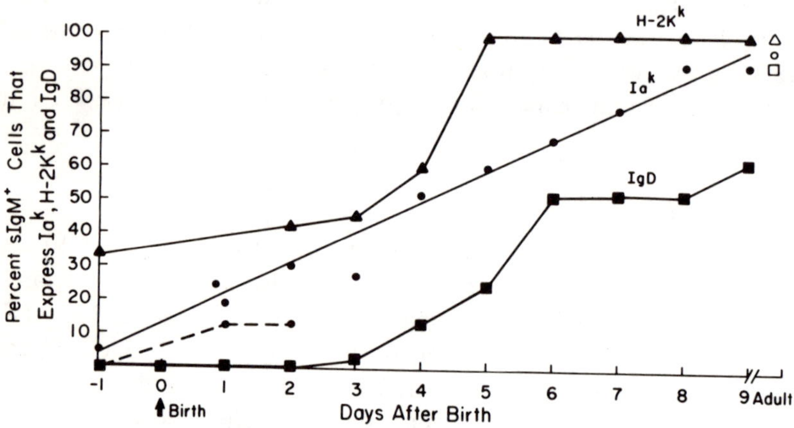

Figure 4. Appearance of Ia, IgD, and H-2K markers on B lymphocytes during development. Note the significant delay between detection of Ia versus IgD markers on B cells. sIgM is found on B lymphocyte membranes prior to either Ia or IgD.

* L. B. Vogler and coworkers: unpublished observations.

surface constituents, and therefore their possible role in regulation of antibody responses.

Mitogen Receptors

The discovery that high concentrations of certain antigens can stimulate multiple clones of B lymphocytes to divide and to differentiate into antibody-secreting plasma cells without the help of T cells led to an intensive investigation of this phenomenon and to many theories on the mechanisms of B-cell triggering.[46-48] Discrimination of mitogen receptors from antigen receptors on B cells has been difficult. It has been demonstrated that from 20 to 40 percent of B lymphocytes in adult mouse spleens can be stimulated by a particular mitogen—for example, lipopolysaccharide (LPS), lipoprotein, and dextrans—to synthesize polyclonal antibodies. All these cells may be assumed to exhibit distinct mitogen receptors, but could not possess specific antigen receptors for these molecules.[49] Strains of mice lacking the ability to respond to LPS have recently been reported to lack B-cell surface receptors for this mitogen; nevertheless, mice of genetic nonresponder strains are capable of recognizing LPS as an antigen when it is coupled to a carrier protein.[50] Moreover nonresponder mice have the capability of making antibodies to the putative LPS receptor on normal murine B lymphocytes, thus providing a useful reagent for future analyses of the LPS receptor.

Most of the currently available information about the ontogeny of mitogen receptors on B cells is indirect and derived from investigation of murine B-lymphocyte responses at various developmental ages. Cultured fetal liver cells require prolonged periods of incubation in vitro after exposure to the B-cell mitogens, LPS, dextran sulfate, and lipoprotein; regardless of the age of the fetal donor, responsive cells in the liver generate antibody-producing cells during a period in culture that corresponds to five to seven days after the birth of the animal.[51] Fetal liver cells adoptively transferred into X-irradiated adult mice also develop responsiveness to different mitogens in a specific order. Response to dextran sulfate could be demonstrated by eight days, to LPS at eleven days, and to PPD at fifteen days after transfer of cells.

Proliferative responses peaked earlier than did antibody production.[52] Other studies indicate that the expression of sIgM coincides with acquisition of the capacity of B cells to respond to LPS stimulation with plasma cell generation.[53-55] Thus expression of a functional receptor for LPS precedes the expression by developing B lymphocytes of several other B-lymphocyte surface receptors—FcR, CR, Ia, IgG, for example.

It has recently been proposed that mouse fetal liver cells as early as fifteen days of gestation may exhibit mitogen receptors.[56] From the foregoing information it would appear that the receptors are not functional, at least in terms of inducing maturation along B-cell lines. It is at least reasonable to suggest that the existence of a receptor during ontogeny may not coincide with its ability to function as a molecule capable of receiving and generating signals that the B lymphocyte is ready to translate into cell proliferation and/or differentiation.

STRUCTURE OF RECEPTORS (Table 1)

Immunoglobulins

Renewed interest is mounting with regard to the synthesis, structure, and insertion of membrane-bound IgM. Early studies of radio-iodinated IgM revealed that it is primarily monomeric (8S versus 19S for the secreted pentameric molecule), bound to a 39,000-45,000 M.W. membrane molecule, and contains carbo-

TABLE 1. BIOCHEMICAL CHARACTERISTICS OF B-CELL SURFACE COMPONENTS

Receptor	Estimated M.W.	Components	Sugars (Percent)	Susceptibility to Pronase	Susceptibility to Trypsin
IgM	180 K	2H 70-80 K 2L 22-23 K	10-15	++	—
IgD	180-200 K	2H 63-70 K 2L 22-23 K	15	++++	++++
γFcR	120 K	2? 65 K	?	++++	—
CR	?	?	?	++++	++++
Ia	55-58 K	34 K + 29 K	+	+++	+++

hydrate residues similar to those of secreted Ig molecules.[57-60] It was postulated that sIgM and secreted IgM have the same sites of synthesis and assembly, with the polymerization of the 8S subunit to 19S molecules occurring shortly before or simultaneously with secretion.[61] Other data suggest that membrane-bound IgM and secreted IgM are derived from different pools.[62]

Recently acquired information indicates that the membrane-associated molecule differs from intracellular and secreted IgM by its possession of an extra carboxy-terminal piece containing a high concentration of hydrophobic amino acids.[63] Moreover, nonsecreting bovine leukemic lymphocytes possess membrane IgM with a molecular weight 10,000 daltons greater than serum IgM.[64] Surface IgM binds detergents far more avidly than serum IgM.* The extra piece of membrane IgM contributes an increased acidic mobility to these molecules, as compared to cytoplasmic "precursors" when both are analyzed on two-dimensional acrylamide gel.[65] Pulse-chase experiments are consistent with conversion of cytoplasmic Ig with intermediate mobilities to surface Ig with more rapid migration in this system. Addition of carbohydrate residues cannot explain these differences. It has not been resolved whether insertion of IgM into specific proreceptor molecules—for example, FcR [66]—can account for the extra amino acid residues associated with Ig isolated from the cell surface.

The structure of the other major cell surface immunoglobulin, IgD, has been even more difficult to analyze due to its remarkable susceptibility to proteolysis and acidic denaturation.[67] Originally, acrylamide gel-electrophoresis patterns indicated that mouse IgD was significantly smaller than its human counterpart. More recent data show that this size variation is attributable to proteolytic artifacts incurred during the isolation procedure.[68] If degradation is minimized, membrane IgD is obtained that migrates more slowly than IgD myeloma proteins on SDS-PAGE, suggesting an extra structure, amino acid or carbohydrate, in membrane-bound IgD molecules of about 2,500 daltons.[69] IgD is attached to cell membranes by its Fc region; treatment with proteolytic enzymes—trypsin, thrombin, or fibrinogen, for example—

* R. M. E. Parkhouse: personal communication.

degrades the F(ab) portion, but leaves the Fc structure relatively intact. Further digestion will cleave the Fc fragment near its carboxy terminus insertion into the B-cell membrane.[70]

Secreted IgD has a carbohydrate content of approximately 15 percent, all sugar residues residing on the δ chain. Since the molecular weight of the δ chain is about 63,000 daltons, amino acids must comprise 53,000 daltons of these chains. It is interesting to note that, while most secreted IgD molecules in humans possess λ light chains ($k/\lambda=0.05$), membrane IgD has a relatively normal distribution of light chains ($k/\lambda=1.5$).[71, 72] Apparently subsets of IgD$^+$ lymphocytes are selected for terminal differentiation to IgD-secreting plasma cells by a mechanism involving the light chain class.

Because a minor proportion of B lymphocytes in blood, spleen, and lymph node cells possess sIgA and sIgG, little information exists about the structure of these immunoglobulins in B-cell membranes. Whether these isotypes at the cell surface possess different carboxy termini than their cytoplasmic or secreted counterparts remains a question for future study.

Receptors for Immunoglobulin Fc Portions (FcR)

The γFcR demonstrates exquisite sensitivity to pronase and phospholipase C digestion, yet it is quite resistant to proteolysis by trypsin and chymotrypsin. These results were interpreted to suggest that the FcR is composed of a phospholipoprotein inserted into the cell membrane.[73] Initial molecular weight determinations of solubilized FcR, isolated by binding to sepharose-linked aggregated human IgG, revealed polypeptides with molecular weights of 65,000, 18,000, and 15,000 daltons. Each molecule could complex with aggregated IgG. The 18,000- and 15,000-dalton polypeptides proved to be degradation products, probably of the 65,000 molecular weight protein.[74] Data from studies using multiple inhibitors of proteolytic enzymes reveal that even the 65,000-dalton molecule might be a degraded component of an FcR with a molecular weight of 120,000 daltons. The larger molecule seems to be a single-chain polypeptide; brief exposure to pronase results in rapid cleavage to smaller proteins with molecular weights of 65,000, 45,000, 30,000, 18,000, and 15,000 daltons.[75] Structural studies of the γFcR have not yet provided insight into the mechanism of binding to the IgG Fc region. Since receptor avidity for Fc varies with the cell type and physical presentation of IgG—for example, soluble versus aggregated versus bound to an insoluble matrix—it is conceivable that separate γFc receptors exist that may regulate distinct functional alterations on different cells, depending on the status of IgG interacting with them. Alternatively, the γFcR on specific cells may all be similar in basic composition yet exhibit novel distributions on cell surfaces.

Complement Component Receptors (CR)

Scant data exist on the structure of the receptors identified for complement components C3b, C3d, and C4. It can be inferred that CR are distinct from Ia, FcR, and sIg, since antisera directed against the latter molecules do not interfere with binding of activated complement components by B cells. Sensitivity of CR to trypsin suggests that CR contains protein that is involved in complement binding.[76,77]

Ia Antigen

Several methods have been used to show that the Ia antigen is distinct from sIg, other H-2 antigens, and FcR in the mouse. Absorption of H-2K or D antigens or Ig from cell lysates by specific antisera do not remove Ia binding to anti-Ia alloantisera.[78] The Ia antigens cap independently of other H-2 markers, sIg, FcR, and CR on B-cell surfaces.[79,80] The developmental sequence of Ia appearance on B cells also can be distinguished from the initial presentation of sIg, H-2 antigens, FcR, and CR, as mentioned earlier.

Surface Ia determinants are glycoprotein molecules having a molecular weight of ~58,000 daltons, which under reducing conditions migrate on SDS-PAGE as two components weighing 33,000 and 25,000 daltons.[81,82] Alloantigenicity resides in the protein portion of the molecule, since proteases and not glycosidases remove reactivity with alloantisera. It is noteworthy that all Ia determinants encoded by a specific subregion of the I locus appear to reside on one molecule after sequential immunoprecipitation. These studies indicate that distinct molecules are encoded by the I-A, I-B, and I-C subregions.[83, 84] More recent information indicates that Ia alloantigens in mice are composed of two chains, α and β, with molecular weights that vary according to both subregion and haplotype—$\alpha = 34$ or 31×10^3 daltons; $\beta = 26$ or 29×10^3 daltons. During detergent lysis of murine B lymphocytes, α and β chains appear to associate through disulfide linkages, which can be prevented by inclusion of alkylating agents in the lysis buffer.[85]

Using alloantisera from multiparous women or exhaustively adsorbed xenoantisera against human lymphoblastoid-cell lines, several groups have demonstrated the existence on human B lymphocytes of molecules with Ia-like qualities.[86, 87] These molecules possess molecular weights of 34,000 and 29,000 daltons under nonreducing conditions, and can be cleaved by exogenous or endogenous proteases to 30,000 and 23,000 dalton glycoproteins. Such antigens appear primarily on B lymphocytes and cells of the monocyte/macrophage series, but can also be seen under certain conditions on the surfaces of "null-", dendritic-, T-, and skin Langerhan-cells.[88, 89]

Mitogen Receptors

A. Coutinho and colleagues reasoned that antibodies may bind mitogens via sites on the antibody molecule very similar to the binding site on a mitogen receptor.[90] They used antibodies to the antigen-binding region (idiotype) of a dextran-binding murine myeloma (J558) to detect the dextran mitogen receptor on B lymphocytes. These receptors were shown to be distinct from sIg by capping experiments, and were present on ~20 percent of mature splenic-B lymphocytes. Anti-idiotype to the dextran-binding myeloma could apparently bind avidly to this receptor and trigger polyclonal antibody responses to the same extent as exposure to the mitogen itself. This finding, coupled with the recent demonstration that the T-cell-antigen receptor bears idiotype markers of the V_H region,[91-93] led to intriguing possibilities of regulation of immunoglobulin synthesis and modulation of B-cell clonal diversity by anti-idiotypic antibodies.[94]

FUNCTIONAL ROLE OF B-LYMPHOCYTE RECEPTORS

In this section we will discuss functional roles for each class of B-lymphocyte receptor described in preceding sections, realizing that this area of B-lymphocyte analysis still remains incomplete and controversial. In fact, for many of the recognizable components of B-cell surfaces, evidence for specificity of binding and functional consequences is still lacking. Conversely, B-lymphocyte surface components may bind mitogens or other ligands without demonstrable functional consequences.[95]

Surface Immunoglobulins (sIg)

The discovery of immunoglobulins on the surface of lymphocytes generated an explosion of information and theories concerning their function in regulating B-lymphocyte proliferation and differentiation into antibody-secreting plasma cells. Several investigators have demonstrated the ability of sIg molecules to bind specific antigens, and each B lymphocyte synthesizes antibody molecules of identical antigen-binding and idiotype specificities, independent of the isotype(s) that appear on its surface.[96,97] Several investigators have proposed that binding of antibody to sIg is a necessary initial step for induction of plasma cell differentiation under normal conditions.

Perturbation of sIgM by cross-linking antibodies or by multivalent antigens results in distinct changes in B-cell development and function. The effects of sIgM cross-linking by multivalent ligands vary depending on the stage in the life history of the B cell at which this occurs. R. G. Mage and S. Drey first showed that exposure of unborn or newborn rabbits heterozygous for kappa light-chain allotypes—b^4b^5 rabbits, for example—to antibodies directed toward the paternal allotype—anti-b_5, for example—inhibits the subsequent development of B cells expressing immunoglobulin molecules of the suppressed allotype.[98] Initial studies in this laboratory showed that when anti-μ was injected into chick embryos *in ovo* [99] or into neonatal mice,[100] abortion of the development of sIg^+ cells of all isotypes followed; pre-B cells are unaffected by this treatment.[101] Thus it appeared reasonable to propose that $sIgM^+$ B lymphocytes are the progenitors of B cells committed to synthesis of each of the Ig classes. This concept was further strengthened by in vitro experiments in which neonatal liver cells or adult spleen cells stimulated with the polyclonal B-cell mitogen, LPS, and treated with anti-μ were unable to differentiate to plasma cells of any class.[102,103] The dose of anti-μ required for suppression of cells at various developmental stages demonstrated a very rapid loss of susceptibility to anti-μ suppression by more mature B-lymphocytes (Figure 5). The ease

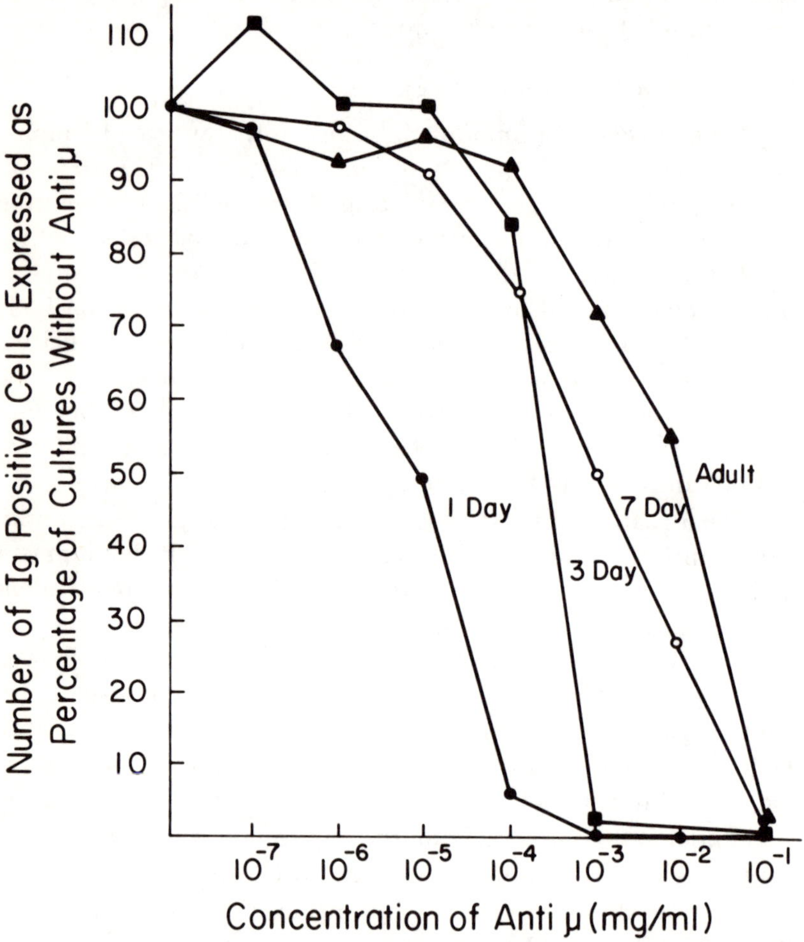

Figure 5. Suppression by anti-μ of LPS-induced B-lymphocyte differentiation at various developmental stages. Lymphocytes from neonatal mice can be inhibited with concentrations of anti-μ well below those necessary for suppression of adult B cells.

of functional modulation of B lymphocytes from newborn mice after treatment with anti-Ig parallels tolerance-induction requirements at various developmental stages. Newly formed B cells are made tolerant to very small doses of multivalent antigens in the absence of carrier-primed T cells, whereas adult spleen cells require 10,000-fold greater concentrations of the same antigen to produce tolerant cells.[104-106] The kinetics of dose and developmental susceptibility of B cells to antigenic tolerance induction directly parallel the data from anti-μ suppressed spleen cells in culture. In the former case, of course, the suppression is limited to the clones of B cells bearing antibodies to the particular antigen used for tolerance induction, whereas the suppression is panclonal when anti-μ antibodies are used. Another means through which selective clonal abortion can be effected is by treatment of neonates with antibodies to antigen-complimentarity determinants of a particular antibody, that is anti-idiotype antibodies.[107, 108]

In contrast to suppression of immature B lymphocytes with antibodies to IgM determinants, mature B cells can be induced to proliferate after exposure to anti-IgM. Either intact or $F(ab')_2$ fragments of antibodies reactive with IgM determinants can stimulate B cells.[109] Induction appears to depend more on the developmental stage of the lymphocyte; splenocytes from older mice are more readily triggered than cells from young mice, <6 months old. Only anti-μ, and not anti-γ, -δ, or -α, stimulates mouse B cells, and there appears to be no correlation with surface density of IgM or IgD.[110] Thus, depending on the maturational status of the B cells, anti-IgM can induce significantly different patterns of expression.

After exposure to either multivalent antigens or anti-Ig—intact or $F(ab')_2$ fragments—sIg begins to migrate rapidly toward one pole of the cell. This phenomenon, called capping, is energy-dependent and involves association of surface molecules with the cytoskeleton.[111-113] Two mechanisms have recently been proposed to account for the capping phenomenon. The first mechanism occurs after binding of a cross-linking ligand to a non-Ig receptor—anti-Ia or anti-H2K, for example; this event stimulates cell motility, resulting in increased membrane fluidity. The receptor-ligand complex remains anchored as the cell moves, and thus appears to aggregate at the pole away from the moving edge. Thus drugs that interfere with cell movement (colchicine, cytochalasin B, Ca^{++}, local quaternary anesthetics) will prevent capping in this system. The other mechanism is seen in lymphocytes after exposure to anti-Ig. Here cross-linking of ligands activates the cytoskeletal microfilaments and microtubules, with subsequent active migration of surface Ig to one pole, usually at the uropod. Capping is then followed by cell movement. This process appears to be less affected by agents that disrupt the cytoskeleton, since only combinations of colchicine plus cytochalasin B prevent capping of sIg, each drug alone being ineffective in altering redistribution of sIg. Motile B lymphocytes can cap sIg spontaneously without a cross-linking ligand, but these molecules then redistribute themselves after motility ceases.[114, 115] The propensity for spontaneous movement of surface molecules in motile lymphocytes creates special problems in determining whether sIg is distributed randomly on the cell surface.[116]

The rate of Ig receptor movement also seems to vary on B cells after exposure to antigen or anti-Ig. Molecules migrate in a discontinuous manner with apparent contractile jerks,[117] which contrasts distinctly with the continuous lipid flow model proposed by M. S. Bretscher.[118] Thus reorganization of sIgM by anti-μ or antigens stimulates dramatic changes in cell movement and expression; redistribution of sIg may provide signals necessary for antibody message translation, directly or in combination with a second receptor (enhanced by redistribution), or merely serve to focus antigen away from a second site, which acts to induce lymphocyte transformation.

Since the discovery of IgD on B-lymphocyte cell surfaces, the biological role this molecule plays during B-cell development and triggering remains a mystery.[119-122] Like IgM, this surface immunoglobulin can bind antigens, is usually detected on cells also bearing IgM (but not on all $sIgM^+$ cells), and possesses the same V region as the IgM with which it is coexpressed on a B cell's surface.[123-125] Two major immunological questions have been: Can one define a receptor function for sIgD related to B-cell triggering?; Does it possess some activity lacking in sIgM, or does it cooperate with sIgM or other receptors required for B-lymphocyte triggering?

As mentioned earlier, surface IgD is acquired relatively late in ontogeny, that is, after sIgM, sIgG, and sIgA expression, and is usually present coincidentally with other isotypes (Figure 3). Since immune tolerance can be induced far more readily using immature B cells ($\mu^+\delta^-$) rather than mature cells ($\mu^+\delta^+$), one postulated role for sIgD is a protective mechanism against tolerance induction. The utility of this function would be as follows: during development, fetal B lymphocytes are exposed to a wide variety of self antigens. Prolonged reactivity to such stimulants would interfere with subsequent development of the organism. Thus development of sIgM$^+$ B cells reacting against autoantigens would be aborted by the tolerance mechanism. After birth, exposure to a wide variety of environmental agents occurs. Tolerance to these antigens would be detrimental to existence and, as mentioned earlier, it is difficult to produce at this age. Many B lymphocytes during this developmental period have acquired both sIgM and sIgD.

Another possible role of sIgD might be to functionally recognize certain types of antigens. Thus neonatal spleen cells can respond to some antigens that do not require T cell help for plasma-cell generation (T-independent or TI), but are unable to be stimulated by other groups of TI antigens and antigens that need specific T-cell factors for B-cell responses (T-dependent or TD). Adult spleen B cells can be induced by both sets of antigens (TI and TD). Since sIgD is absent on neonatal cells and appears on lymphocytes during progression toward adulthood, it is conceivable that this Ig enables B cells to respond to specific classes of immunogens. A strain of mice with an X-linked deficiency in immune responsiveness to certain TI and TD antigens (CBA/N) possesses B cells with a surface phenotype typical of immature B lymphocytes, that is, a paucity of sIgD. The patterns of antigen reactivity observed in such mice bolster the correlation between sIg expression and response pattern to specific antigens.[126, 127]

Treatment of mouse spleen cells with anti-δ antisera affects subsequent plasma cell generation in a manner unique from exposure to anti-μ. Responses to several TI antigens [128,129] and polyclonal mitogens [130] are not reduced after incubation with anti-δ. Antibody production against TD antigens is severely reduced, however, when anti-δ is included in tissue cultures, again suggesting distinct roles for these two surface immunoglobulins.

Recently it has been proposed that sIgM and sIgD give complementary signals in the elicitation of an antibody response. Thus IgM receptors may elicit a signal for B-cell proliferation or tolerance induction, and IgD receptors may provide a signal for differentiation into a mature antibody-secreting plasma cell.[131]

Another controversial feature of sIgD concerns its presence on the surface of memory B cells committed to IgG synthesis. Various investigators have reported that memory cells may possess sIgM, sIgG, or sIgD.[132-137] Few investigations have examined whether multiple isotypes exist on memory lymphocytes, and have not considered the role of interactions between Ig receptor molecules on such cells. Thus it is still quite conceivable that some memory cells bear multiple isotypes that are systematically lost after triggering. Data supporting this contention are derived from sequential examinations of B-lymphocyte sIg after stimulation with polyclonal mitogens or antigens. Several investigators have shown either rapid or gradual loss of sIgD on previously δ^+ cells after exposure to a mitogen, while γ and α isotype markers either appear *de novo* or are unchanged.[138-140]

A novel proposed role for sIgD concerns its ability to elicit an anti-idiotype response.[141] After binding antigen the variable region of IgD would be removed from the cell surface, either by shedding or due to its extreme sensitivity to proteolysis. Antigen-antibody complexes could then generate an anti-idiotype response that would serve to regulate Ig production by recognizing cell-surface Ig remaining on the relevant cell.

Major differences in experimental design exist between most of the studies on the role of sIg in immune responses. Thus the disparate results obtained by various investigators may be related to the experimental protocol employed—immunization procedures, cell selection methods, assays for cell products, for example. It is also apparent that these studies do not coincidentally consider the function of other receptors that arise in parallel during development with the Ig isotype being examined. These variable factors create problems in assigning specific roles for sIg receptors.

FcR

Firm data concerning the precise role of the FcR in B-lymphocyte immune responses are thus far lacking. It is known that the γFcR on B cells is associated in its spatial distribution or mobility with both sIgM and Ia antigens. When sIgM is capped the FcR is drawn along with it. If the FcR is capped first, however, sIg does not appear to move coincidentally.[142] This intriguing relationship was found to occur primarily with sIgM and FcR, since antisera to IgD did not redistribute FcR.[143]

It was also noted that treatment of mouse lymphocytes with anti-Ia antisera prevented binding of IgG by Fc receptors,[144] while pretreatment of cells with aggregated IgG did not significantly interfere with subsequent binding of anti-Ia antibodies.[145] Antisera directed against other cell-surface constituents (Lyb4 antigen, MBLA, H-2K, and D and mouse Ig) also diminish FcR binding to antibody-coated erythrocytes, probably by steric hindrance. From these observations it was speculated that because of its close spatial relationship with Ia molecules, cocapping with sIg, and the postulated role of Ia antigens in receiving T-cell helper factors, FcR may act as an intermediate signal generator that relays information from T helper factor-Ia complexes and antigen-antibody receptors to a second messenger inside the cell.[146]

The FcR on B cells could also serve to dampen the antibody response, since incubation of mouse spleen cells in vitro with aggregated IgG diminishes subsequent plasma-cell generation without significantly affecting proliferation. Thus, after antigen stimulation and antibody production, antigen-antibody complexes may bind to the FcR and act as a feedback control mechanism for further Ig synthesis.[147]

Finally, the FcR could function as a proreceptor to bind sIg molecules to the cell membrane.[148] Since most B cells are $sIgM^+$, the need for the presence of γFcR on these cells is not easily explained. Populations of B cells do, however, exist with μ-, α-, and ϵ-FcR;[149-151] δFcR has not yet been detected.

Complement Receptor (CR)

Studies of the function of the complement receptor in triggering B lymphocyte responses again reveal a potpourri of conflicting evidence and opinions. Summarizing a large volume of information, the CR may function to augment certain T-dependent antibody responses. The CR could facilitate the frequency of a cell's encounters with specific antigens by binding Ag-Ab-C, in addition to antigen binding by surface Ig. Since it is known that C3 causes release of immune complexes from the CR, the subsequent loss of these complexes may permit B cells to respond to signals initiated at other receptors—sIg, Ia, Fc, mitogen.[152,153] Exogenous C3 does not appear to be necessary for B-cell triggering by polyclonal mitogens or T-independent antigens, and no correlation has been demonstrated between the ability of a mitogen to activate the complement pathway with its B cell-stimulating properties. Supporting the idea that the CR may be required for responses to specific classes of antigens is the data indicating that CR^- B lymphocytes are capable of responding to certain TI antigens and mitogens, but not to TD antigens.[154] Serum, depleted of C3 by anti-C3 antibodies, could support induction of murine spleen cells by TI antigens, but not by certain TD antigens.[155] Specific presentation of antigens in each group, however, may have determined responsiveness of B lymphocytes; thus the CR may facilitate T-B interactions only to soluble TD antigens by augmenting the stability of antigen-sIg unions.[156] Furthermore, there appears to be inhibition of complement binding by macrophages after treatment with anti-H-2 antisera,[157] which suggests that the CR can interact with other surface receptors known to be involved in B-cell responsiveness to TD antigens.

Ia Antigens

Extensive data support the contention that the Ir genes of the H-2 region play a crucial role in regulation of the immune response to a variety of antigens.[158,159] Products of the I loci are detected serologically on the surface of several cells involved in immune recognition, and they appear to be intimately associated with factors known to regulate B-T interactions and antigen responsiveness. T cells activated with specific carrier molecules produce factors that can either help or suppress antibody production to haptens coupled to the carrier; several of the factors produced by T cells have been shown to bear Ia determinants in both mice and humans.[160-163] Binding of these factors by Ia antigens may induce a signal to trigger B lymphocytes subsequent to antigen binding. Interactions with both receptors, then, would be a prerequisite for normal responsiveness to an immunogen; each signal separately might produce partial responses of either: a) cell growth (amplification); or b) cell differentiation and antibody synthesis; but neither is sufficient for complete induction.

Mitogens

A mitogen can be defined as a molecule that can generate a proliferative response in a population of cells. In the lymphoid system such compounds may stimulate B cells, T cells, or both.[164] Simple binding of a mitogen to a cell does not indicate that the cell bears a receptor that is able to transmit a proliferative signal. For example, both B and T lymphocytes bind the mitogens Con A and PHA, yet only T cells are triggered to divide after exposure to these molecules. It can be assumed that B cells lack functional mitogen receptors for the above lectins.

The controversy over the role of mitogen receptors in B-lymphocyte immune responses can be detailed as follows. One proposal suggests that the mitogen receptor exists as the sole surface molecule regulating ultimate B-cell triggering for antibody production. According to this view, other surface receptors behave in a passive manner to either focus antigen, sIg and CR, for example, or interfere with interactions at the mitogenic site (Fc). T-independent antigens would be the prototype molecules with mitogenic capacities.[165,166] In low concentrations these compounds could be bound to antigen receptors of appropriate clones of B cells, and focused to provide a proliferative and differentiated signal in the absence of T cells via interaction with a mitogen receptor. This concept was extended to suggest that all B-lymphocyte triggering was the result of one nonspecific stimulus, provided either by antigen, in the case of TI antigens, or T-cell factors for TD antigens.

The arguments against the one nonspecific signal mechanism for B-cell induction by substances with mitogenic capabilities are quite strong; samples of the opposing evidence are:

1. With several TI antigens, mitogenicity and generation of primary antibody responses are completely separate; polymers need not demonstrate immunogenic or mitogenic properties to act as carriers for the generation of primary antihapten responses.[167]

2. Polymyxin-B treatment of the B-cell mitogen, LPS, severely reduces its mitogenic activity while not interfering with its capacity to behave as a TI antigen.[168]

3. Generation of memory B cells to TD antigens can occur after exposure to antigen in the absence of T cells; [169] addition of T cells or a T-replacing factor (TRF) permitted differentiation to proceed.[170,171]

4. The presence of multiple isotypes on B cells responsive to various mitogens and antigens suggests a functional role for the molecules as active receptor molecules.[172]

5. Treatment of B lymphocytes capable of responding to LPS with anti-μ abolishes polyclonal antibody induction, suggesting that sIgM possesses a function independent of the mitogen receptor for B-cell stimulation.[173,174]

6. Anti-Ia antisera or blockage of FcR also severely diminished LPS induction of B lymphocytes.[175,176]

7. Finally, it has been shown that mitogens, like antigens, require presentation on Ia-positive cells in order to trigger responses of immunocompetent cells.[177]

These examples, plus numerous others, support the contention that a minimum of two signals are necessary to generate a secondary antibody response, and probably a primary one, too. The mitogen receptor could be comparable to a T cell-factor receptor in providing auxiliary or complementary information to a cell during its activation.

Recent evidence suggesting that the dextran receptor bears the idiotypic determinant of antidextran antibody permits speculation that these moieties are involved in a feedback network with T-cell factors and receptors to modulate immune responses. It has been proposed that the T-cell receptor for antigen possesses determinants comparable to the V_H region of specific antibodies. Anti-idiotype antibodies would then be able to interact with T-cell receptors and mitogen receptors to either enhance or diminish antibody production.[178]

Lymphocyte B (Lyb) Antigens

Several Lyb antigens have been detected on murine B lymphocyte membranes by using alloantisera developed in mice lacking these markers. These entities appear to define subpopulations of B lymphocytes, comparable to both FcR and CR. A functional role for one of these antigens, Lyb3, has been described recently.[179, 180] Anti-Lyb3 reacts with a subset of splenic B lymphocytes in mature mice. When B cells are treated with this antiserum, plus suboptimal doses of a T-dependent antigen, an enhanced antibody response is obtained. This effect is not demonstrable in $(CBA/N \times Balb/C)F_1$ male mice that lack the Lyb3 determinant. Anti-Lyb3 did not function by polyclonally activating B cells, nor were T cells required for its stimulative capacity. Thus Lyb3 may function to receive a second signal, which, in conjunction with specific antigens, generates B-lymphocyte differentiation. In this regard it may play a role similar to FcR and CR on B-cell membranes.

CONCLUSIONS

In this section we will attempt to consolidate the available information on expression and function of B-cell receptors and propose molecular mechanisms by which they may alter phenotypic expression. It is clear that subsets of B lymphocytes can be defined by the presence or absence of various receptors, and that these cells behave differently when subjected to similar conditions of growth and stimulation. Ontogenic studies demonstrate that multiple receptors become detectable and functional during the same stages of development, and that B cells rarely possess or lack only one specific receptor. Dissecting the contributions each of these molecules makes toward regulation of B-lymphocyte growth and differentiation thus becomes a very difficult task.

Certain pieces of information facilitate speculations about regulatory interactions of the surface constituents of B lymphocytes:

1. No evidence available contradicts the observation that B lymphocytes triggered to differentiate to plasma cells must possess sIg of at least one isotype.

2. A great deal of information indicates that receptors can migrate freely within the plane of the membrane, and, after binding a specific ligand, proceed to aggregate in certain patterns (patching or capping). This aggregation may involve receptor cross-linking, binding to other cell surface or internal molecules, for example, other receptors and actin, or stimulation of cell movement.

3. After binding to receptors, many receptor-ligand complexes can be internalized; entrance into the cell may involve capping of molecules or direct endocytosis of smaller complexes.

4. Certain hormones, both steroid and polypeptide, can bind to receptors present either in the cytosol, internal membranes (RER, SER, Golgi apparatus), or nucleus.

5. Many changes in cell metabolism—including variations in cGMP and cAMP levels, phosphorylation of membrane lipids, a burst of protein synthesis coupled with increased numbers of active free polysomes, and altered patterns of histone phosphorylation and acetylation—occur in B cells after exposure to a number of hormones, lectins, or antigens.

The first step in B-lymphocyte activation, then, would be to bind a specific antigen by Ia. Upon interaction, stimulation of associated adenyl cyclase would occur, resulting in a proliferative signal transmitted via changes in nuclear protein composition. Cross-linking of sIgM by anti-μ would initiate capping and internalization of IgM-anti-μ microfilaments. Breakdown products of this complex might then specifically interfere with enzymes involved in further cellular differentiation, for example, guanidyl cyclase.

A second signal, supplied by either a specific T-cell helper factor, nonspecific T-cell factor, or polyclonal mitogen would then be necessary to initiate differentiation by B lymphocytes to plasma cells. Activation of guanidyl cyclase associated with the receptors for these factors might alter phosphorylation of enzymes and proteins that would stimulate transcription and translation of molecules necessary for expression of specialized cell products. The mitogen receptor would necessarily be associated with both adenyl and guanidyl cyclase, since both cell proliferation and differentiation can be induced by polyclonal B-cell mitogens (Figure 6).

The Ia antigens are postulated to associate with specific T-cell factors. This receptor, unlike mitogen and nonspecific T-cell helper factors, would only interact with guanidyl cyclase, and thus be unable to send proliferative signals to the cell.

The role of other receptors on B cells, such as the ones for

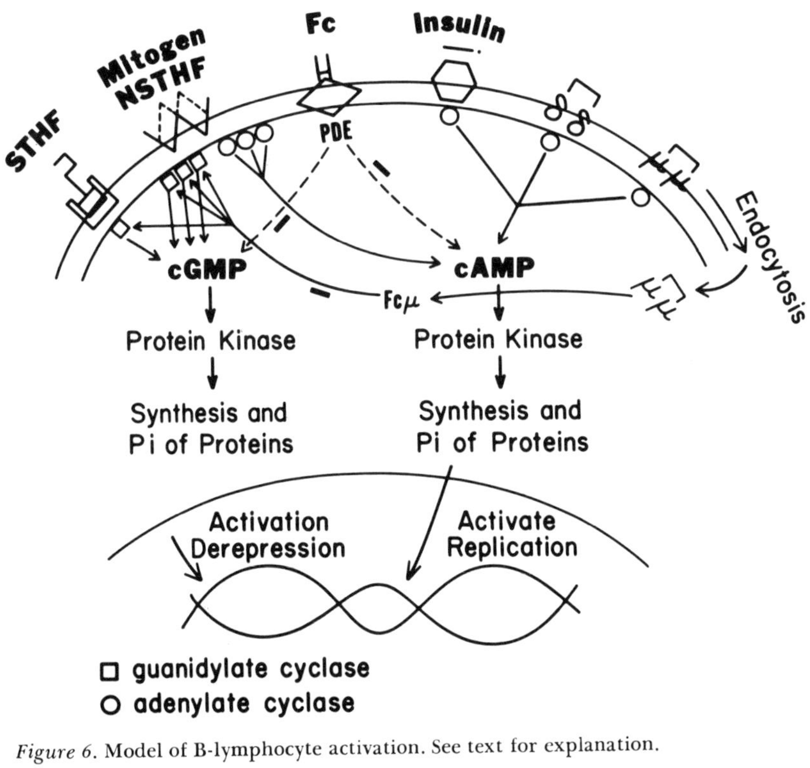

Figure 6. Model of B-lymphocyte activation. See text for explanation.

Fc and complement components, would be that of positive or negative modulators of events at the sIg, mitogen, or T-cell factor receptors. Both the Fc and complement receptors could focus any passively shed (Ag-Ab-C3) complexes on the cell surface and activate either adenyl or guanidyl cyclase. Alternatively, they may interact with enzymes that destroy cyclic nucleotides (phosphodiesterases) and thereby dampen activation processes. Since anti-Ia antibodies interfere with binding at Fcγ receptors, an inhibitory mechanism via FcR could be blocked by T-cell factors that have complexed to the Ia receptors.

The role of hormones in B-cell activation is still uncertain, but one could also speculate that they, too, cooperate in either amplifying or dampening signals initiated by other receptors through alteration of cyclic nucleotides or enzymes involved in stimulating protein synthesis and energy metabolism in B cells.

We have not mentioned mechanisms of binding between receptors and ligands, but multiple means of both specific and nonspecific interactions exist. Some of these include noncovalent protein-protein, protein-carbohydrate, carbohydrate-carbohydrate, and lipid-lipid interactions. The strength of these bonds would depend on a number of physicochemical considerations too involved to discuss here. Covalent alterations might involve cell-surface enzymes, such as exoproteases, exoglycosylases, and glycosyl transferases, which can bind numerous mitogens and factors known to be glycoproteins.

It can now be appreciated that the large number of receptors on the B lymphocyte surface play important roles in regulating the expression and terminal differentiation of a cell that constitutes a major sector of the host defense system. Thus a complex system of checks and balances and fine tuning of activation by multiple components seems only logical. With this recognition, it is clear that we have no more than barely scratched the surface of the B cell.

NOTES

1. J. J. T. Owen, M. D. Cooper, and M. C. Raff, "*In Vitro* Generation of B Lymphocytes in Mouse Foetal Liver, A Mammalian 'Bursa Equivalent'," *Nature* 249 (1974): 361-63.

2. M. C. Raff et al., "Early Production of Intracellular IgM by B Lymphocyte Precursors in Mouse," *Nature* 259 (1976): 224-26.

3. Ibid.

4. M. C. Raff et al., "Differences in Susceptibility of Mature and Immature Mouse B-Lymphocytes to Anti-Immunoglobulin-Induced Immunoglobulin Suppression *In Vitro*," *Journal of Experimental Medicine* 142 (1975): 1052-64.

5. F. Melchers, H. Von Boehmer, and R. A. Phillips, "B Lymphocyte Subpopulations in the Mouse. Organ Distribution and Ontogeny of Immunoglobulin Synthesizing and Mitogen-Sensitive Cells," *Transplantation Review* 25 (1975): 26-58.

6. P. D. Burrows et al., "Pre-B Cells: Bone Marrow Persistence in Anti-μ Suppressed Mice, Conversion to B Lymphocytes, and Recovery After Destruction by Cyclophosphamide," *Journal of Immunology* 120 (1978): 1526-31.

7. J. Andersson et al., "Clonal Growth and Maturation to Immunoglobulin Secretion *in vitro* of Every Growth-Inducible B Lymphocyte," *Cell* 10 (1977): 27-34.

8. Y. J. Rosenberg and C. R. Parish, "Ontogeny of the Antibody-Forming Cell Line in Mice. IV. Appearance of Cells Bearing Fc Receptors, Complement Receptors and Surface Immunoglobulin," *Journal of Immunology* 118 (1977): 612-17.

9. T. A. Andrew and J. J. T. Owen, "Studies on the Earliest Sites of B-Cell Differentiation in the Mouse Embryo," *Developmental and Comparative Immunology* 2 (1978): 339-46.

10. W. E. Gathings, A. R. Lawton, and M. D. Cooper, "Immunofluorescent Studies of the Development of Pre-B Cells, B Lymphocytes and Immunoglobulin Isotype Diversity in Humans," *European Journal of Immunology* 7 (1977): 804-10.

11. F. Melchers, "Murine Embryonic B-Lymphocyte Development in Placenta," submitted for publication in *Nature*.

12. G. J. V. Nossal and B. L. Pike, "Evidence for the Clonal Abortion Theory of B-Lymphocyte Tolerance," *Journal of Experimental Medicine* 141 (1975): 904-17.

13. ——————, "Studies on the Differentiation of B Lymphocytes in the Mouse," *Immunology* 25 (1973): 33-45.

14. E. R. Abney et al., "Sequential Expression of Immunoglobulin on Developing Mouse B Lymphocytes: A Systematic Survey That Suggests a Model for the Generation of Immunoglobulin Isotype Diversity," *Journal of Immunology* 120 (1978): 2041-49.

15. P. G. Spear et al., "Characterization of Splenic Lymphoid Cells in Fetal and Newborn Mice," *Journal of Experimental Medicine* 138 (1973): 557-73.

16. G. J. V. Nossal and B. L. Pike, "Differentiation of B Lymphocytes from Stem Cell Precursors," *Advances in Experimental Medicine and Biology* 29 (1972): 11-18.

17. Gathings, Lawton, and Cooper, "Immunofluorescent Studies" (See note 10).

18. J. M. Davie et al., "Ontogeny of Immunoglobulin-Bearing Lymphocytes and DNP-Specific Antigen-Binding Cells in Guinea Pigs," *Journal of Immunology* 112 (1974): 2202-09.

19. Abney et al., "Expression of Immunoglobulin" (See note 14).

20. A. R. Lawton, J. F. Kearney, and M. D. Cooper, "Control of Expression of C-Region Genes During Development of B Cells," in *Progress in Immunology. III. Proceedings of the Third International Congress on Immunology,* ed. T. E. Mandel et al., (New York: Elsevier North-Holland, Inc., 1977): 171-82.

21. A. R. Lawton and M. D. Cooper, "Modification of B-Lymphocyte Differen-

tiation by Anti-Immunoglobulins," in *Contemporary Topics in Immunobiology,* vol. 3, ed. M. D. Cooper and N. L. Warner (New York: Plenum Publishing Company, 1974): 193-225.

22. C. W. Pierce, S. M. Solliday, and R. Asofsky, "Immune Responses *in vitro.* IV. Suppression of Primary γM, γG and γA Plaque-Forming Cell Responses in Mouse Spleen Cell Cultures by Class-Specific Antibody to Mouse Immunoglobulins," *Journal of Experimental Medicine* 135 (1972): 675-97.

23. S. K. Pierce and N. R. Klinman, "The Allogeneic Bisection of Carrier-Specific Enhancement of Monoclonal B-Cell Responses," *Journal of Experimental Medicine* 142 (1975): 1165-79.

24. J. M. Davie and W. E. Paul, "Role of T Lymphocytes in the Humoral Immune Response. I. Proliferation of B Lymphocytes in Thymus-Deprived Mice," *Journal of Immunology* 113 (1974): 1438-45.

25. D. H. Sherr, M. R. Szewczuk, and G. W. Siskind, "Ontogeny of B-Lymphocyte Function. V. Thymus Cell Involvement in the Functional Maturation of B-Lymphocytes from Fetal Mice Transferred into Adult Irradiated Hosts," *Journal of Experimental Medicine* 147 (1978): 196-206.

26. E. Rabellino et al., "Immunoglobulins on the Surface of Lymphocytes. I. Distribution and Quantitation," *Journal of Experimental Medicine* 133 (1971): 156-67.

27. C. L. Sidman and E. R. Unanue, "Development of B Lymphocytes. I. Cell Populations and a Critical Event During Ontogeny," *Journal of Immunology* 114 (1975): 1730-35.

28. A. Basten et al., "A Receptor for Antibody on B Lymphocytes. I. Method of Detection and Functional Significance," *Journal of Experimental Medicine* 135 (1972): 610-26.

29. A. Basten, N. L. Warner, and T. Mandel, "A Receptor for Antibody on B Lymphocytes. II. Immunochemical and Electron Microscopy Characteristics," *Journal of Experimental Medicine* 135 (1972): 627-42.

30. H. B. Dickler, "Lymphocyte Receptors for Immunoglobulin," *Advances in Immunology* 24 (1976): 167-214.

31. H. B. Dickler and H. G. Kunkel, "Interaction of Aggregated γ-Globulin with B Lymphocytes," *Journal of Experimental Medicine* 136 (1972): 191-96.

32. S. S. Frøland, F. Wisløff, and T. E. Michaelsen, "Human Lymphocytes with Receptors for IgG. A Population of Cells Distinct from T- and B-Lymphocytes," *International Archives of Allergy and Applied Immunology* 47 (1974): 124-38.

33. Basten et al., "Receptor. Detection and Significance" (See note 28).

34. Basten, Warner, and Mandel, "Receptor. Characteristics" (See note 29).

35. L. Forni and B. Pernis, "Interactions Between Fc Receptors and Membrane Immunoglobulins on B Lymphocytes," in *Membrane Receptors of Lymphocytes,* ed. M. Seligmann, J. L. Preud'homme, and F. M. Kourilsky (New York: American Elsevier Publishing Company, 1975): 193-201.

36. C. Bianco, R. Patrick, and V. Nussenzweig, "A Population of Lymphocytes Bearing a Membrane Receptor for Antigen-Antibody-Complement Complexes," *Journal of Experimental Medicine* 132 (1970): 702-20.

37. C. Bianco and V. Nussenzweig, "Theta-Bearing and Complement-Receptor Lymphocytes Are Distinct Populations," *Science* 173 (1971): 154-55.

38. Sidman and Unanue, "Cell Populations" (See note 27).

39. M. C. Gelfand et al., "Ontogeny of B Lymphocytes. II. Relative Rates of Appearance of Lymphocytes Bearing Surface Immunoglobulin and Complement Receptors," *Journal of Experimental Medicine* 139 (1974): 1125-41.

40. D. H. Katz and B. Benacerraf, "Genetic Control of Lymphocyte Interactions and Differentiation," in *The Role of Products of the Histocompatibility Gene Complex in Immune Responses,* ed. D. H. Katz and B. Benacerraf (New York: Academic Press, 1976): 355-85.

41. J. L. Press et al., "Ia Antigens on B Cells: Relationships to B-Cell Precursor Function and to Surface Immunoglobulin," in *Membrane Receptors of Lymphocytes,* ed. M. Seligmann, J. L. Preud'homme, and F. M. Kourilsky (Amsterdam: North-Holland Publishing Company, 1975): 247-62.

42. J. L. Press, "Splenic Expression of Ia Alloantigens During Murine Neonatal Development," *Federation Proceedings* 34 (1975): 945.

43. U. Hämmerling et al., "The Ontogeny of Murine B Lymphocytes. I. Induction of Phenotypic Conversion of Ia^- to Ia^+ Lymphocytes," *Journal of Immunology* 115 (1975): 1425-31.

44. U. Hämmerling, A. F. Chin, and J. Abbott, "Ontogeny of Murine B Lymphocytes: Sequence of B-Cell Differentiation from Surface-Immunoglobulin-Negative Precursors to Plasma Cells," *Proceedings of the National Academy of Sciences* 73 (1976): 2008-12.

45. J. F. Kearney et al., "Ontogeny of Ia and IgD on IgM-Bearing B Lymphocytes in Man," *Journal of Experimental Medicine* 146 (1977): 297-301.

46. J. Watson, E. Trenkner, and W. Cohn, "The Use of Bacterial Lipopolysaccharides to Show That Two Signals Are Required for the Induction of Antibody Synthesis," *Journal of Experimental Medicine* 138 (1973): 699-714.

47. A. Coutinho and G. Möller, "Immune Activation of B Cells: Evidence for 'One Nonspecific Triggering Signal' Not Delivered by the Ig Receptor," *Scandinavian Journal of Immunology* 3 (1974): 133-46.

48. ——————, "The Self-Nonself Discrimination: A One-Signal Mechanism," *Scandinavian Journal of Immunology* 4 (1975): 99-102.

49. Andersson et al., "Clonal Growth" (See note 7).

50. J. Watson and R. Riblett, "Genetic Control of Responses to Bacterial Lipopolysaccharides in Mice. II. A Gene That Influences a Membrane Component Involved in the Activation of Bone Marrow-Derived Lymphocytes by Lipopolysaccharides," *Journal of Immunology* 114 (1975): 1462-68.

51. F. Melchers, "Immunoglobulin Synthesis and Mitogen Reactivity: Markers for B-Lymphocyte Differentiation," in *Development of Host Defenses,* ed. M. D. Cooper and D. H. Dayton (New York: Raven Press, 1977): 11-24.

52. E. Gronowicz, A. Coutinho, and G. Möller, "Differentiation of B Cells: Sequential Appearance of Responsiveness to Polyclonal Activators," *Scandinavian Journal of Immunology* 3 (1974): 413-21.

53. J. F. Kearney and A. R. Lawton, "B-Lymphocyte Differentiation Induced by Lipopolysaccharide. I. Generation of Cells Synthesizing Four Major Immunoglobulin Classes," *Journal of Immunology* 115 (1975): 671-76.

54. ——————, "B-Lymphocyte Differentiation Induced by Lipopolysaccharide. II. Response of Fetal Lymphocytes," *Journal of Immunology* 115 (1975): 677-81.

55. F. Melchers, J. Andersson, and R. A. Phillips, "Ontogeny of Murine B Lymphocytes: Development of Ig Synthesis and of Reactivities to Mitogens and

Anti-Ig Antibodies," *Cold Spring Harbor Symposia on Quantitative Biology* 41 (1976): 147-58.

56. A. Coutinho, L. Forni, and B. Blomberg, "Shared Antigenic Determinants by Mitogen Receptors and Antibody Molecules to the Same Thymus-Independent Antigen," *Journal of Experimental Medicine* 148 (1978): 862-70.

57. E. S. Vitetta, S. Baur, and J. W. Uhr, "Cell-Surface Immunoglobulin. II. Isolation and Characterization of Immunoglobulin from Mouse Splenic Lymphocytes," *Journal of Experimental Medicine* 134 (1971): 242-64.

58. E. S. Vitetta and J. W. Uhr, "Cell-Surface Immunoglobulin. IX. A New Method for the Study of Synthesis, Intracellular Transport and Exteriorization in Murine Splenocytes," *Journal of Experimental Medicine* 139 (1974): 1599-1620.

59. J. J. Marchalonis, R. E. Cone, and J. C. Atwell, "Isolation and Partial Characterization of Lymphocyte Surface Immunoglobulins," *Journal of Experimental Medicine* 135 (1972): 956-71.

60. E. Premkumar, P. A. Singer, and A. R. Williamson, "A Human Lymphoid Cell Line Secreting Immunoglobulin G and Retaining Immunoglobulin M in the Plasma Membrane," *Cell* 5 (1975): 87-92.

61. R. M. E. Parkhouse and B. Askonas, "Immunoglobulin M Biosynthesis. Intracellular Accumulation of 7S Subunits," *Biochemistry Journal* 115 (1969): 163-69.

62. Y-S. Choi, "Biosynthesis of Membrane Bound Ig and Secretion of Ig by Chicken Lymphoid Cells," *Biochemistry* 15 (1976): 1037-42.

63. P. B. Williams, R. T. Kubo, and H. M. Grey, "μ-Chains from a Nonsecretor B-Cell Line Differ from Secreted μ-Chains at the C-Terminal End," *Journal of Immunology* 121 (1978): 2435-39.

64. J. Haimovich, "Membrane Immunoglobulins of Bovine Lymphocytes," *Immunochemistry* 14 (1977): 337-40.

65. J. Goding, "Characterization of Intracellular, Membrane and Secreted IgM," in *Proceedings of an International Conference on B Lymphocytes in the Immune Response,* ed. M. D. Cooper et al., (New York: Elsevier North-Holland, 1979): forthcoming.

66. R. Ramasamy, A. Munro, and C. Milstein, "Possible Role for the Fc Receptor on B Lymphocytes," *Nature* 249 (1974): 573-74.

67. A. Bourgois, E. R. Abney, and R. M. E. Parkhouse, "Mouse Immunoglobulin Receptors on Lymphocytes: Identification of IgM and IgD Molecules by Tryptic Cleavage and a Postulated Role for Cell Surface IgD," *European Journal of Immunology* 7 (1977): 210-13.

68. R. M. E. Parkhouse and M. D. Cooper, "A Model for the Differentiation of B Lymphocytes with Implications for the Biological Role of IgD," *Immunological Review* 37 (1977): 105-26.

69. H. L. Spiegelberg, "The Structure and Biology of Human IgD," *Immunological Review* 37 (1977): 3-24.

70. S. M. Goyert, T. E. Hugh, and H. L. Spiegelberg, "Sites of 'Spontaneous' Degradation of IgD," *Journal of Immunology* 118 (1977): 2138-42.

71. B. Pernis, G. Chiappino, and D. S. Rowe, "Cells Producing IgD Immunoglobulins in Human Spleen," *Nature* 211 (1966): 424-25.

72. D. S. Rowe et al., "Immunoglobulin D as a Lymphocyte Receptor," *Journal of Experimental Medicine* 138 (1973): 965-72.

73. A. Basten et al., "A Receptor for Antibody on B Lymphocytes. III. Rela-

tionship of the Receptor to Immunoglobulin and Ia Determinants," *International Archives of Allergy and Applied Immunology* 50 (1976): 309-21.

74. L. Rask et al., "Isolation and Properties of a Murine Spleen Cell Fc Receptor," *Nature* 257 (1975): 231-33.

75. A. Bourgois, E. R. Abney, R. M. E. Parkhouse, "Structure of Mouse Fc Receptor," *European Journal of Immunology* 7 (1977): 691-95.

76. Bianco, Patrick, and Nussenzweig, "Antigen-Antibody-Complement Complexes" (See note 36).

77. Bianco and Nussenzweig, "Theta-Bearing Lymphocytes" (See note 37).

78. S. E. Cullen et al., "Evidence for More Than One Ia Antigenic Specificity on Molecules Determined by the I-A Subregion of the Mouse Major Histocompatibility Complex," *Journal of Immunology* 116 (1976): 549-53.

79. E. R. Unanue et al., "The Presence of I-Region-Associated Antigens on B Cells in Molecules Distinct from Immunoglobulin and H-2K and H-2D," *Proceedings of the National Academy of Sciences* 71 (1974): 5014-16.

80. J. W. Goding, E. White, and J. J. Marchalonis, "Partial Characterization of Ia Antigens on Murine Thymocytes," *Nature* 257 (1975): 230-31.

81. Cullen et al., "Evidence for Antigenic Specificity" (See note 78).

82. B. D. Schwartz and S. E. Cullen, "The Chemistry of Mouse and Guinea Pig Ia Antigens," in *The Role of Products of the Histocompatibility Gene Complex in Immune Responses,* ed. D. H. Katz and B. Benacerraf (New York: Academic Press, 1976): 691-701.

83. Cullen et al., Evidence for Antigenic Specificity" (See note 78).

84. J. Klein, "The Major Histocompatibility Complex of the Mouse," *Science* 203 (1979): 516-21.

85. R. G. Cook et al., "Structural Studies on the Murine Ia Alloantigens. II. Molecular Weight Characterization of the Products of the I-A and I-E/C Subregions," *Journal of Immunology* 121 (1978): 2257-62.

86. R. J. Winchester et al., "Recognition by Pregnancy Serums of Non-HL-A Alloantigens Selectively Expressed on B Lymphocytes," *Journal of Experimental Medicine* 141 (1975): 924-29.

87. J. L. Strominger et al., "Isolation and Structure of Products of the Human Histocompatibility Gene Complex," in *The Role of Products of the Histocompatibility Gene Complex in Immune Responses,* ed. D. H. Katz and B. Benacerraf (New York: Academic Press, 1976): 621-43.

88. J. A. Frelinger et al., "Evidence for the Expression of Ia (H-2-Associated) Antigens on Thymus-Derived Lymphocytes," *Journal of Experimental Medicine* 140 (1974): 1273-84.

89. G. J. Hämmerling et al., "Tissue Distribution of Ia Antigens: Ia Spermatozoa, Macrophages and Epidermal Cells," *Immunogenetics* 1 (1975): 428.

90. Coutinho, Forni, and Blomberg, "Shared Antigenic Determinants" (See note 56).

91. H. Binz and H. Wigzell, "Shared Idiotypic Determinants on B and T Lymphocytes Reactive Against the Same Antigenic Determinants. I. Demonstration of Similar or Identical Idiotypes on IpG Molecules and T-Cell Receptors with Specificity for the Same Alloantigens," *Journal of Experimental Medicine* 142 (1975): 197-211.

92. ————, "Shared Idiotypic Determinants on B and T Lymphocytes

Reactive Against the Same Antigenic Determinants. V. Biochemical and Serological Characteristics of Naturally Occurring Soluble Antigen-Binding T-Lymphocyte Derived Molecules," *Scandinavian Journal of Immunology* 5 (1976): 559-71.

93. K. Eichmann and K. Rajewsky, "Induction of T and B Cell Immunity by Anti-Idiotypic Antibody," *European Journal of Immunology* 5 (1975): 661-66.

94. N. K. Jerne, "Towards a Network Theory of the Immune System," *Annales de l'Institut Pasteur* 125C (1974): 373-89.

95. G. Janossy and M. Greaves, "Functional Analysis of Murine and Human B-Lymphocyte Subsets," *Transplantation Review* 24 (1975): 177-236.

96. F. Salsano et al., "Same Idiotype of B-Lymphocyte Membrane IgD and IgM. Formal Evidence for Monoclonality of Chronic Lymphocytic Leukemia Cells," *Scandinavian Journal of Immunology* 3 (1974): 841-46.

97. S. M. Fu, R. J. Winchester, and H. G. Kunkel, "Similar Idiotypic Specificity for the Membrane IgD and IgM of Human B Lymphocytes," *Journal of Immunology* 114 (1975): 250-52.

98. R. G. Mage and S. Dray, "Persistent Altered Phenotypic Expression of Allelic γG-Immunoglobulin Allotypes in Heterozygous Rabbits Exposed to Isoantibodies in Fetal and Neonatal Life," *Journal of Immunology* 95 (1965): 525.

99. P. W. Kincade, A. R. Lawton, and M. D. Cooper, "Restriction of Surface Immunoglobulin Determinants to Lymphocytes of the Plasma Cell Line," *Journal of Immunology* 106 (1971): 1421-23.

100. A. R. Lawton et al., "Suppression of Immunoglobulin Class Synthesis in Mice. I. Effects of Treatment with Antibody to μ Chain," *Journal of Experimental Medicine* 135 (1972): 277-97.

101. Burrows et al., "Pre-B Cells" (See note 6).

102. J. Andersson, W. W. Bullock, and F. Melchers, "Inhibition of Mitogenic Stimulation of Mouse Lymphocytes by Anti-Mouse Immunoglobulin Antibodies. I. Mode of Action," *European Journal of Immunology* 4 (1974): 715-22.

103. J. F. Kearney, M. D. Cooper, and A. R. Lawton, "B-Lymphocyte Differentiation Induced by Lipopolysaccharide. III. Suppression of B-Cell Maturation by Anti-Mouse Immunoglobulin Antibodies," *Journal of Immunology* 116 (1976): 1664-68.

104. Nossal and Pike, "Clonal Abortion Theory" (See note 12).

105. E. S. Metcalf and N. R. Klinman, "*In vitro* Tolerance Induction of Neonatal Murine B Cells," *Journal of Experimental Medicine* 143 (1976): 1327-40.

106. ——————, "*In vitro* Tolerance Induction of Bone-Marrow Cells: A Marker for B-Cell Maturation," *Journal of Immunology* 118 (1977): 2111-16.

107. D. S. Strayer et al., "Neonatal Tolerance Induced by Antibody Against Antigen Specific Receptors," *Science* 186 (1974): 640-42.

108. A. Augustin and H. Cosenza, "Expression of New Idiotypes Following Neonatal Idiotypic Suppression of a Dominant Clone," *European Journal of Immunology* 6 (1976): 497-501.

109. S. Sell and P. G. H. Gell, "Studies on Rabbit Lymphocytes *in vitro*. I. Stimulation of Blast Transformation with an Antiallotype Serum," *Journal of Experimental Medicine* 122 (1965): 423-39.

110. H. L. Weiner, J. W. Moorhead, and H. N. Claman, "Anti-Immunoglobulin Stimulation of Murine Lymphocytes. I. Age Dependency of the Proliferative Response," *Journal of Immunology* 116 (1976): 1656-61.

111. R. B. Taylor et al., "Redistribution and Pinocytosis of Lymphocyte Surface Immunoglobulin Molecules Induced by Anti-Immunoglobulin Antibody," *Nature (New Biology)* 233 (1971): 225-29.

112. G. F. Schreiner and E. R. Unanue, "Membrane and Cytoplasmic Changes in B Lymphocytes Induced by Ligand-Surface Immunoglobulin Interaction," *Advances in Immunology* 24 (1976): 37-165.

113. G. C. Gabiani et al., "Actin and Tubulin Co-Cap with Surface Immunoglobulins in Mouse B Lymphocytes," *Nature* 269 (1977): 697-98.

114. Schreiner and Unanue, "Membrane and Cytoplasmic Changes" (See note 112).

115. ————, "Capping and the Lymphocyte: Models for Membrane Reorganization," *Journal of Immunology* 119 (1977): 1549-51.

116. A. S. Perelson, "Spatial Distribution of Surface Immunoglobulin on B Lymphocytes," *Experimental Cell Research* 112 (1978): 309-21.

117. M. A. Feitelson and R. F. Ashman, " 'Stop and Go' Kinetics of Receptor Movements Induced by Anti-Ig or Antigen on Individual Lymphocytes," *Cellular Immunology* 34 (1977): 310-18.

118. M. S. Bretscher, "Directed Lipid Flow in Cell Membranes," *Nature* 260 (1976): 21-25.

119. Rowe et al., "Immunoglobulin D" (See note 72).

120. J. A. Van Boxel et al., "IgD-Bearing Human Lymphocytes," *Journal of Immunology* 109 (1972): 648-51.

121. U. Melcher et al., "Cell-Surface Immunoglobulin. X. Identification of an IgD-Like Molecule on the Surface of Murine Splenocytes," *Journal of Experimental Medicine* 140 (1974): 1427-31.

122. E. R. Abney and R. M. E. Parkhouse, "Candidate for Immunoglobulin D Present on Murine B Lymphocytes," *Nature* 252 (1974): 600-02.

123. Salsano et al., "Evidence for Monoclonality" (See note 96).

124. Fu, Winchester, and Kunkel, "Idiotypic Specificity" (See note 97).

125. B. Pernis, J. C. Brouet, and M. Seligmann, "IgD and IgM on the Membrane of Lymphoid Cells in Macroglobulinemia. Evidence for Identity of Membrane IgD and IgM Antibody in a Case with Anti-IgG Receptors," *European Journal of Immunology* 4 (1974): 776-78.

126. D. E. Mosier, I. Scher, and W. E. Paul, "*In vitro* Responses of CBA/N/Mice: Spleen Cells of Mice with an X-Linked Defect That Precludes Immune Responses to Several Thymus-Independent Antigens Can Respond to TNP-Lipopolysaccharide," *Journal of Immunology* 117 (1976): 1363-69.

127. D. E. Mosier et al., "Surface Immunoglobulin D as a Functional Receptor for a Subclass of B Lymphocytes," *Immunological Reviews* 37 (1977): 89-104.

128. F. S. Ligler et al., "Inactivation of Antigen-Responsive Clones with Antisera Specific for IgM or IgD," *Journal of Immunology* 120 (1978): 1139-42.

129. J. C. Cambier et al., "Blocking of Primary *in vitro* Antibody Responses to Thymus-Independent and Thymus-Dependent Antigens with Antiserum Specific for IgM or IgD," *Proceedings of the National Academy of Sciences* 75 (1978): 432-35.

130. J. F. Kearney et al., "B-Cell Differentiation Induced by Lipopolysaccharide. V. Suppression of Plasma Cell Maturation by Anti-μ: Mode of Action and Characteristics of Suppressed Cells," *Journal of Immunology* 120 (1978): 158-66.

131. J. R. Kettman et al. "The Role of Receptor IgM and IgD in Determining Triggering and Induction of Tolerance in Murine B Cells," *Immunological Reviews* 43 (1979): 69-95.

132. E. R. Abney et al., "Immunoglobulin M Receptors on Memory Cells of Immunoglobulin G Antibody-Forming Cell Clones," *European Journal of Immunology* 6 (1976): 443-50.

133. S. Strober, "Maturation of B Lymphocytes in Rats. III. Two Subpopulations of Memory B Cells in the Thoracic Duct Lymph Differ by Size, Turnover Rate and Surface Immunoglobulin," *Journal of Immunology* 117 (1976): 1288-94.

134. I. Zan-Bar, S. Strober, and E. S. Vitetta, "The Relationship Between Surface Ig Isotype and Immune Function of Murine B Lymphocytes. I. Surface Ig Isotypes on Primed B Cells in the Spleen," *Journal of Experimental Medicine* 145 (1977): 1188-1205.

135. K. Okumura et al., "Demonstration That IgG Memory Is Carried by IgG Bearing Cells," *European Journal of Immunology* 6 (1976): 467-72.

136. R. L. Coffman and M. Cohn, "The Class of Surface Ig on Virgin and Memory B Lymphocytes," *Journal of Immunology* 118 (1977): 1806-15.

137. S. J. Black et al., "Expression of IgD by Murine Lymphocytes. Loss of Surface IgD Indicates Maturation of Memory B Cells," *Journal of Experimental Medicine* 147 (1978): 984-96.

138. M. D. Cooper et al., "Generation of Immunoglobulin Class Diversity in B Cells: A Discussion with Emphasis on IgD Development," *Annales de l'Institut Pasteur* 127C (1976): 573-81.

139. A. Bourgois et al., "Surface Immunogobulins of Lipopolysaccharide-Stimulated Spleen Cells. The Behavior of IgM, IgD and IgG," *European Journal of Immunology* 7 (1976): 151-53.

140. J. L. Preud'homme, "Loss of Surface IgD by Human B Lymphocytes During Polyclonal Activation," *European Journal of Immunology* 7 (1977): 191-93.

141. Bourgois, Abney, and Parkhouse, "Identification of Molecules" (See note 67).

142. A. K. Abbas and E. R. Unanue, "Inter-Relationships of Surface Immunoglobulins and Fc Receptors on Mouse B Lymphocytes," *Journal of Immunology* 115 (1975): 1665–71.

143. Forni and Pernis, "Fc Receptors and Immunoglobulins" (See note 35).

144. H. B. Dickler and D. H. Sachs, "Evidence for Identity or Close Association of the Fc Receptor of B Lymphocytes and Alloantigens Determined by the IR Region of the H-2 Complex," *Journal of Experimental Medicine* 140 (1974): 779-96.

145. Basten et al., "Receptor for Antibody" (See note 73).

146. H. B. Dickler et al., "Studies on the Nature of the Relationship Between Ia Antigens and Fc Receptors on Murine B Lymphocytes," *Journal of Immunology* 119 (1977): 348-54.

147. G. G. B. Klaus and A. K. Abbas, "Antigen-Receptor Interactions in the Induction of B-Lymphocyte Unresponsiveness," *Current Topics in Microbiology and Immunology* 78 (1977): 31-68.

148. Ramasamy, Munro, and Milstein, "Fc Receptor on B Lymphocytes" (See note 66).

149. Dickler, "Lymphocyte Receptors" (See note 30).

150. Ferrarini et al., "Receptors for IgM on Certain Human B Lymphocytes," *Journal of Immunology* 119 (1977): 1525-29.

151. W. J. Pichler and S. Broder, "Fc-IgM and Fc-IgG Receptors on Human Circulating B Lymphocytes," *Journal of Immunology* 121 (1978): 887-90.

152. G. W. Miller, P. H. Saluk, and V. Nussenzweig, "Complement-Dependent Release of Immune Complexes from the Lymphocyte Membrane," *Journal of Experimental Medicine* 138 (1973): 495-507.

153. G. W. Miller and V. Nussenzweig, "A New Complement Function: Solubilization of Antigen-Antibody Aggregates," *Proceedings of the National Academy of Sciences* 72 (1975): 418-22.

154. G. K. Lewis et al., "Murine B-Cell Subpopulations Responsive to T-Dependent and T-Independent Antigens," *Journal of Experimental Medicine* 144 (1976): 382-97.

155. M. B. Pepys, "Role of Complement in the Induction of Immunologcial Responses," *Transplantation Reviews* 32 (1976): 93-120.

156. C. R. Parish and A. B. Chilcott, "Functional Significance of the Complement Receptors on B Lymphocytes," *Cellular Immunology* 20 (1975): 290-303.

157. M. Schlesinger and M. Chaouat, "The Association of H-2 Antigens and EAC Receptors on the Surface of Peritoneal Cells," *European Journal of Immunology* 5 (1975): 27-32.

158. Katz and Benacerraf, "Lymphocyte Interactions" (See note 40).

159. B. Benacerraf and H. O. McDevitt, "Histocompatibility-Linked Immune Response Genes," *Science* 175 (1972): 273-79.

160. M. J. Taussig et al., "Antigen-Specific T-Cell Factor in Cell Cooperation. Mapping Within the I Region of the H-2 Complex and Ability to Cooperate Across Allogenic Barriers," *Journal of Experimental Medicine* 142 (1975): 694-700.

161. T. L. Delovitch and H. O. McDevitt, "*In vitro* Analysis of Allogeneic Lymphocyte Interaction. I. Characterization and Cellular Origin of an Ia-Positive Helper Factor-Allogeneic Effect Factor," *Journal of Experimental Medicine* 146 (1977): 1019-32.

162. T. Tokuhisa et al., "An Antigen-Specific I Region Gene Product That Augments the Antibody Response," *Journal of Immunology* 120 (1978): 414-21.

163. F. B. Mudawwar, E. J. Yunis, and R. S. Geha, "Antigen-Specific Helper Factor in Man," *Journal of Experimental Medicine* 148 (1978): 1032-43.

164. Janossy and Greaves, "Murine and B-Lymphocyte Subsets" (See note 95).

165. Coutinho and Möller, "Immune Activation of B Cells" (See note 47).

166. ———, "Self-Nonself Discrimination" (See note 48).

167. G. G. B. Klaus, G. Janossy, and J. H. Humphrey, "The Immunological Properties of Haptens Coupled to Thymus-Independent Carrier Molecules. III. The Role of the Immunogenicity and Mitogenicity of the Carrier in the Induction of Primary IgM Anti-Hapten Responses," *European Journal of Immunology* 5 (1975): 105-11.

168. D. M. Jacobs and D. C. Morrison, "Dissociation Between Mitogenicity and Immunogenicity of TNP-Lipopolysaccharide, a T-Independent Antigen," *Journal of Experimental Medicine* 141 (1975): 1453-58.

169. J. W. Schrader, "The Role of T Cells in IgG Production; Thymus-Dependent Antigens Induce B-Cell Memory in the Absence of T Cells," *Journal of Immunology* 114 (1975): 1665-69.

170. R. W. Dutton, "Separate Signals for the Initiation of Proliferation and Differentiation in the B Cell Response to Antigen," *Transplantation Proceedings* 23 (1975): 66-77.

171. A Schimpl and E. Wecker, "A Third Signal in B Cell Activation Given by TRF," *Transplantation Reviews* 23 (1975): 176-88.

172. Parkhouse and Cooper, "Differentiation of Lymphocytes" (See note 68).

173. Andersson, Bullock, and Melchers, "Inhibition of Stimulation" (See note 102).

174. Kearney, Cooper, and Lawton, "B-Lymphocyte Differentiation" (See note 103).

175. J. L. Ryan et al., "Inhibition of Lymphocyte Mitogenesis by Immobilized Antigen-Antibody Complexes," *Journal of Experimental Medicine* 142 (1975): 814-26.

176. J. E. Niederhuber et al., "Effects of Anti-Ia Sera on Mitogenic Responses. II. Differential Expression of the Ia Marker on Phytohemagglutinin- and Concanavalin A-Reactive T Cells," *Journal of Experimental Medicine* 143 (1976): 379-81.

177. S. Habu and M. C. Raff, "Accessory Cell Dependence of Lectin-Induced Proliferation of Mouse T Lymphocytes," *European Journal of Immunology* 7 (1977): 451-57.

178. Coutinho, Forni, and Blomberg, "Antigenic Determinants" (See note 56).

179. B. Huber, R. K. Gershon, and H. Cantor, "Identification of a B-Cell Surface Structure Involved in Antigen-Dependent Triggering: Absence of This Structure on B Cells from CBA/N Mutant Mice," *Journal of Experimental Medicine* 145 (1977): 10-20.

180. R. E. Cone et al., "Molecular Identification of a Surface Structure on B Cells (Lyb-3) and Its Relationship to B Cell Triggering," *Journal of Immunology* 120 (1978): 1733-40.

POLYMORPHIC MAJOR TRANSPLANTATION ANTIGENS AND CELL-MEDIATED IMMUNITY TO INTRACELLULAR PARASITES *

Rolf M. Zinkernagel †

Cell-mediated immunity (CMI) plays a major role in protecting higher vertebrate hosts from infection by intracellular parasites. Other defense mechanisms, specific and nonspecific, are also involved, but CMI often seems to be the crucial factor in early and rapid clearance of parasites.[1,2] During the last few years it has become clear that thymus-derived lymphocytes (T cells) are specific not only for foreign, that is, parasite antigens expressed on surfaces of infected cells, but for self antigens expressed by the cells.[3] The self antigens involved have been defind as products of the major histocompatibility gene complex (MHC),‡ which also codes for the major transplantation antigens involved in graft rejection.[4-6]

* Some experiments summarized herein have been supported by United States Public Health Service grants A1-13779 and A1-00273. This paper, submitted in November 1978, is Publication No. 1664 from the Department of Immunology, Scripps Clinic and Research Foundation.

† The author wishes to thank Ms. Phyllis Minick and Ms. Annette Parson for their excellent editorial and secretarial help.

‡ The major histocompatibility gene complex of mice (*H-2*) is located on the seventeenth chromosome and is about 0.5 recombination units (centimorgans) long, which corresponds to about the length of the genome of *E. coli*. The H-2 complex is subdivided into *K, I, S,* and *D* regions; K and D regions corresponding to the human A and B regions of *HLA* code for serologically defined major transplantation antigens that are present on all cells of an individual; the murine *I* region, in humans the *D* region, codes for Ia antigens expressed on certain subsets of lymphocytes and for immune response genes regulating antibody response; the (*S*) region codes for complement factors and other serum proteins

Interestingly enough, T cells are restricted to different MHC products according to function. T cells involved in nonlytic functions, such as cooperation with B cells, delayed-type hypersensitivity, or activation of macrophages, are specific for *H-2I* (*HLA-D*) regions, whereas cytotoxic T cells are specific for *H-2K, D* (*HLA-A, B*) regions. Major transplantation antigens therefore seem to comprise an array of receptors for cell-specific differentiation signals that are triggered by antigen-specific T cells.[7]

The association of T cell functions with MHC products may have evolved under selective pressure within creatures continuously invaded by intracellular parasites; eventually the survivors developed a mobile police force to protect themselves and their descendents. Cytotoxic T cells can destroy acutely virus-infected cells via H-2K,D (or HLA-A,B) during the virus's eclipse phase, and thus can eliminate the virus. Alternatively, nonlytic T cells can activate phagocytic cells, which then digest intracellular bacteria much more efficiently.

T cells differentiate specificity for self according to the MHC of the thymus where they mature.[8, 9] Thus differentiation of T cells' specificity for self is influenced more by the thymic MHC than by the genome of the maturing lymphohemopoietic stem cells. Full immunocompetence is possible, however, only when some of the lymphohemopoietic cells confronting T cells "post-thymically" are histocompatible with the thymus that selected the T cells' specificity. In addition, only antigen presented on cells of lymphohemopoietic origin may be immunogenic, and thus trigger effector T cells; antigen-presenting cells and thymus must therefore be histocompatible for T cells to become sensitized. No one knows whether lymphohemopoietic cells influence T cell maturation by generating diversity of the receptor repertoire specific for foreign antigens or by expanding the small numbers of committed T cells.[10]

Along with the imposition of restricted specificity, the thymic MHC also apparently selects the capacity of T cells to respond to antigens that are under the control of MHC-coded immune response (*Ir*) genes.[11, 12] Furthermore, MHC regions that code for the restricting element also code for *Ir* genes regulating responsiveness of these T cells.[13-15] These two findings suggest that gene products of restricting genes and *Ir* coding genes are either closely linked or identical.

In theoretical terms this work offers considerable promise for clarifying T cell differentiation and recognition. The practical applications lie in revising the premise for reconstituting immunodeficient patients and enhancing our understanding of the mechanisms leading to MHC disease associations.[16,17] In this regard a brief review of our basic assumptions is appropriate. First, T cells protect a host from intracellular parasites via MHC-coded differentiation signal receptors in an immune-specific fashion, and destroy infected host cells accompanied by inflammatory processes; second, MHC-coded restricting differentiation signal receptors influence directly or indirectly T-cell responsiveness. Based on these two assumptions, susceptibility to and maintenance of disease can be viewed as resulting not only from the balance between immunoprotective *versus* immunodestructive effects of T cells, but from the MHC-dictated and restriction-specificity-dependent responsiveness of T cells.[18]

The outcome of the susceptibility of resistance to disease may fall into four categories. High responsiveness results in relatively *increased* resistance to infectious *cytopathogenic virus,* but *decreased* resistance to *poorly cytopathogenic* viruses, which tends to tip the balance of immune protection toward *autoaggression,* since T cell-mediated tissue damage is favored by the wider spread of virus. In contrast, low responsiveness results in relatively *great susceptibility to acute cytopathogenic virus* infection, and therefore has been eliminated by selection, but possibly *decreased the susceptibility to virus-induced autoaggressive diseases.*[19] For some time chronic virus infections have been suspected as the cause of so-called autoimmune diseases. In fact most of the associations between MHC and disease relate to autoaggressive (or autoimmune) disease, not to acute infections. Resistance or susceptibility to acute virus infections would influence survival of a species dramatically, since acute infections can occur well before an individual is reproductively mature. Natural selection would tend to eliminate these types of low responders. In contrast, the more chronic autoimmune types of diseases tend to become life-limiting much later, after an individual reproduces; natural selection therefore, need not discard susceptibility of this type, and MHC-associated diseases may become manifest.

NOTES

1. A. L. Notkins, ed., *Viral Immunology and Immunopathology* (New York: Academic Press, 1976).

2. G. Möller, ed., *Transplantation Review*, vol. 19 (Copenhagen: Munksgaard, 1976).

3. R. M. Zinkernagel and P. C. Doherty, "Activity of Sensitized Thymus-Derived Lymphocytes in Lymphocytic Choriomeningitis Reflects Immunological Surveillance Against Altered Self Components," *Nature* 251 (1974): 547-48.

4. P. C. Doherty and R. M. Zinkernagel, "H-2 Compatibility Is Required for T-Cell Mediated Lysis of Target Cells Infected with Lymphocytic Choriomeningitis Virus," *Journal of Experimental Medicine* 141 (1975): 502-07.

5. ————, "Enhanced Immunological Surveillance in Mice Heterozygous at the H-2 Gene Complex," *Nature* 256 (1975): 50-52.

6. R. M. Zinkernagel and P. C. Doherty, "Major Transplantation Antigens Virus and Specificity of Surveillance T Cells. The 'Altered Self' Hypothesis," *Contemporary Topics in Immunobiology* 7 (1977): 179-220.

7. R. M. Zinkernagel, "Role of the H-2 Gene Complex in Cell-Mediated Immunity to Infectious Disease," *Transplantation Proceedings* IX (1977): 1835-38.

8. R. M. Zinkernagel et al., "On the Thymus in the Differentiation of 'H-2 Self-Recognition' by T Cells: Evidence for Dual Recognition?" *Journal of Experimental Medicine* 147 (1978): 882-96.

9. R. M. Zinkernagel, "Thymus and Lymphohemopoietic Cells: Their Role in T-Cell Maturation in Selection of T Cells' H-2-Restriction-Specificity and in H-2-Linked *Ir* Gene Control," *Immunological Reviews,* in press.

10. Ibid.

11. H. von Boehmer, W. Haas, and N. K. Jerne, "Major Histocompatibility Complex-Linked Immune Responsiveness Is Acquired by Lymphocytes of Low-Responder Mice Differentiating in Thymus of High-Responder Mice," *Proceedings of the National Academy of Sciences* 75 (1978): 2439-42.

12. R. M. Zinkernagel et al., "In Irraditation Chimeras, K or D Regions of the Chimeric Host, Not of the Donor Lymphocytes, Determine Immune Responsiveness of Antiviral Cytotoxic T Cells," *Journal of Experimental Medicine* 148 (1978): 805-10.

13. B. Benacerraf and H. O. McDevitt, "Histocompatibility-Linked Immune Response Genes," *Science* 175 (1972): 273-79.

14. B. Benacerraf and R. Germain, "The Immune Response Genes of the Major Histocompatibility Complex," *Immunological Reviews* 38 (1978).

15. P. C. Doherty et al., Cytotoxic T-Cell Responses in Mice Infected with Influenza and Vaccinia Viruses Vary in Magnitude with H-2 Genotype," *Journal of Experimental Medicine* 148 (1978): 534.

16. R. M. Zinkernagel et al., "*Ir* Genes in H-2 Regulate Generation of Anti-

Viral Cytotoxic T Cells: Mapping to K or D and Dominance of Unresponsiveness," *Journal of Experimental Medicine* 148 (1978): 592.

17. Zinkernagel, "Thymus and Lymphohemopoietic Cells" (See note 9).
18. Zinkernagel et al., "Irradiation Chimeras" (See note 12).
19. Zinkernagel, "Thymus and Lymphohemopoietic Cells" (See note 9).

III. HORMONES AND RECEPTORS

CURRENT CONCEPTS OF THE INSULIN RECEPTOR AND ITS FUNCTION

Michael S. Kappy, Maxine A. Lesniak,
Jesse Roth, and Robert S. Bar *

INTRODUCTION

The concepts of hormones and of receptors were both introduced in the late nineteenth century. In the past seventy years, however, the study of hormones has advanced while the study of receptors has lagged.

This dominance of hormone over receptor and target cells in the study of the endocrine system allowed endocrine disorders to be viewed as diseases of hormones. It later became clear that in many endocrine disorders that had been viewed as hormone deficiency states the hormone was present in supernormal concentrations and the biological effect was deficient because the target cell was defective. Thus progress in the study of the receptor and the relationship between hormone and receptor is advancing, as will be demonstrated in this chapter on the insulin receptor.

Disorders of Glucose Metabolism

It has long been known that glucose stimulates the β-cells of the pancreas to release insulin, which acts on target cells to promote glucose utilization and storage. Only glucose could be measured, however, so the system was oversimplified—hyperglycemia

---

* Dr. Kappy is with the Division of Pediatric Endocrinology, Department of Pediatrics, the Johns Hopkins Hospital; Ms. Lesniak and Dr. Roth are with the Diabetes Branch, National Institute of Arthritis, Metabolism, and Digestive Diseases, National Institutes of Health; and Dr. Bar is with the Department of Internal Medicine, Veterans Administration Hospital, University of Iowa. Only Dr. Roth attended the conference.

was due to insulin deficiency, and hypoglycemia was due to insulin excess. The dramatic response of diabetics to insulin administration and the similarity of patients with islet cell tumors to symptoms of diabetics who received too much insulin strengthened this simple formulation. With the introduction of methods to measure insulin levels in blood, especially the radioimmunoassay, it became clear that only a minority of patients conform to this scheme.[1]

Diabetics who are truly insulin-requiring are indeed insulin-deficient, but the majority of hyperglycemic patients are not absolutely insulin-dependent and have normal or supernormal concentrations of circulating insulin (Table 1). Further, the insulin in these patients is biologically intact. Conversely, in those patients with hypoglycemia due to islet-cell tumors, insulin levels were found to be inappropriately elevated, which conformed to expectations, while in many patients with low normal glucose or frank hypoglycemia, plasma insulin was noted to be low or unmeasurable (Table 1).

Thus the radioimmunoassay for insulin, while it confirmed

TABLE 1. CONCORDANCE AND DISCONCORDANCE OF CIRCULATING HORMONE CONCENTRATIONS AND THE BIOLOGICAL STATE

	Hormone Concentration Subnormal or Unmeasurably Low	Hormone Concentration Normal or Elevated
	Concordant	Discordant
Biological Response Deficient	—Insulin-dependent (juvenile type) diabetes (~20%)	—Noninsulin-dependent (adult type) diabetics (~80%) —Growth hormone excess —Glucocorticoid excess —Uremia —Syndromes of extreme insulin resistance
	Discordant	Concordant
Biological Response Excessive	—Nonislet cell tumors —Anorexia nervosa —Growth hormone deficiency —Glucocorticoid deficiency	—Islet cell tumors —Infants of diabetic mothers

that hormone levels were concordant with biological observations in some patients, raised a paradox in most conditions—the hormone levels were discordant with the biological observations (Table 1). The introduction of methods to measure the receptor has resolved many of these apparent paradoxes. In many disorders of glucose metabolism where hormone levels were discordant with the clinical state, it was shown that the receptor was altered and reflected the clinical state.

The Structure of the Insulin Receptor

The structure of the receptor for insulin is just beginning to be elucidated. It is an integral membrane protein—it cannot be removed from the membrane without the use of detergents and requires the continued presence of detergents to remain soluble. It contains carbohydrates, has a mass in the range of 10^5 to 10^6 daltons, and appears to be composed of several subunits, each in the range of 10^4 to 10^5 daltons.[2]

Cells for Study of Insulin Receptors in Humans

Although partially purified preparations of insulin receptors are available, they are not easily applied to the study of clinical situations.[3-6] For most human studies the insulin receptors on circulating monocytes and isolated fat cells have proven to be the most useful. Human monocytes and fat cells have identical characteristics of insulin binding (Table 2), which are also shared by insulin receptors on liver and muscle, the tissue most relevant for overall glucose homeostasis in vivo.

In fat cells, both the binding of insulin to its receptors and the biological effects of insulin have been measured. Insulin had no known metabolic effect on the human monocyte, although insulin had been shown to modulate two specific immune functions of murine monocyte-like cells, that is, macrophages.[7, 8] Recent studies have shown that insulin does stimulate hexose metabolism in human monocytes.[9]

The human monocyte has proven to be an especially useful source of insulin receptors for study of disease states in man. Monocytes can be obtained safely, rapidly, and in large quantity by

TABLE 2. COMMON PROPERTIES OF INSULIN RECEPTORS IN HUMAN ADIPOCYTES AND IN HUMAN MONOCYTES

Specificity of binding
Competition by modified insulins is directly proportional to their biological potencies
Rapid and reversible binding of insulin
Finite number of binding sites (i.e., binding is saturable)
pH dependence of binding
Sharp optimum at 7.8-8.0
Temperature dependence of binding
Steady-state binding greater at lower temperatures
Curvilinear Scatchard plots
Consistent with negative cooperativity
Acceleration of the dissociation of receptor-bound ^{125}I-insulin by unlabeled insulin
Kinetic equivalent of negative cooperativity
Affinity of receptor for insulin

routine venipuncture, thus allowing repeated study of a given patient. (Even though the monocytes represent a minority of the total cells in this preparation [10-30 percent], they account for the majority of insulin binding [$\simeq$ 90 percent].)[10] Most important, in several disease states changes in the insulin receptors on monocytes correlate with the state of overall insulin sensitivity, suggesting that the monocyte receptor may accurately mirror changes in other, more relevant receptors for insulin—liver and muscle, for example.

More recently, erythrocytes have been used to study insulin binding to receptors. This is especially useful for infants and children because only a few milliliters of blood are needed, in contrast to the 50-100 ml. needed for studies of monocytes. In each pathological and physiological state it remains to be determined how closely the insulin receptor of each of these tissues reflects change in liver (and muscle).

Methods of Measuring Receptor

Measurements of insulin binding to cells use methodology that is similar to that used in other competitive binding assays.[11] ^{125}I-labeled insulin is incubated with cells—monocytes, erythrocytes, or adipocytes—and various concentrations of unlabeled insulin under specified conditions. After steady state or equilibrium is attained, the ^{125}I-labeled hormone bound to the cells is separated from the free hormone, usually by centrifugation. Data from such a competition assay can then be used to derive receptor concentration (binding capability of the total cell population) and receptor affinity (Figure 1). Receptor concentration is determined from the Scatchard plot, which plots bound/free radioactivity as a function of the concentration of bound hormone.[12] A typical Scatchard plot for insulin binding is concave upward (Figure 1, lower left). Extrapolation of the curve to the horizontal axis gives an intercept value equivalent to the maximum binding capacity, from which receptor concentration is calculated as follows:

Binding sites per cell = maximum binding capacity (expressed in moles/liter) X Avogadro's number X (1/cell concentration expressed per liter).

The curvilinear Scatchard plot for insulin binding had been thought to represent two or more orders of independent sites, each with its fixed affinity for insulin. P. De Meyts and J. Roth have, however, presented data suggesting that the insulin receptors represent a single class of sites displaying site-site interactions of the negatively cooperative type;[13, 14] that is, the affinity of the insulin receptors progressively decreases as more receptor sites are occupied by insulin, thereby yielding the curvilinear plot. These investigators have developed the average affinity profile, a graphic analysis to depict the changing affinity with increasing receptor occupancy by insulin[15] (Figure 1, lower right). Thus at low physiological concentrations of insulin most receptor sites are empty, and the receptors are in their highest affinity state. With increasing receptor occupancy by insulin the affinity of the remaining receptors decreases progressively. Although this interpretation has been questioned, we feel that extensive equilibrium and kinetic data support the negative cooperativity model.

Biological Relevance of Insulin Receptor Measurements

To establish the specificity of the binding to the receptors, several species of insulin and insulin derivatives were compared for their biological potency in stimulating glucose oxidation in isolated rat adipocytes (Figure 2, A), and the same insulin prepara-

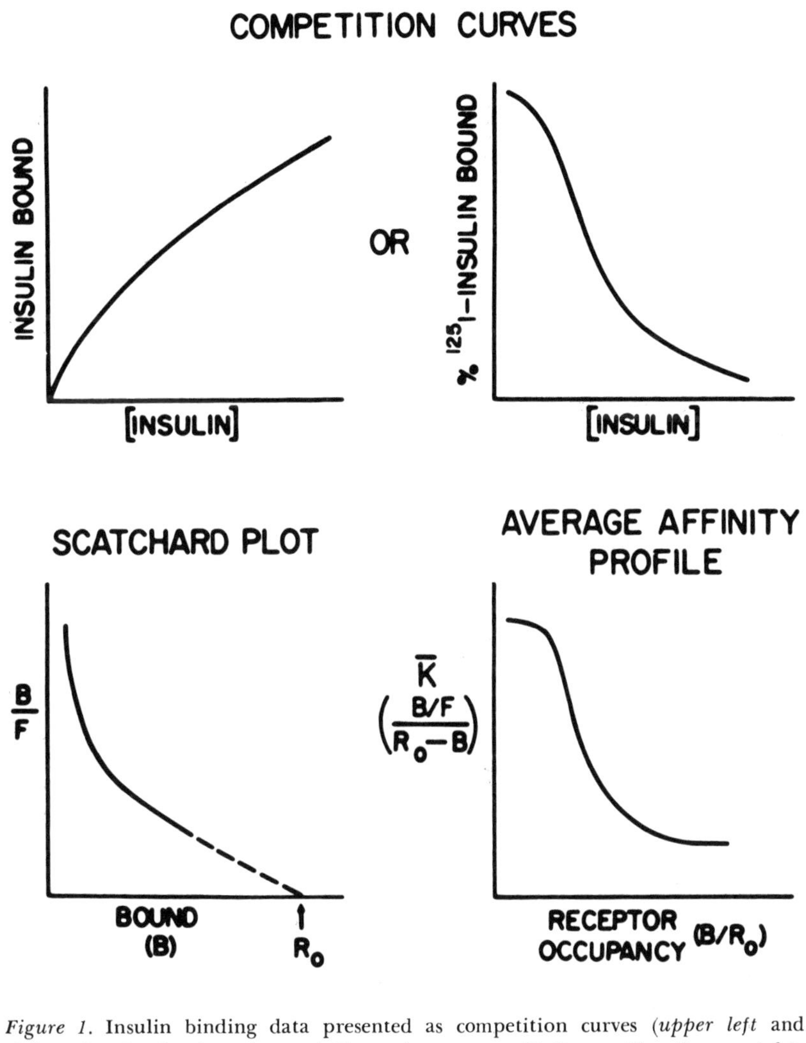

Figure 1. Insulin binding data presented as competition curves (*upper left* and *right*), Scatchard plot (*lower left*), and average affinity profile (*lower right*). K = average affinity; B/F = bound/free hormone; R_o = total receptor concentrations; B = concentration of bound hormone.

SOURCE: P. De Meyts, "Cooperative Properties of Hormone Receptors in Cell Membranes," *Journal of Supramolecular Structure* 4 (1976): 241-58.

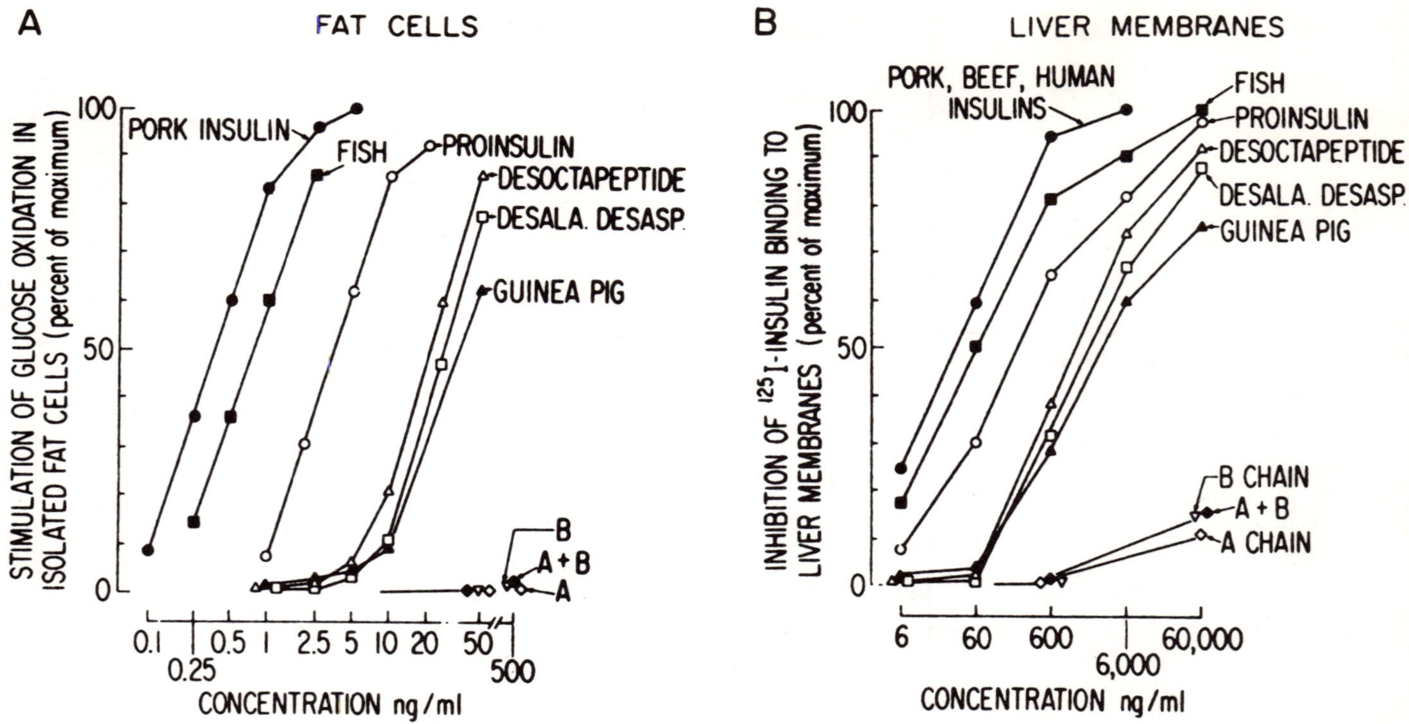

Figure 2. A, Preparations of insulin and their ability to stimulate ^{14}C glucose oxidation in isolated fat cells. B, Several preparations of insulin and their ability to inhibit binding of ^{125}I-insulin to liver membranes.

SOURCE: P. Freychet, J. Roth, and D. M. Neville, Jr., "Insulin Receptors in the Liver: Specific Binding of ^{125}I-Insulin to the Plasma Membrane and Its Relation to Insulin Bioactivity," *Proceedings of the National Academy of Sciences* 68 (1971): 1833-37.

tions were compared for their ability to compete with ^{125}I-labeled pork insulin in binding to receptors on purified plasma membranes of rat liver (Figure 2, B). Notice that the rank order of potencies is the same for both stimulation of glucose oxidation and for binding to receptors. Thus the capacity of each insulin to stimulate biological processes is predicted by its relative affinity for the insulin receptors.[16, 17] Similar studies have been performed to validate the specificity of insulin binding to sites on each cell.

Regulation of Receptors

In the case of the receptor for insulin, numerous biologically relevant events produce changes in the receptor (Table 3); growth, differentiation, and environment can have profound effects. Total calories, dietary composition, timing of meals, and exercise have major effects at the level of the receptor, with changes in concentration and/or affinity. The homologous ligand itself acts to regulate both concentration and affinity of its own receptor; regulation of affinity appears to be a physicochemical process intrinsic to the receptor that can be reproduced fully with detergent-solubilized receptors, whereas regulation of receptor concentration is a regulatory event that requires cellular processes beyond the receptor. Studies in several in vitro systems have convincingly demonstrated that insulin can directly regulate its own receptor.[18-21] When cultured lymphoblastoid cells were grown in the presence of insulin the concentration of their insulin receptors was reduced; the decrement in receptor concentration was directly proportional to the concentration of ambient insulin. In these studies the insulin-induced loss of receptors appeared to be mediated through an

TABLE 3. "BIOLOGICALLY RELEVANT" REGULATORS OF AFFINITY AND/OR CONCENTRATION OF INSULIN RECEPTORS

1. Exercise	7. Age
2. Meals	8. c-AMP
3. Diet	9. pH
4. Insulin	10. Ketones
5. Other hormones	11. Sulfonylureas
6. Growth and development	12. Antireceptor antibodies

increased degradation rate of insulin receptors, a process that required active protein synthesis. The relationship between receptor concentration and ambient insulin is shown in Figure 3.

Elevated concentrations of hormone may act not only to decrease binding to receptor, but, at one or more postreceptor sites, to desensitize the target cell. Such postreceptor desensitization may be important in extreme hyperinsulinemia.

The ability of a hormone to regulate its own receptor appears to be a generalized property of hormones. Growth hormone (GH), thyrotropin-releasing hormone (TRH), and the β-adrenergic agents all decrease either the affinity or concentration of their respective receptors.[22-24] In contrast, prolactin increases the concentration of its own receptors.[25] Such control mechanisms, if operative under physiological conditions, not only have importance in our understanding of normal and pathologic states, but may provide a rational basis for therapy with exogenous hormones, insulin-requiring diabetics, for example.

Other (heterologous) hormones and drugs can affect target cell receptors for insulin and thereby alter target-cell sensitivity to insulin. In addition, the insulin receptor is extremely sensitive to pH [26,27] (Figure 4), and the change in pH from 7.4 to 6.8, as might occur with diabetic ketoacidosis, might well be associated with a major decrease in insulin binding due to an alteration in receptor affinity. In addition, acidosis can have profound effects on receptor concentration.[28] Given that the receptor is at the interface between the outside and inside of the cell, it is reasonable, albeit retrospectively, that it is responsive to signals from both the external and internal environments.

Role of Receptor in Insulin Degradation

The insulin receptor also acts to regulate the degradation of insulin. In vivo there are at least two mechanisms for insulin degradation, both at the level of the cell. (For the majority of peptide hormones in vivo no degradation occurs in plasma.) One mode of degradation does not require binding to receptor.[29, 30] Another major mechanism of insulin degradation, however, requires binding of hormone to receptors.[31, 32] It is believed that

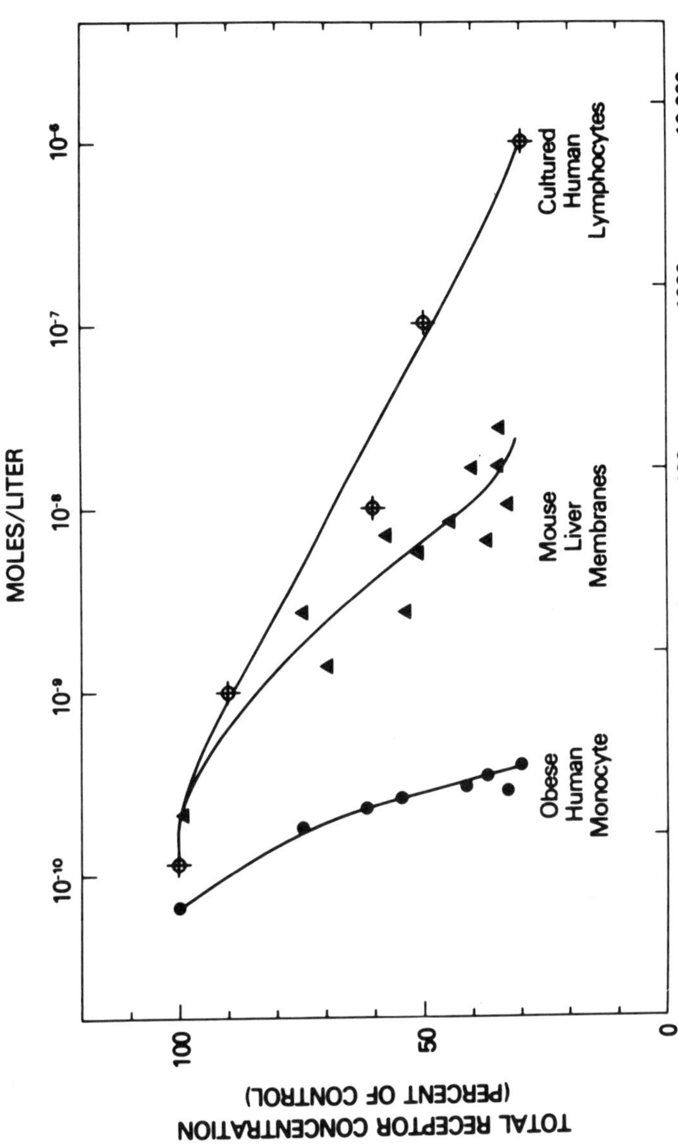

Figure 3. Insulin receptor concentrations as a function of ambient insulin concentration in the obese human, obese mice, and cultured human lymphocytes.

SOURCE: R. S. Bar and J. Roth, "Insulin Receptor Status in Disease States in Man," *Archives of Internal Medicine* 137 (1977): 474-81. Copyright 1977, American Medical Association.

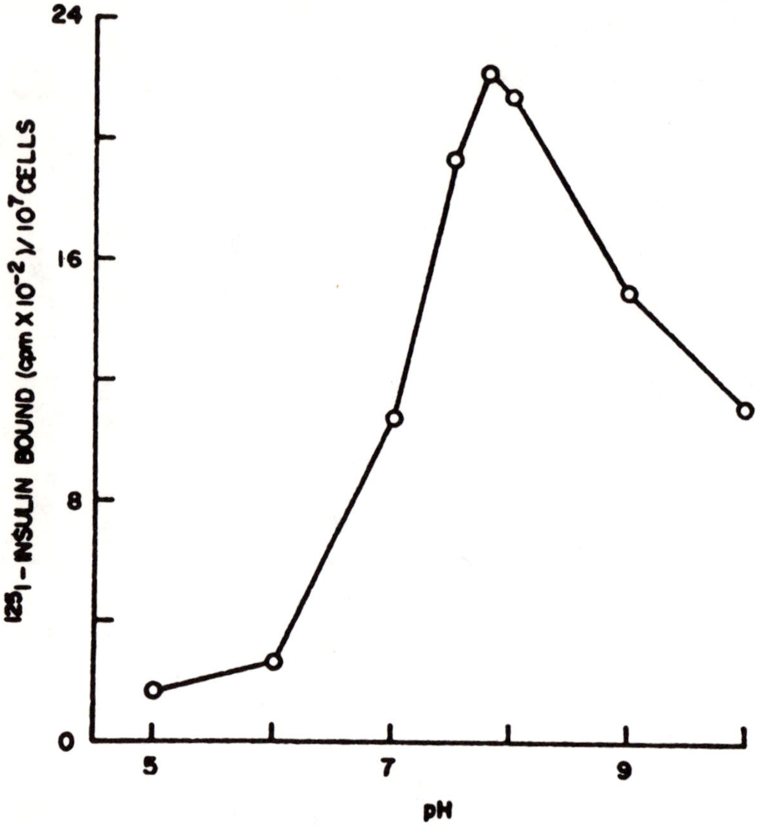

Figure 4. Effect of pH on insulin binding to receptor. Specific binding of ^{125}I-insulin to receptors on cultured human lymphocytes (IM-9 line) is plotted as a function of the pH of the medium. This sharp pH dependence of binding is characteristic of all insulin receptors.

SOURCE: J. R. Gavin, III et al., "Characteristics of the Human Lymphocyte Insulin Receptor," *Journal of Biological Chemistry* 248 (1973): 2202-07.

the receptor does not carry out the degradation but delivers the hormone to a cellular site where the hormone gets degraded. Thus the insulin-receptor interaction regulates degradation of both hormone and receptor, regulates receptor concentration and affinity, cross-linking and translocation of hormone-receptor complexes, and postreceptor events; cause and effect relationships among these functions are as yet uncertain.

INSULIN RECEPTORS IN DISORDERS OF GLUCOSE TOLERANCE AND INSULIN SENSITIVITY

Obesity in Man

Insulin receptors have been evaluated extensively in obesity, the most common hormone-resistant state in humans.[33-39] The binding of ^{125}I-labeled insulin to either monocytes or adipocytes in some obese patients is entirely normal, whereas in other patients there is impairment of insulin binding that can be very marked (Figure 5). Scatchard plots of such data indicate that the decrease in insulin binding is due entirely to a decrease in receptor concentration (Figure 6); the remaining receptors have normal affinity for insulin. When the binding data on an individual patient are compared with the clinical state of that patient, it is the obese patient with the lowest binding and the greatest receptor loss who has the most severe insulin resistance. Obese patients who show no evidence of insulin resistance have normal binding of insulin and normal concentrations of insulin receptors on their monocytes and adipocytes. The relationship between insulin receptor concentration and clinical insulin resistance—reflected in vivo in the basal levels of circulating insulin—is depicted for several obese patients in Figure 7.[40] The inverse relationship between receptor concentration and ambient levels of insulin in vivo was maintained after chronic low-calorie diet, which resulted in restoration of sensitivity to insulin.[41] After insulin-resistant obese patients dieted for several weeks their plasma insulin levels fell and insulin binding to monocytes increased (Figure 8). The increased binding was due entirely to a rise in receptor concentration. In these studies the improved clinical sensitivity to insulin and the increase in receptor concentrations occurred before a substantial amount of weight was lost, suggesting that adiposity, per se, is not the direct cause of the insulin resistance, but that diet may be directly involved in the pathogenesis of the resistant state.

With mild to moderate degrees of hyperinsulinemia and insulin resistance, the receptor defect appears entirely to account for the hormone resistance. With more severe hyperinsulinemia and resistance, postreceptor defects have been detected, in addi-

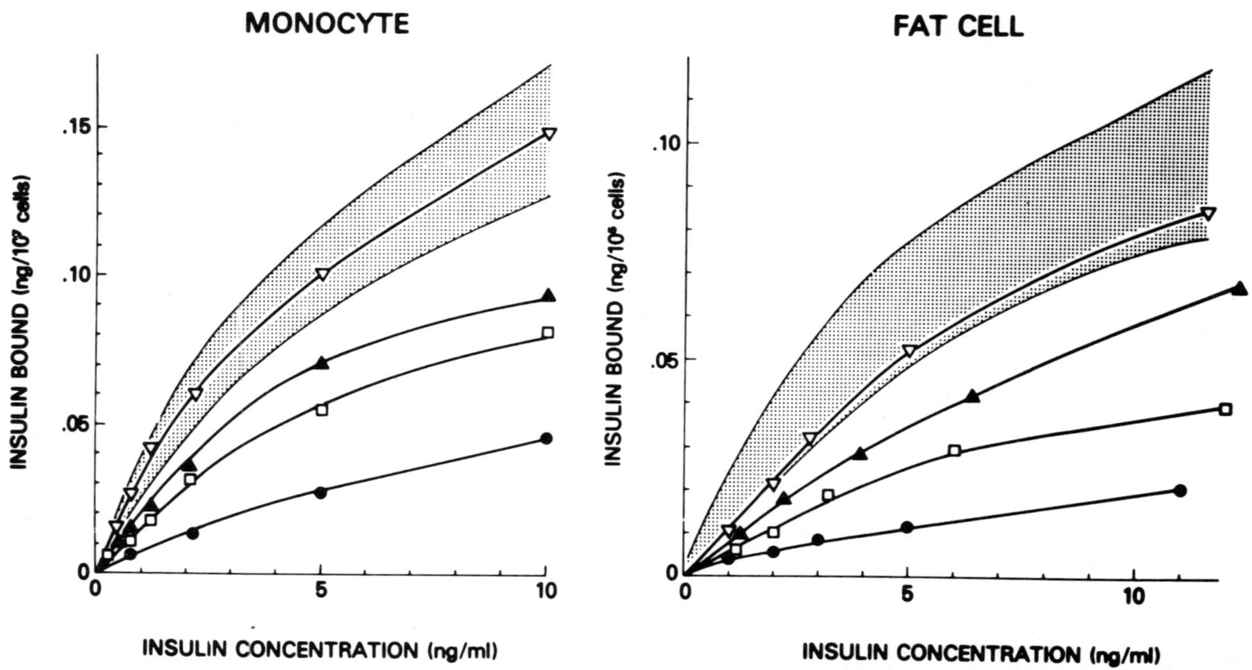

Figure 5. Insulin binding to circulating monocytes from four obese patients (*left*), and insulin binding to isolated fat cell from four different obese patients. Bound insulin is plotted as a function of the amount of insulin in the incubation assay. Shaded areas are the ranges of normal (mean ± 1 S.D.).

SOURCE: R. S. Bar et al., "Regulation of Insulin Receptors in Normal and Abnormal Physiology in Humans," *Advances in Internal Medicine* 24 (1979): 23-52.

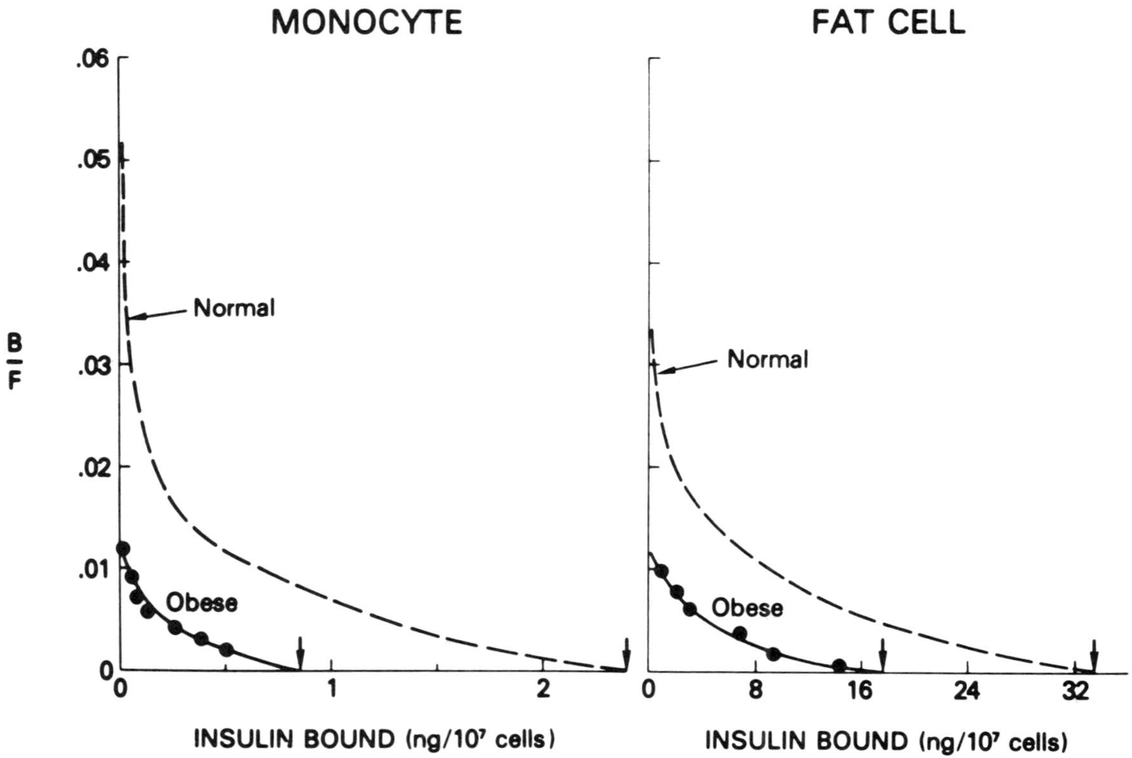

Figure 6. Scatchard plots of insulin binding from the two obese patients in Figure 5 with the lowest insulin binding. Dashed lines represent mean data from several studies in thin, normal subjects. Inverted arrows (abscissa intercepts) designate the maximal binding capacities and permit calculation of receptor concentrations. B/F = bound/free hormone.

SOURCE: R. S. Bar et al., "Regulation of Insulin Receptors in Normal and Abnormal Physiology in Humans," *Advances in Internal Medicine* 24 (1979): 23-52.

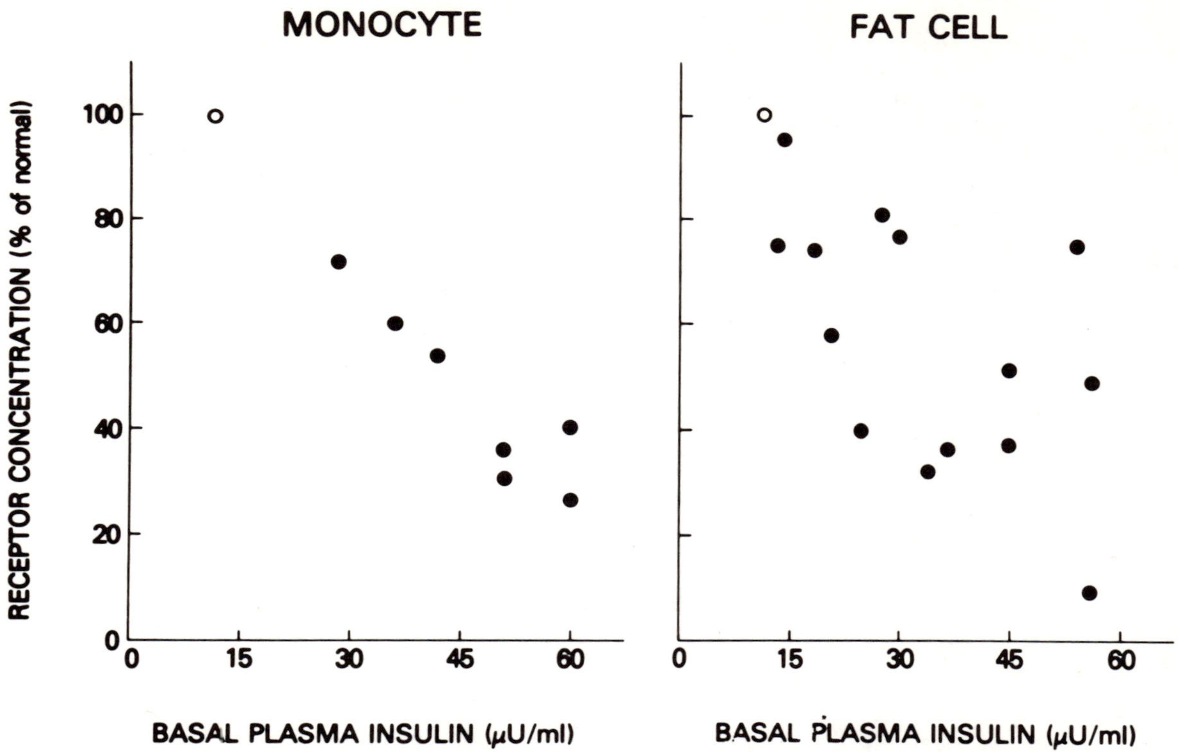

Figure 7. Correlation between basal (fasting) levels of insulin and receptor concentrations on monocytes (*left*) and fat cells (*right*) from a group of obese patients. Open circles mean receptor concentrations in a large group of thin normal subjects.

SOURCE: R. S. Bar et al., "Regulation of Insulin Receptors in Normal and Abnormal Physiology in Humans," *Advances in Internal Medicine* 24 (1979): 23-52.

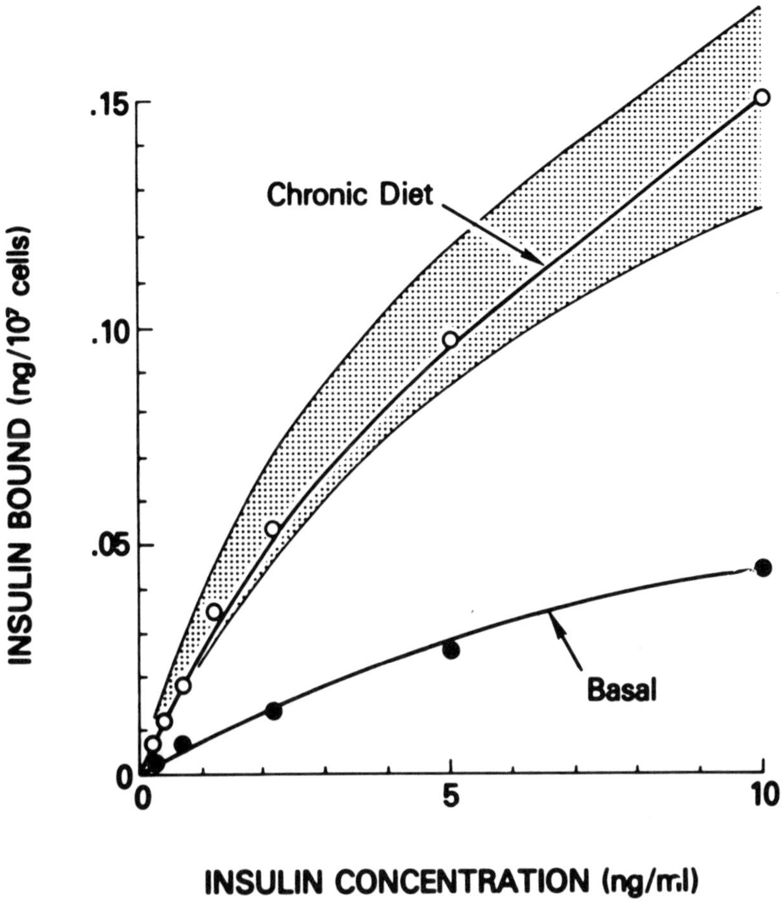

Figure 8. The effect of low-calorie, chronic diet on insulin binding to monocytes from an insulin-resistant obese patient. The shaded area is the normal range (mean ± 1. S.D.).

SOURCE: R. S. Bar and J. Roth, "Insulin Receptor Status in Disease States in Man," *Archives of Internal Medicine* 137 (1977): 474-81. Copyright 1977, American Medical Association.

tion to receptor defects (see the following sections on ob/ob mice), which may add to the insulin resistance.[42]

Obesity in Rodents

The alterations in obese rodents are even more marked than in humans. The ob/ob mice, in which obesity is inherited as a

Mendelian recessive, have severe hyperinsulinemia, extreme resistance to insulin in vivo and in vitro, and a marked decrease in the concentration of receptors for insulin[43] (Figure 9). The decrease in receptor concentrations has been demonstrated in multiple tissues, including the liver,[44-46] skeletal muscle,[47, 48] fat,[49] heart,[50] and thymus lymphocytes.[51] The defect is present when measured in whole cells as well as in crude preparations of membranes and in highly purified plasma membranes.[52] The residual receptors are normal in all respects, including their affinity for hormone.[53] The defect is specific for the insulin receptor; other cell membrane components, including other receptors, enzymes, and structural proteins, are unaltered.

Restriction of food intake in the ob/ob mouse to the level of that of normal (thin) littermates resulted in weight loss, reduction of plasma insulin concentrations, and near restoration of receptor concentrations[54] (Figure 9). Acute fasting of the obese animal for twenty-four hours results in a sharp fall in the circulating insulin concentration and a doubling of the receptor concentration. Thin littermates who are fasted for comparable periods show an increase in receptor concentration of about 20 percent (in contrast, thin humans who fast show no change in receptors in up to seventy-two hours of fasting). When the ob/ob animals are fasted, but their insulin concentrations are maintained at high levels by injections of long-acting insulin, the receptor concentrations remain low. When the obese mice are fasted, but insulin replacement is withheld until the last hour or two of the fast, however, receptor concentrations were at the level observed in animals that were fasted but were given no insulin.[55]

The db/db mouse, another model of recessively inherited obesity, shows similar degrees of hyperinsulinemia, insulin resistance, and receptor deficiency[56] (Figure 9). When normal mice are given gold thioglucose or other gold salts, cells in the central nervous system are destroyed, which often results in chronic overeating. As these animals overeat and gain weight, the plasma insulin concentrations in the basal state rise and receptor concentrations decrease. If, after these animals have become obese, food intake is restricted, the process is reversed, with a progressive fall

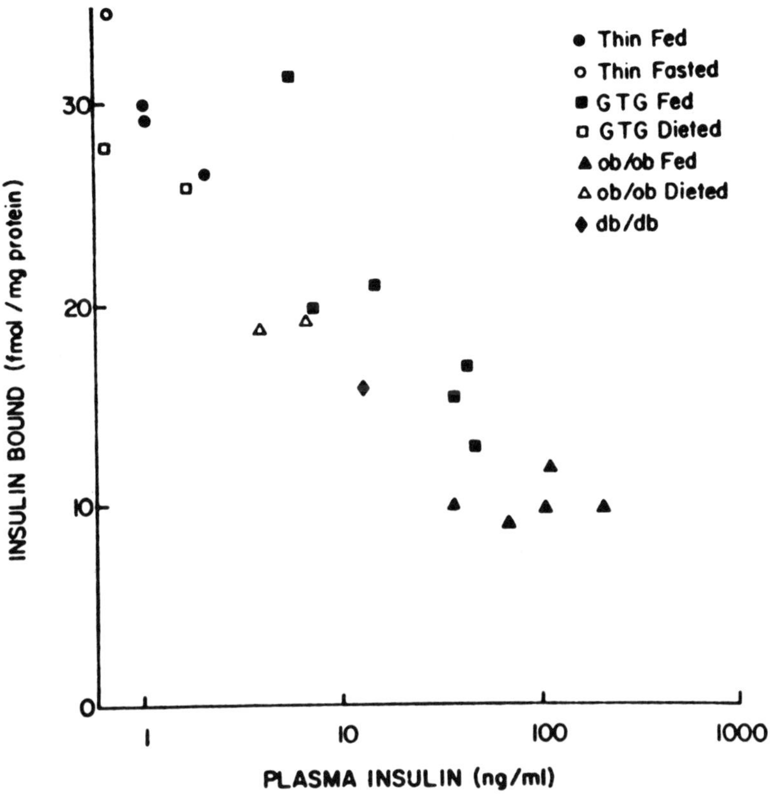

Figure 9. Relationship of receptor concentration to basal plasma insulin concentration. This figure summarizes a large number of studies on mice with two forms of inherited obesity (ob/ob and db/db), as well as on a group of normal mice that have become obese following treatment with gold thioglucose (GTG). The obese mice are compared to their thin littermates under a variety of experimental manipulations of feeding. Insulin receptor concentrations obtained from studies of ^{125}I-insulin binding to highly purified plasma membranes from liver are plotted as a function of the plasma insulin concentration in the basal state at the time of the experiment. It should be emphasized that this inverse correlation is related to the chronic level, not to the acute level, of the ambient insulin.

SOURCE: A. H. Soll et al., "Insulin Receptor Deficiency in Genetic and Acquired Obesity," *Journal of Clinical Investigation* 56 (1975): 769-80.

in weight and in plasma insulin, associated with restoration of receptor concentrations to normal.

As with the two forms of inherited obesity, the receptor concentration can be predicted by the magnitude of the elevation of the plasma insulin; in all cases the residual receptors are normal and are reduced only in number. With severe degrees of hyperinsulinemia 75-80 percent of the receptors are lost; further increases in plasma insulin concentration produce no further decreases in receptor concentration.[57-59] It should be noted that in severe hyperinsulinemia and insulin resistance, as in the ob/ob mouse, there appear to be postreceptor defects, in addition to the defects at the level of the receptor.[60, 61]

The data presented thus far are consistent with the following pathogenesis of insulin resistance in obesity: overeating → hyperglycemia → hyperinsulinemia → decrease in insulin receptor concentration → insulin resistance. It should be pointed out, however, that the existing data are equally consistent with alternative sequences of events. Given that excess calories and excess carbohydrate or fat, as well as inactivity (all of which are present in the obese state), can each produce defects at the level of insulin binding to receptors, it may be difficult to establish the sequence of events without more detailed study. In at least one model of obesity and insulin resistance—produced by electrolytic lesions in the hypothalamus—hyperinsulinemia may be the earliest defect detected, before the commencement of overeating or inactivity.[62]

Diabetes Mellitus in Adults

Studies of receptors have been performed in both thin and obese adult-onset (insulin-independent) diabetics. The majority of adult-onset diabetics are obese and insulin resistant and demonstrate the inverse relationship between receptor concentration and basal levels of insulin.

J. M. Olefsky and coworkers studied a group of thin, insulin-dependent diabetics and found the same inverse relationship between receptor concentration and ambient levels of insulin.[63] In these thin diabetics Olefsky also noted improvement in insulin binding to receptor after treatment with sulfonylureas. Acutely, in vivo and in vitro, sulfonylureas cause release of insulin from the β-cells of the pancreas, but the chronic use of sulfonylureas is not associated with any demonstrable increase in circulating insulin. The studies by Olefsky, which showed that receptors are increased, without any change in hormone concentration, in patients who respond to the drug but not in nonresponders, led him to conclude that sulfonylureas act by way of their effect on receptors. Similar results have been obtained in normal rodents.

Acromegaly

A regulatory role of insulin on its own receptor concentration has been suggested in acromegaly, another clinical state characterized by hyperinsulinemia and insulin resistance. As in obesity, an inverse relationship was found between ambient levels of basal (fasting) insulin and the concentrations of insulin receptors on circulating monocytes (Figure 10). In acromegaly, however, there were also associated striking changes in receptor affinity, which may play a key role in determining whether glucose tolerance is severely disturbed.[64]

Uremia

Uremia is often associated with glucose intolerance, hyperinsulinemia, and insulin resistance, but the molecular mechanisms are complex and difficult to isolate. In a uremic model in rodents, insulin receptors were decreased and the magnitude of the receptor deficiency was correlated with the magnitude of the hyperinsulinemia.[65]

Effects of Glucocorticoid on Receptor Status in Rodents[66-70]

Glucocorticoid excess is typically associated with insulin resistance, with or without glucose intolerance, and with a subnormal affinity of insulin for its receptors (Figure 11). Conversely, glucocorticoid deficiency is associated with a decrease in blood glucose and supernormal sensitivity to the effects of insulin. Insulin receptors in rats that are glucocorticoid deficient have an increase in affinity[71] (Figure 11). In both cases receptor concentrations show little or no change. This was a bit unexpected because,

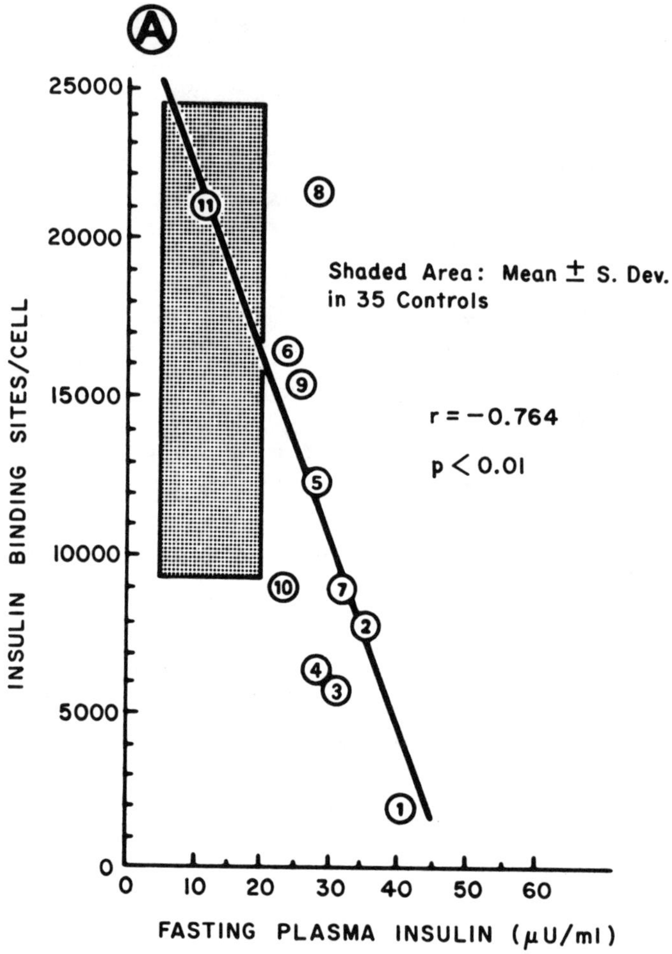

Figure 10. Total receptor concentration, expressed as the number of insulin binding sites per cell, is plotted as a function of the fasting plasma insulin concentration for each of eleven acromegalic patients. The shaded area represents the mean ± 1 S.D. for the normal subjects.

SOURCE: M. Muggeo et al., "Insulin Resistance of Acromegaly: Evidence for Two Alterations in the Insulin Receptor on Circulating Monocytes," *Journal of Clinical Endocrinology and Metabolism* 48 (1979): 17-25.

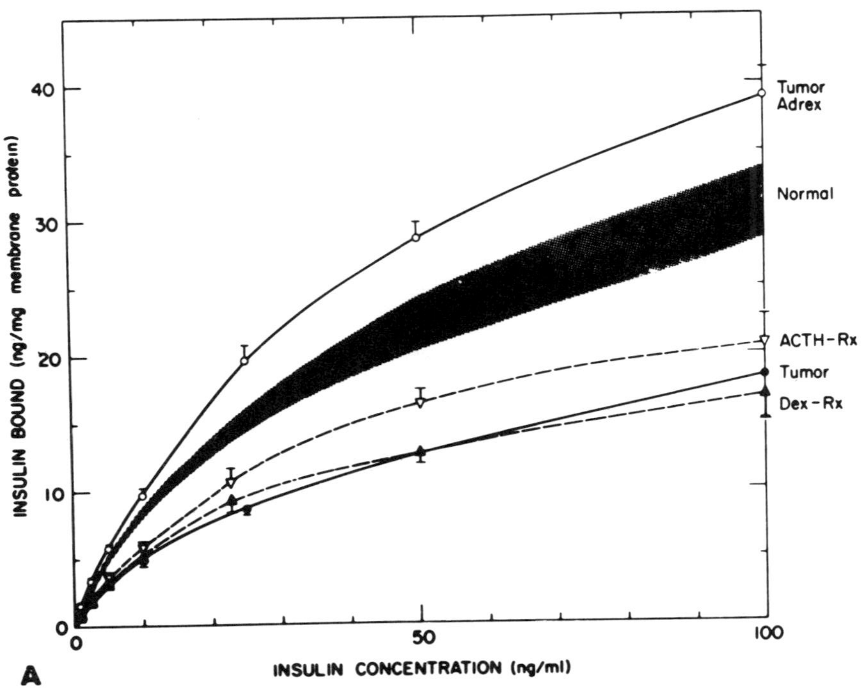

	Cortico-Sterone µg/100ml	Glucose mg/100ml	Insulin ng/ml
Tumor Adrex	4	120	1.6
Normal	5	130	1.5
ACTH-Rx	95	122	4.0
Tumor	53	110	10.0
Dex-Rx	-	192	7.5

B

Figure 11. Effects of glucocorticoid deficiency and excess on insulin binding. Normal rats were treated with ACTH ("ACTH-Rx") or dexamethasone (Dex-Rx). The curve labeled "Tumor" represents rats bearing tumors that hypersecrete ACTH; that labeled "Tumor Adrex" represents the same animals following adrenalectomy. (Although these tumors also hypersecrete growth hormone and prolactin, the effect on the insulin receptor in these animals appears to reflect only the hyper- and hypoadrenal states, with little or no effect from the other two hormones.) The depression of insulin binding due to glucocorticoid excess and the elevated level of insulin binding in adrenalectomized animals on careful analysis appears to be largely or entirely due to changes in receptor affinity, with little or no changes in receptor concentration.

SOURCE: J. Roth, "Receptors for Peptide Hormones," in *Endocrinology,* vol. 3, ed. L. DeGroot (New York: Grune and Stratton, 1979): 2037-54.

with glucocorticoid excess, the severity of the hyperinsulinemia and insulin resistance is of the same magnitude as in acromegaly and obesity. Thus, in the case of steroid excess, the regulatory role of hyperinsulinemia and hypoinsulinemia in regulating the concentration of its own receptors appears to be nullified, and some mechanism other than hyperinsulinemia appears to be the primary regulator of the insulin receptor. The molecular mechanisms for this nullification and for the alteration in receptor affinity are as yet unknown.

RECEPTOR AUTOANTIBODIES

Extreme Insulin Resistance Associated with Acanthosis Nigricans

There are patients who have extreme insulin resistance, extreme hyperinsulinemia, and severe defects in insulin binding to receptors, associated typically with the skin lesion of acanthosis nigricans.[72] Thus far all cases have been sporadic, with no familial clusters. The characteristic features of these patients are shown in Table 4.

The patients with Type A syndrome are female children or young adults who have no known cause of insulin resistance but do have problems with masculinization and hirsutism. On fasting for forty-eight to seventy-two hours the plasma insulin falls substantially, but is still elevated; over this interval there is no change in receptor concentration or affinity. The mechanisms for these defects is unknown, as is the sequence of events that leads to the receptor defect, hyperinsulinemia, and insulin resistance.

In another subgroup, designated Type B, there are multiple features suggesting an autoimmune process, although only a minority of the patients fit into the realm of previously defined autoimmune disorders. By definition, all Type B patients have circulating antibodies that reduce the binding of insulin to its own receptors. The autoantibodies are polyclonal, primarily IgG, and are specific for the insulin receptor.[73] In an individual patient the titer of autoantibodies correlates with both the degree of clinical resistance to insulin and the level of impairment of insulin binding to the patient's own monocytes.

TABLE 4. SYNDROME OF EXTREME INSULIN RESISTANCE WITH ACANTHOSIS NIGRICANS

1. Extreme elevation of endogenous plasma insulin concentrations—basal and stimulated
2. Extreme resistance to exogenous insulin
3. Absence of any previously known cause of insulin resistance, including obesity and lipoatrophy
4. Glucose tolerance ranges from normal to severe hyperglycemia with moderate ketoacidosis
5. Marked impairment of insulin binding to its receptor on freshly obtained cells (monocytes)
6. Three or more clinical subtypes:
   Type A. No immunological features
      Young females—children and adolescents
      Disordered sexual function—amenorrhea; hirsutism; enlargement of clitoris and/or labiae; polycystic ovaries
      Receptor concentration decreased; receptor affinity normal
      Accelerated early growth; no autoimmune features
   Type B. Autoimmune features
      Older; mostly females
      Immunological features—elevated erythrocyte sedimentation rate; leukopenia; alopecia; arthralgias; nephritis; antinuclear antibodies; some patients have well defined autoimmune disease (e.g. systemic lupus)
      Antireceptor antibodies—bind to insulin receptor; block insulin binding; impair insulin action; mimic insulin action
      Clinical course—insulin resistance may be replaced by hypoglycemia; nephritis and nephrotic syndrome associated with proliferation of cellular receptors with low affinity for insulin; remissions (spontaneous or drug induced) and death in one case
   Type C. Very similar to Type A except that insulin binding to receptor is normal

In addition to blocking the binding of insulin to its receptors, several of these autoantibodies produce insulin-like effects in adipocyte, muscle, and other cell preparations in vitro. These acute, insulin-like effects are transient and are followed by cellular adaptations that lead to a state of cellular insensitivity to insulin; thus, both in vitro and in vivo, chronic exposure to the autoantibodies results in insulin resistance.[74-77]

Autoantibodies to the insulin receptor have also been detected in patients with ataxia telangiectasia, a genetic disorder characterized by progressive neuromuscular disease, diverse abnormalities of the immune system, and insulin resistance. The autoantibodies are low molecular weight IgM, and in general are of lower titer than in most Type B patients.[78, 79] Both ataxia and Type B patients have similar qualitative abnormalities of insulin binding to their own monocytes (Figure 12). Maximal insulin binding is decreased in both syndromes, and receptor affinity is markedly decreased—in contrast to the obese patients; receptor concentration is either normal or increased.

Several lines of evidence indicate that the underlying recep-

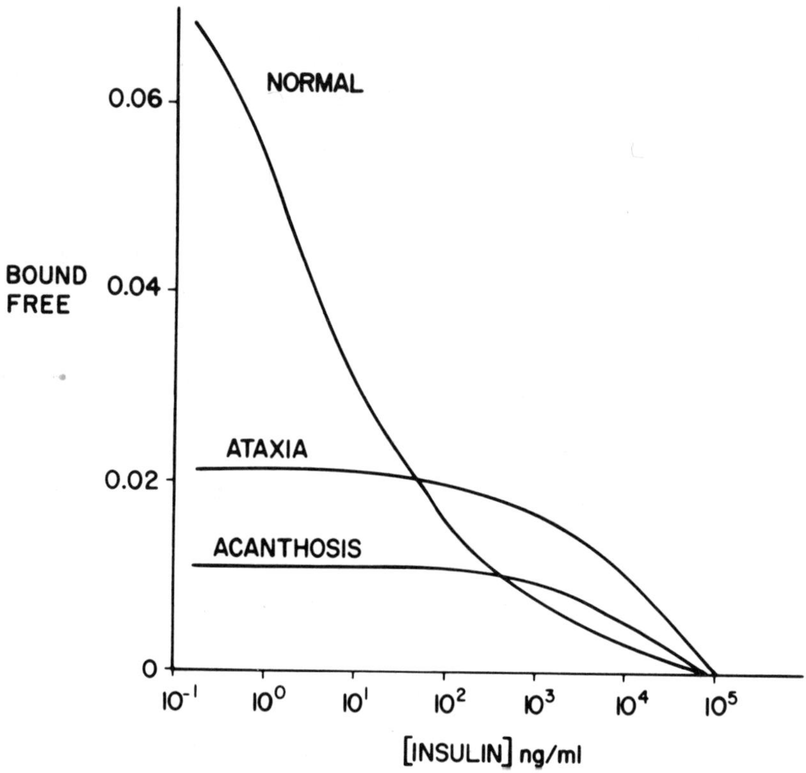

Figure 12. Insulin binding to circulating monocytes in a normal individual, a patient with insulin resistance and ataxia telangiectasia, and a patient with the syndrome of acanthosis nigricans with insulin resistance, type B. Note the decreasing binding and the "flat" shape of the binding curve in the ataxia and type B patients, consistent with a marked decrease in receptor affinity.

SOURCE: R. S. Bar and J. Roth, "Altered Insulin Receptors, Insulin Resistance and Diabetes Mellitus," in *Secondary Diabetes: The Spectrum of Diabetic Syndromes,* ed. S. Podolsky and M. Viswanathan (New York: Raven Press, forthcoming).

tor in these two syndromes is normal, and that the autoantibodies induce the observed binding abnormalities and clinical resistance to the action of insulin.[80] Thus exposure of cultured cells or normal monocytes to sera from Type B or ataxia patients also induces marked decreased in receptor affinity. Elution of the antibodies from normal cells or from the patient's own cells restores receptor affinity to normal. Furthermore, receptors on cultured lymphocytes and fibroblasts from affected patients have normal insulin binding. Finally, when adipocytes are chronically exposed to the autoantibodies of Type B patients, the cells develop decreased binding of insulin and become refractory to the effects of insulin.

Similar antibodies have been found in the blood of the NZO mouse, a strain that has autoimmunity and insulin-resistant diabetes.[81]

Lipoatrophic Diabetes

Lipoatrophic diabetes includes a heterogeneous group of patients with glucose tolerance and insulin resistance. Except for the familial tendency and the lipoatrophy, which may be congenital, they resemble the Type A and C patients. Insulin binding to receptors is quite variable, being low, normal, or elevated. Even within an individual family the binding pattern may differ.[82]

POSTRECEPTOR DEFECTS

Not all states of insulin resistance are accompanied by changes in insulin receptors. There are multiple examples of studies in vitro and in animals in which receptor events do not account for the observed cellular effects of insulin.[83] Often we can infer the existence of a defect at a site beyond the receptor (more properly, beyond binding). For example, we recently reported on a young woman with severe insulin resistance and clinical features identical with type A patients who had insulin resistance and acanthosis nigricans.[84] Characterization of insulin receptors in this patient, however, demonstrated normal receptors— concentration, affinity, specificity, and cooperative interactions—thereby indicating the presence of postreceptor defect(s) as the cause of her insulin resistance. M. Kobayashi and colleagues have since reported a patient with leprechaunism who had severe insulin resistance and normal receptors.[85] Cultured fibroblasts from this patient demonstrated normal insulin receptors but an impaired stimulation of glucose transport by insulin, further implicating a postreceptor (or postbinding) defect.

Additional examples of the dissociation of receptor events from the biological effects of insulin have been reported in animal studies.[86] Fasted obese mice, for example, have increased binding of insulin to adipocyte receptors, whereas insulin stimulation of glucose oxidation in the same cells is decreased. Such findings provide evidence that events subsequent to the binding of insulin can predominate and, in some instances, counteract the observed receptor change. With the future development of methods to assess the specific transmembrane and intracellular pathways of insulin action, it is likely that multiple examples of postreceptor defects will be found.

Cautions in Interpretation

The triad at steady state of 1) elevated concentration of biologically active hormone, 2) resistance to the effect of hormone at the target cell, and 3) decrease in hormone binding to receptors can be reached by any of multiple routes. The primary defect or earliest defect in the cycle could be hormone excess or receptor deficiency or distal defect in hormone action. Any one would yield the same triad at steady state, since excess hormone desensitizes the target cell (at the receptor and/or steps beyond receptor), and a resistant target cell (at the receptor or beyond) produces hypersecretion of hormone. Thus elucidation of the sequence of events requires that the steady state be perturbed and that variables be dealt with individually when possible.

DISORDERS OF RECEPTOR DESIGN (SPECIFICITY SPILLOVER)

Disease Manifestations Due to Degeneracies of Hormone Receptor Specificity

As a first approximation, all cells in the body are exposed to the same concentration of all hormones. For a system to have specificity, the target cell must have specific receptors that recognize each particular hormone. Ideally, each hormone should be unique and have its own unique receptor; often each recognition unit is not unique. Among the steroid and peptide hormones there are families of hormones within which members have structural similarities, presumably because they evolved from a common ancestor. Likewise, overlaps exist with receptors. Hormones have high affinity for their own receptors, but, in addition, may also have some finite affinity for the receptors of a structurally related hormone.

Normally the spillover of one hormone on to the receptors of another related hormone is so slight that the one hormone causes no significant biological effects via the receptor for the other hormone. In pathological states, where the concentration of one hormone is elevated, disease manifestations will obviously result from excessive effect mediated through its own receptors. In addition, when the concentration of that hormone becomes sufficiently high, its weak affinity for another class of receptor can now produce significant biological effects through the spillover pathway. Two examples from the insulin family will be presented.

The Family of Insulin-Like Peptides [87-98]

In addition to insulin and its precursors, such as proinsulin, there exists in the circulation a series of closely related insulin-like growth-promoting substances. The best characterized are somatomedins A and C, NSILA-s (nonsuppressible insulin-like activity soluble in acid ethanol, which is composed of two similar peptides, insulin-like growth factor [IGF] I and II),[99] and multiplication-stimulating activity (MSA). Other related peptides not covered here include NSILA-p or (NSILP), somatomedin B, nerve growth factor, and relaxin. These insulin-like peptides are similar to insulin with molecular weights in the range of insulin and proinsulin. They do not bind to anti-insulin antibodies, however, and therefore do not react in the radioimmunoassay for insulin. They produce all the metabolic effects of insulin, and at high concentrations are capable of maximally stimulating insulin-sensitive pathways, that is, they are full agonists. They are about 1 percent as potent as insulin on a molar basis, biologically and in their reactivity for the insulin receptor, suggesting strongly that their insulin-like properties are exercised through the insulin receptor.[100-103]

These peptides also promote cell growth, stimulating cell division in numerous test systems. When they exercise their growth-promoting activities they do so by binding to another set of receptors, which, in contrast to the insulin receptor, often bind insulin-like growth factors better than they bind insulin and proinsulin.

Insulin and proinsulin also stimulate cell growth, but are less potent than the insulin-like growth factors; the growth-promoting properties of insulin and proinsulin are probably not mediated via the insulin ("metabolic") receptor, but through the receptors that prefer insulin-like factors ("growth" receptors). Thus both insulin and insulin-like growth factors interact with their own receptors, but, in addition, have weak but significant reactivities for one or more closely related classes of receptors (Figure 13, A, B, and C). At low concentrations each produces its own characteristic effects, whereas at very high concentrations, in addition to producing its own effects, each can produce effects mediated through the other receptors (Figure 14). It should be emphasized that these receptors are all highly specific. They discriminate totally between insulins and all noninsulin peptides, binding only insulin and insulin-like peptides; among the insulins, each receptor class has a different order of affinities.

This overall formulation is supported by recent studies by C. R. Kahn and G. L. King and their coworkers using antireceptor antibodies, which block hormone binding to the insulin receptor but not to the growth factor receptors.[104] The antibody in its bivalent form is also insulinomimetic, that is, it can mimic all of the metabolic responses of insulin acutely, but notably fails to affect cell growth. Further, the monovalent form of the antibody, which blocks insulin binding but is not insulinomimetic (it is a pure antagonist), blocks the metabolic response to both insulin and a growth factor (MSA), while it leaves totally unaffected the growth-promoting effects of both insulin and the growth factor.

Hypoglycemia Associated with Nonislet Cell Tumors

As shown in Table 5, tumors associated with fasting hypoglycemia may be of islet cell or nonislet cell origin.[105-107] Clinically, the syndromes are indistinguishable from one another, and both are capable of causing profound hypoglycemia. With the advent of the radioimmunoassay for insulin, it became clear that patients with islet cell tumors had inappropriate elevation of the plasma insulin, whereas those with nonislet cell tumors rarely, if ever, did. With the development of a specific radioreceptor assay for NSILA-s it has been found that about 30-50 percent of patients with nonislet cell tumors have an inappropriate elevation of the plasma NSILAs, where patients with the islet cell tumors do not.[108-110]

In patients with nonislet cell tumors that produce NSILA-s in excess, the free (unbound) concentration of the insulin-like growth factor(s) becomes sufficiently high so that the insulin-like (metabolic) effects are equal to the bioactivity produced by high insulin levels in patients with islet cell tumors [111-112] (Table 6). We conclude that in these patients the hypoglycemic effect of the insulin-like growth factor is being exercised through a spillover at the insulin receptor.

Insulin-Like Effects of Somatotropin

While somatotropin pituitary growth hormone acts to antagonize the effects of insulin, human growth hormone can produce an acute insulin-like effect.[113] With the introduction of the radioimmunoassay for insulin it became clear that this acute in-

TABLE 5. TUMORS ASSOCIATED WITH FASTING HYPOGLYCEMIA

Tumors	Hypoglycemia	Inappropriate Elevation of Plasma Insulin	Elevation of Plasma NSILA-s
Islet cell	Yes	Yes	No
Nonislet cell	Yes	No	Yes
Mesenchymal			
Hepatoma			
Adrenocortical carcinoma			
Other			

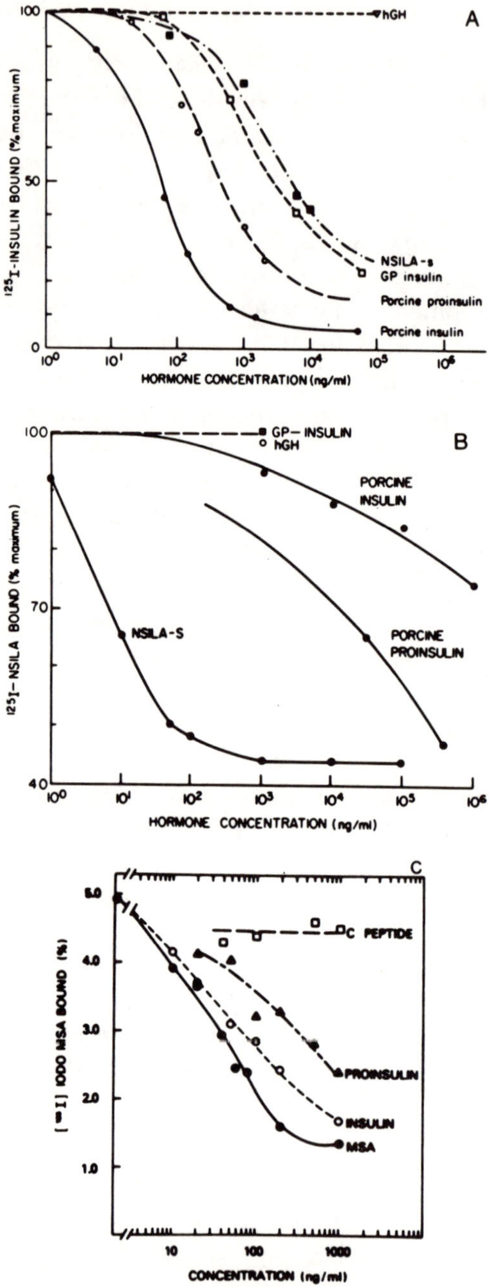
A
hGH
NSILA-s
GP insulin
Porcine proinsulin
Porcine insulin
^{125}I-INSULIN BOUND (% maximum)
HORMONE CONCENTRATION (ng/ml)
B
GP– INSULIN
hGH
PORCINE INSULIN
NSILA-S
PORCINE PROINSULIN
^{125}I–NSILA BOUND (% maximum)
HORMONE CONCENTRATION (ng/ml)
C
C PEPTIDE
PROINSULIN
INSULIN
MSA
IODO MSA BOUND (%)
CONCENTRATION (ng/ml)

Figure 13. Characterization of the specificity of three insulin-related receptors. A, *at top,* shows the characteristic specificity of the insulin receptor, which is the same for all tissues of all vertebrate species studied. With ^{125}I-insulin as a tracer, pork insulin binds better than pork proinsulin, which is more potent than guinea pig insulin and NSILA-s (IGF I and II). Almost 100 different insulin-related substances have been tested, and in nearly all cases the ability of the material to bind to the insulin receptor is predicted by its biological potency, measured as in vitro metabolic responses. B, *center,* shows the specific NSILA-s receptor on the plasma membrane of rat liver. With ^{125}I-NSILA-s as tracer, NSILA-s and MSA both react strongly; pork proinsulin and pork insulin react much less well. C, *bottom,* shows another more typical growth-factor type of receptor. The insulin-like growth factors bind modestly, better than insulin or proinsulin, and the binding affinity overall is lower than that observed for either the insulin receptor or the NSILA-s specific receptor. From the data available it appears that there are several more growth factor receptors present in different species or in different tissues (not shown here), which may demonstrate an affinity for the growth factors similar to that shown in this figure but in which insulin and proinsulin react much less well.

SOURCES: 13, A, K. Megyesi et al., "Radioreceptor Assay of Plasma NSILA-s in Man: Basal and Stimulated Levels in Normal and Pathologic States," in *Growth Hormone and Related Peptides,* Proceedings of the III International Symposium, Milan (Amsterdam: Excerpta Medica, 1975): 127-40. 13, B, K. Megyesi et al., "Hypoglycemia in Association with Extrapancreatic Tumors: Demonstration of Elevated Plasma NSILA-s by a New Radioreceptor Assay," *Journal of Endocrinology and Metabolism* 38 (1974): 931-34. 13, C, S. P. Nissley et al., "Proinsulin Binds to a Growth Peptide Receptor and Stimulates DNA Synthesis in Chick Embryo Fibroblasts," *Endocrinology* 101 (1979): 708-16.

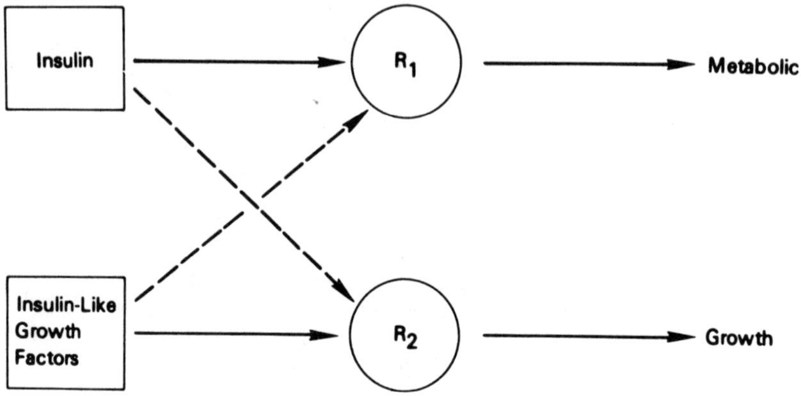

Figure 14. The relationship of insulin and insulin-like growth factors to their receptors. In this schematic outline insulin is shown to react strongly with its own receptors (R_1) to produce the characteristic metabolic responses. The insulin-like growth factors react strongly with their receptors (R_2) to produce growth-promoting effects. In addition, each hormone interacts with receptors for the other, although the heterologous affinities are distinctly lower than the affinities for homologous receptors.

SOURCE: J. Roth, "Receptors for Peptide Hormones," in *Endocrinology,* vol. 3, ed. L. DeGroot (New York: Grune and Stratton, 1979): 2037-54.

sulin-like effect of somatotropin was not due to an acute release of insulin. The studies illustrated in Figure 15 demonstrate that growth hormone, acting through its own receptor, produces an acute rise in NSILA-s, which is accompanied by acute insulin-like effects, namely, a fall in the concentration of free-fatty acids in three of the patients, and a fall in the blood glucose in one of the patients without any change in plasma concentration of insulin itself. These data indicate that somatotropin causes an acute release of one or more insulin-like growth factors, which are not detected in the radioimmunoassay for insulin, but are detected in the radioreceptor assay for NSILA-s and MSA; the growth factors at high concentrations interact with the insulin receptor to produce the insulin-like effect. Thus this represents another example of specificity spillover involving, presumably, the interaction of one or more insulin-like growth factors with the insulin receptor to produce insulin-like metabolic effects.

TABLE 6. THE INSULIN-LIKE ACTIVITY IN THE PLASMA OF PATIENTS WITH ISLET CELL AND NONISLET CELL TUMORS

	Schematic Outline					Representative Values					
						Component Parts				Total Insulin-like Activity (M.W. 5,000-10,000)	
	Plasma [INS]	+	Plasma [NSILA-s]	=	Total Effective Bioactivity	NSILA-s Receptor Assays		Insulin Immuno-assay			
Patient Condition						Insulin-Like Activity of Free NSILA-s		Insulin + Proinsulin		Calculated as Sum of Components Parts	Measured by Radioreceptor Assay with Insulin Receptor
Normal	Normal	+	Normal	=	Normal	10	+	8	=	18 μU/ml.	18 μU/ml.
Islet cell tumor ("insulinoma")	Elevated	+	Normal	=	Elevated	5	+	130	=	135 μU/ml.	55 μU/ml.
Nonislet cell tumor (30-50% of cases)	Normal	+	Elevated	=	Elevated	120	+	15	=	135 μU/ml.	150 μU/ml.

NOTES: The schematic outline (*left*) represents our interpretation of the published data. The representative values (*right*) are derived from data we have obtained from three patients reported elsewhere. Plasma from each of the patients was filtered at neutral pH on Sephadex G-50, and the effluent fractions corresponding to molecular weights of 5,000 to 10,000 daltons were collected. At neutral pH, 90 percent or more of the NSILA-s is bound to one or more plasma proteins, with only a small fraction of the NSILA-s existing in the free form. During gel filtration at neutral pH the bound NSILA-s migrates in the void volume, whereas the free NSILA-s appears in the 5,000 to 10,000 M.W. region. The NSILA-s specific radioreceptor assay was used to estimate the content of free NSILA-s and the results were converted into insulin-like equivalents based on the purity of the standards used in the receptor assay. The insulin, including proinsulin and its intermediates, were measured by radioimmunoassay for insulin. The total insulin-like activity of plasma is shown as the sum of these two assays. Independently, this material was studied in a radioreceptor assay for insulin, which measures substances on the basis of their ability to bind to the insulin receptor, which is proportional to their ability to stimulate metabolic processes. It can be seen that the insulin-like activity measured directly by the radioreceptor assay for insulin gives results that are quite close to those expected by measurements of the sum of the parts independently. In the patient with the islet cell tumor, the insulin content measured by radioimmunoassay may be overestimated because the immunoreactive insulin in plasma of patients with islet cell tumors often is enriched with proinsulin-like components that react more strongly with the antibody in the radioimmunoassay than they do with the insulin receptor.

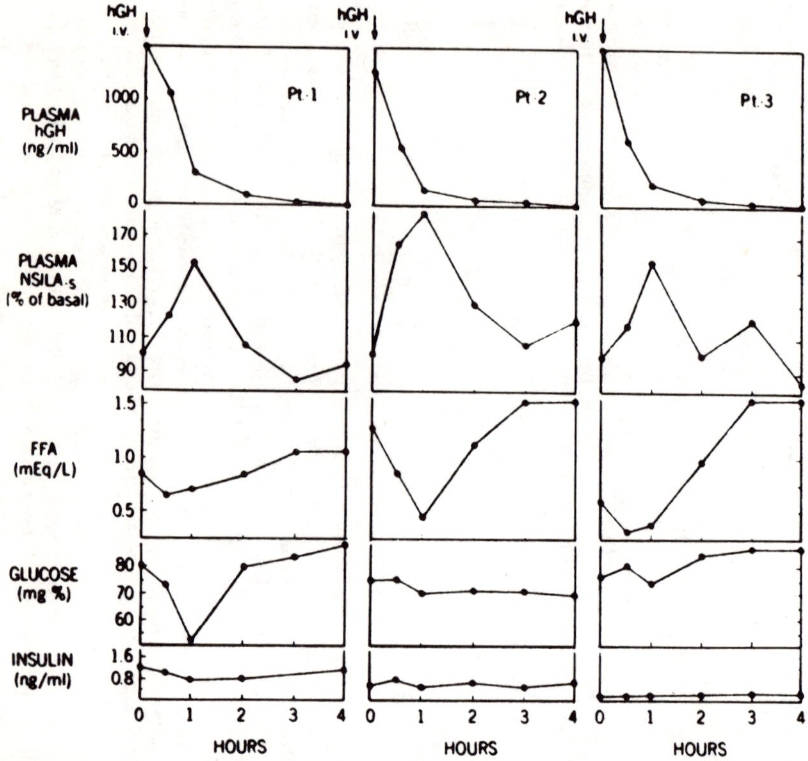

Figure 15. The acute insulin-like effects of growth hormone in vivo. Three hypopituitary patients were given 10 mg. of human growth hormone (hGH) intravenously at a single injection. Plasma concentrations of hGH, which were very high immediately following the injection, fell rapidly and progressively throughout the course of the experiment. Plasma NSILA-s, measured in a specific NSILA-s radioreceptor assay, were subnormal at the beginning of the experiment, but rose rapidly, with a peak at about one hour following the injection, and a fall to normal or near normal levels shortly thereafter. Simultaneous with the rise in plasma NSILA-s was a fall in the plasma concentration of free fatty acids (FFA) in all three patients, and a significant fall in the blood glucose in one patient. Note that plasma insulin concentrations, measured by the insulin radioimmunoassay, were unaltered throughout the experiment.

SOURCE: K. Megyesi et al., "Circulating NSILA-s in Man: Preliminary Studies of Stimuli *in Vivo* and of Binding to Plasma Components," *Journal of Clinical Endocrinology and Metabolism* 41 (1975): 475-84.

Infants of Diabetic Mothers

Mothers with hyperglycemia during late pregnancy often produce babies with a characteristic syndrome.[114] The babies have elevated levels of circulating insulin, presumably due to transplacental passage of glucose and possibly other substances that act as stimuli to the babies' beta cells. They are typically big, with both length and weight being proportionately increased. They have organomegaly, an excess deposition of stored substrates, including fat, glycogen, and protein, and may show hypoglycemia postnatally. We conclude that the excess of stored substrates and the hypoglycemia are due to elevated insulin levels acting on the insulin receptor. On the other hand, we attribute the increased length and macrosomia to the effect of excess insulin acting through growth factor receptors to produce these changes.[115] Thus excess levels of insulin act on the insulin receptor to produce metabolic effects, and cross-react via growth factor receptors to produce effects on somatic growth.

Growth-Promoting Effects of Insulin in Cell Cultures

Insulin at high concentration can substitute for serum in permitting the growth of cells in culture. Recently, J. Bottenstein and colleagues have shown for a wide range of cell lines that insulin is one of the essential ingredients for growth in culture in the absence of serum.[116] While it is possible that insulin in some cases is acting via the insulin receptor, we suggest that when insulin acts as a promoter of cell division and cell growth it typically acts through receptors for insulin-like growth factors.[117]

SUMMARY

In only a small minority of patients with disorders of glucose metabolism is the insulin level the dominant determinant of the clinical state; most often the target cell is the cause of the disorder. With the introduction of methods to measure directly insulin binding to its receptors, it is now clear that in the majority of endocrine disorders the receptor rather than the hormone is exerting the major influence. Moreover, even in those conditions where the hormone level is playing the dominant role, events at the receptor are important modulators of the clinical state.

NOTES

1. R. S. Yalow and S. A. Berson, "Immunology of Endogenous Plasma Insulin in Man," *Journal of Clinical Investigation* 39 (1960): 1157-75.

2. U. Lang, C. R. Kahn, and A. Chrambach, "Characterization of the Insulin Receptor from Human Lymphocytes by Quantitative Polyacrylamide Gel Electrophoresis," *Endocrinology,* in press.

3. P. Cuatrecasas, "Affinity Chromatography and Purification of the Insulin Receptor of Liver Cell Membranes," *Proceedings of the National Academy of Sciences* 69 (1972): 1277-81.

4. B. H. Ginsberg, C. R. Kahn and J. Roth, "The Insulin Receptor of the Turkey Erythrocyte: Purification and Characterization of the Membrane-Bound Receptor," *Biochimica et Biophysica Acta* 443 (1977): 227-42.

5. B. H. Ginsberg, "The Insulin Receptor: Properties and Regulation," in *Biochemical Actions of Hormones,* vol. 4, ed. G. Litwak (New York: Academic Press, 1977): 313-49.

6. L. C. Harrison et al., "The Effect of Solubilization on the Properties of the Insulin Receptor of Human Placental Membranes," *Endocrinology,* in press.

7. R. S. Bar, C. R. Kahn, and H. S. Koren, "Insulin Inhibition of Antibody-Dependent Cytotoxicity and Insulin Receptors in Macrophages," *Nature* 265 (1977): 632-35.

8. R. J. Muschel et al., "Modulation of F_c-Mediated Phagocytosis by Cyclic AMP and Insulin in a Macrophage-Like Cell Line," *Journal of Immunology* 119 (1977): 1813-20.

9. H. Beck-Nielsen and O. Pedersen, "Insulin Binding, Insulin Degradation and Glucose Metabolism in Human Monocytes," *Diabetologia* 17 (1979): 77-84.

10. R. H. Schwartz et al., "A Demonstration That Monocytes Rather than Lymphocytes Are in the Insulin Binding Cells of Human Peripheral Blood Mononuclear Leukocyte Preparations: Implications for Studies of Insulin-Resistant States in Man," *Proceedings of the National Academy of Sciences* 72 (1975): 474-78.

11. S. A. Berson et al., "Immunoassay of Protein and Peptide Hormones," *Metabolism* 13 (1964): 1135-53.

12. G. Scatchard, "The Attraction of Proteins for Small Molecules and Ions," *Annals of the New York Academy of Science* 51 (1949): 660-72.

13. P. De Meyts et al., "Insulin Interactions with Its Receptors: Experimental Evidence for Negative Cooperativity," *Biochemical and Biophysical Research Communications* 55 (1973): 154-61.

14. P. De Meyts, "Cooperative Properties of Hormone Receptors in Cell Membranes," *Journal of Supramolecular Structure* 4 (1976): 241-58.

15. P. De Meyts and J. Roth, "Cooperativity in Ligand Binding: A New Graphic Analysis," *Biochemical and Biophysical Research Communications* 66 (1975): 1118-26.

16. P. Freychet, J. Roth, and D. M. Neville, Jr., "Mono-iodoinsulin: Demonstration of Its Biological Activity and Binding to Fat Cells and Liver Membranes," *Biochemical and Biophysical Research Communications* 43 (1971): 400-08.

17. ———, "Insulin Receptors in the Liver: Specific Binding of ^{125}I-Insulin to

the Plasma Membrane and Its Relation to Insulin Bioactivity," *Proceedings of the National Academy of Sciences* 68 (1971): 1833-37.

18. J. R. Gavin, III et al., "Insulin-Dependent Regulation of Insulin Receptor Concentrations: A Direct Demonstration in Cell Culture," *Proceedings of the National Academy of Sciences* 71 (1974): 84-88.

19. F. C. Kosmakos and J. Roth, "Cellular Basis of Insulin-Induced Loss of Insulin Receptors," *Endocrinology* 98 (1976): 69A.

20. W. G. Blackard, P. S. Guzelian, and M. E. Small, "Down Regulation of Insulin Receptors in Primary Cultures of Adult Rat Hepatocytes in Monolayer," *Endocrinology* 103 (1978): 548-53.

21. D. M. Mott, B. V. Howard, and P. H. Bennett, "Stoichiometric Binding and Regulation of Insulin Receptors on Human Diploid Fibroblasts Using Physiologic Insulin Levels," *Journal of Biological Chemistry* 254 (1979): 8762-67.

22. M. A. Lesniak and J. Roth, "Regulation of Receptor Concentration by Homologous Hormone: Effect of Human Growth Hormone on Its Receptor in IM-9 Lymphocytes," *Journal of Biological Chemistry* 251 (1976): 3720-29.

23. P. M. Hinkle and A. H. Tashjian, Jr., "Thyrotropin-Releasing Hormone Regulates the Number of Its Own Receptors in the GH_3 Strain of Pituitary Cells in Culture," *Biochemistry* 14 (1975): 3845-51.

24. J. R. Mickey, R. Tate, and R. J. Lefkowitz, "Sensitivity of Adenylate Cyclase and Decreased β-Adrenergic Receptor Binding after Chronic Exposure to (—)-Isoproterenol *in Vitro*," *Journal of Biological Chemistry* 250 (1975): 5727-29.

25. B. I. Posner, P. A. Kelly, and H. G. Friesen, "Induction of a Lactogenic Receptor in Rat Liver: Influence of Estrogen and the Pituitary," *Proceedings of the National Academy of Sciences* 71 (1974): 2407-10.

26. J. R. Gavin, III et al., "Characteristics of the Human Lymphocyte Insulin Receptor," *Journal of Biological Chemistry* 248 (1973): 2202-07.

27. P. Cuatrecasas, "Properties of the Insulin Receptor of Isolated Fat Cells," *Journal of Biological Chemistry* 23 (1971): 7265-74.

28. H. Beck-Nielsen et al., "A Subgroup of Juvenile Diabetics Characterized by an Insulin Receptor Defect," Paper presented at the 10th Congress of the International Diabetes Federation, Vienna, September 1979.

29. A. J. Zelesnik and J. Roth, "Demonstration of the Insulin Receptor *in Vivo* in Rabbits and Its Possible Role as a Reservoir for the Plasma Hormone," *Journal of Clinical Investigation* 61 (1978): 1363-74.

30. P. Freychet, C. R. Kahn, and D. M. Neville, Jr., "Insulin Interactions with Liver Plasma Membranes: Independence of Binding of the Hormone and Its Degradation," *Journal of Biological Chemistry* 247 (1972): 3953-61.

31. S. Terris and D. F. Steiner, "Binding and Degradation of ^{125}I-Insulin by Rat Hepatocytes," *Journal of Biological Chemistry* 250 (1975): 8389-98.

32. C. R. Kahn and K. L. Baird, "The Fate of Insulin Bound to Adipocytes: Evidence for Compartmentalization and Processing," *Journal of Biological Chemistry* 253 (1978): 4900-06.

33. G. V. Marinetti, L. Schlatz, and K. Reilly, "Hormone-Membrane Interactions," in *Insulin Action: Proceedings of a Symposium,* ed. I. B. Fritz (New York: Academic Press, 1972).

34. J. A. Archer, P. Gorden, and J. Roth, "Defect in Insulin Binding to Recep-

tors in Obese Man: Amelioration with Calorie Restriction," *Journal of Clinical Investigation* 55 (1975): 166-74.

35. J. M. Armatruda, J. N. Livingston, and D. H. Lockwood, "Insulin Receptor: Role in Resistance of Human Obesity to Insulin," *Science* 188 (1975): 264-66.

36. J. M. Olefsky, "Insulin Binding to Adipocytes and Circulating Monocytes from Obese Patients," *Journal of Clinical Investigation* 57 (1976): 1165-72.

37. R. S. Bar et al., "Fluctuations in the Affinity and Concentration of Insulin Receptors on Circulating Monocytes of Obese Patients: Effects of Starvation, Refeeding and Dieting," *Journal of Clinical Investigation* 58 (1976): 1123-35.

38. L. C. Harrison, F. I. R. Martin, and R. A. Melick, "Correlation between Insulin Receptor Binding in Isolated Fat Cells and Insulin Sensitivity in Obese Human Subjects," *Journal of Clinical Investigation* 58 (1976): 1435-41.

39. R. DeFronzo et al., "Dissociation of Insulin Binding to Monocytes and Insulin Action in Intact Man during Starvation and Refeeding," *Clinical Research* 25 (1977): 494A.

40. R. S. Bar et al., "Regulation of Insulin Receptors in Normal and Abnormal Physiology in Humans," *Advances in Internal Medicine* 24 (1979): 23-52.

41. Ibid.

42. O. G. Kolterman, J. R. Insel, and J. M. Olefsky, "Correlation between Decreased Cellular Insulin Receptors and Insulin Resistance in Human Obesity," Paper presented at the 10th Congress of the International Diabetes Federation, Vienna, September 1979.

43. A. H. Soll et al., "Insulin Receptor Deficiency in Genetic and Acquired Obesity," *Journal of Clinical Investigation* 56 (1975): 769-80.

44. Ibid.

45. C. R. Kahn, D. M. Neville, Jr., and J. Roth, "Insulin-Receptor Interaction in the Obese-Hyperglycemic Mouse," *Journal of Biological Chemistry* 248 (1973): 2442-50.

46. C. R. Kahn et al., "Insulin-Receptor Defect in Insulin Resistance: Studies in the Obese-Hyperglycemic Mouse," *Biochemical and Biophysical Research Communications* 48 (1972): 135-42.

47. P. Freychet, "Interactions of Polypeptide Hormones with Cell Membrane Specific Receptors: Studies with Insulin and Glucagon," *Diabetologia* 12 (1976): 83-100.

48. J. Olefsky, V. C. Bacon, and S. Baver, "Insulin Receptors of Skeletal Muscle: Specific Insulin Binding Sites and Demonstration of Decreased Number of Sites in Obese Rats," *Metabolism* 25 (1976): 179-91.

49. P. Freychet et al., "Impairment of Insulin Binding to the Fat Cell Membranes in the Obese Hyperglycemic Mouse," *FEBS Letters* 25 (1972): 339-42.

50. M.-E. Forgue and P. Freychet, "Insulin Receptors in the Heart Muscle: Demonstration of Specific Binding Sites and Impairment of Insulin Binding in the Plasma Membrane of the Obese Hyperglycemic Mouse," *Diabetes* 24 (1975): 715-23.

51. A. H. Soll et al., "Thymic Lymphocytes in Obese (ob/ob) Mice: A Mirror of the Insulin Receptor Defect in Liver and Fat," *Journal of Biological Chemistry* 249 (1974): 4127-31.

52. Kahn, Neville, and Roth, "Obese Hyperglycemic Mouse" (See note 45).

53. A. H. Soll, C. R. Kahn, and D. M. Neville, Jr., "Insulin Binding to Liver Plasma Membranes in the Obese Hyperglycemic (ob/ob) Mouse: Demonstration of

a Decreased Number of Functionally Normal Receptors," *Journal of Biological Chemistry* 250 (1975): 7402-07.

54. Soll et al., "Genetic and Acquired Obesity" (See note 43).

55. Ibid.

56. Ibid.

57. Ibid.

58. Y. LeMarchand-Brustel, B. Jeanrenaud, and P. Freychet, "Insulin Binding and Effects in the Isolated Soleus Muscle of Lean and Obese Mice," *American Journal of Physiology* 234 (1978): E348-58.

59. Y. LeMarchand-Brustel, P. Freychet, and B. Jeanrenaud, "Longitudinal Study in the Establishment of Insulin Resistance in Hypothalamic Obese Mice," *Endocrinology* 102 (1978): 74-85.

60. LeMarchand-Brustel, Jeanrenaud, and Freychet, "Lean and Obese Mice" (See note 58).

61. LeMarchand, Freychet, and Jeanrenaud, "Hypothalamic Obese Mice" (See note 59).

62. H.-R. Berthoud and B. Jeanrenaud, "Acute Hyperinsulinemia and Its Reversal by Vagotomy after Lesions of the Ventromedial Hypothalamus in Anesthetized Rats," *Endocrinology* 105 (1979): 146-51.

63. J. M. Olefsky and G. M. Reaven, "Insulin Binding in Diabetes: Relationships with Plasma Insulin Levels and Insulin Sensitivity," *Diabetes* 26 (1977): 680-88.

64. M. Muggeo et al., "Insulin Resistance of Acromegaly: Evidence for Two Alterations in the Insulin Receptor on Circulating Monocytes," *Journal of Clinical Endocrinology and Metabolism* 48 (1979): 17-25.

65. V. Soman and P. Felig, "Glucagon and Insulin Binding to Liver Membranes in a Partially Nephrectomized Uremic Rat Model," *Journal of Clinical Investigation* 60 (1977): 224-32.

66. C. R. Kahn, "Membrane Receptors for Polypeptide Hormones," in *Methods in Membrane Biology,* vol. 3, ed. E. D. Korn (New York: Plenum Publishing Company, 1975): 81-146.

67. J. M. Olefsky, "Effect of Dexamethasone on Insulin Binding, Glucose Transport, and Glucose Oxidation of Isolated Rat Adipocytes," *Journal of Clinical Investigation* 56 (1975): 1499-1508.

68. J. M. Olefsky et al., "The Effects of Acute and Chronic Dexamethasone Administration on Insulin Binding to Isolated Rat Hepatocytes and Adipocytes," *Metabolism* 24 (1975): 517-27.

69. V. G. Bennett and P. Cuatrecasas, "Insulin Receptor of Fat Cells in Insulin-Resistance Metabolic States," *Science* 176 (1972): 805-06.

70. C. R. Kahn et al., "Alterations in Insulin Binding Induced by Changes *in Vivo* in the Levels of Glucocorticoids and Growth Hormones," *Endocrinology* 103 (1978): 1054-66.

71. Ibid.

72. C. R. Kahn et al., "The Syndromes of Insulin Resistance and Acanthosis Nigricans: Insulin-Receptor Disorders in Man," *New England Journal of Medicine* 294 (1976): 739-45.

73. J. S. Flier et al., "Antibodies That Impair Insulin Receptor Binding in an Unusual Diabetic Syndrome with Severe Insulin Resistance," *Science* 190 (1975): 63-65.

74. Kahn et al., "Disorders in Man" (See note 72).

75. Flier et al., "Unusual Diabetic Syndrome" (See note 73).

76. C. R. Kahn et al., "Effects of Autoantibodies to the Insulin Receptor in Isolated Adipocytes. Studies of Insulin Binding and Insulin Action," *Journal of Clinical Investigation* 60 (1977): 1094-1106.

77. F. A. Karlsson et al., "Desensitization of the Insulin Receptor at an Early Post-Receptor Step by Prolonged Exposure to Anti-Receptor Antibody," *Proceedings of the National Academy of Sciences* 76 (1979): 809-13.

78. R. S. Bar et al., "Extreme Insulin Resistance in Ataxia Telangiectasia: Defect in Affinity of Insulin Receptors," *New England Journal of Medicine* 298 (1978): 1164-71.

79. L. C. Harrison et al., "Insulin Binding Defects Induced by a Serum Globulin Factor in Ataxia Telangiectasia," *Clinical Research* 27 (1979): 252A.

80. M. Muggeo et al., "The Underlying Insulin Receptor in Patients with Anti-Receptor Autoantibodies: Demonstration of Normal Binding and Immunologic Properties," *Journal of Clinical Endocrinology and Metabolism* 49 (1979): 110-19.

81. L. C. Harrison and A. Itin, "A Possible Mechanism for Insulin Resistance and Hyperglycemia in NZO Mice," *Nature* 279 (1979): 334-36.

82. H. Wachslicht-Rodbard et al., "Evidence for Heterogeneity in the Pathogenesis of Lipoatrophic Diabetes as Demonstrated by ^{125}I-Insulin Binding in Circulating Cells," *Clinical Research* 27 (1979): 379A.

83. LeMarchand-Brustel, Freychet, and Jeanrenaud, "Hypothalamic Obese Mice" (See note 59).

84. R. S. Bar et al., "Insulin Resistance, Acanthosis Nigricans and Normal Insulin Receptors in a Young Woman: Evidence for a Post-Receptor Defect," *Journal of Clinical Endocrinology and Metabolism* 47 (1978): 620-26.

85. M. Kobayashi et al., "Insulin Resistance (IR) Due to a Post-Receptor Defect," *Diabetes* 27, suppl. 2 (1978): 84A.

86. D. H. Lockwood et al., "Effect of Fasting on Insulin Binding and Glucose Transport by Fat Cells," *Diabetes* 25, suppl. 1 (1976): 321.

87. C. R. Kahn et al., "Receptors for Peptide Hormones: New Insights into the Pathophysiology of Disease States in Man," *Annals of Internal Medicine* 86 (1977): 205-19.

88. J. J. Van Wyk et al., "The Somatomedins: A Family of Insulin-Like Hormones under Growth Control," *Recent Progress in Hormone Research* 30 (1974): 259-95.

89. W. H. Daughaday et al., "Somatomedin: Proposed Designation for Sulfation Factor," *Nature* 235 (1972): 107.

90. K. Megyesi et al., "Pathophysiology of Non-Suppressible Insulin-Like Activity," in *Endocrinology,* vol. 2, Proceedings of the V International Congress of Endocrinology, Excerpta Medica International Congress, series no. 403, ed. V. H. T. James (Amsterdam: Excerpta Medica, 1976): 173-77.

91. E. Rinderknecht and R. Humbel, "Polypeptides with Non-Suppressible Insulin-Like and Cell Growth Promoting Activities in Human Serum: Isolation, Chemical Characterization, and Some Biological Properties of Forms I and II," *Proceedings of the National Academy of Sciences* 73 (1976): 2365-69.

92. N. C. Dulak and H. M. Temin, "A Partially Purified Polypeptide Fraction

from Rat Liver Cell Conditioned Medium with Multiplication Stimulating Activity for Embryo Fibroblasts," *Journal of Cell Physiology* 81 (1973): 153-60.

93. M. M. Rechler et al., "Purified Human Somatomedin A and Rat Multiplication Stimulating Activity: Mitogens for Cultured Fibroblasts That Cross-React with the Same Growth Peptide Receptors," *European Journal of Biochemistry* 82 (1978): 5-12.

94. J. Zapf et al., "Specific Binding of Non-Suppressible Insulin-Like Activity to Chicken Embryo Fibroblasts and to a Solubilized Receptor," *Archives of Biochemistry and Biophysics* 168 (1975): 630-37.

95. M. M. Rechler et al., "Identification of a Receptor for Somatomedin-Like Polypeptides in Human Fibroblasts," *Journal of Clinical Endocrinology and Metabolism* 44 (1977): 820-31.

96. M. M. Rechler, J. M. Podskalny, and S. P. Nissley, "Characterization of the Binding of Multiplication Stimulating Activity (MSA) to a Receptor for Growth Polypeptides in Chick Embryo Fibroblasts," *Journal of Biological Chemistry* 252 (1977): 3898-3910.

97. ———, "Interaction of Multiplication Stimulation Activity with Chick Embryo Fibroblasts Demonstrates a Growth Receptor," *Nature* 259 (1976): 134-36.

98. J. B. Baseman and N. S. Hayes, "Differential Effect of Hormones on Macromolecular Synthesis and Mitosis in Chick Embryo Cells," *Journal of Cell Biology* 67 (1975): 92-97.

99. Rinderknecht and Humbel, "Polypeptides" (See note 91).

100. Kahn et al., "Peptide Hormones" (See note 87).

101. Van Wyk et al., "Somatomedins" (See note 88).

102. Daughady et al., "Sulfation Factor" (See note 89).

103. Megyesi et al., "Non-Suppressible Insulin-Like Activity" (See note 90).

104. G. L. King et al., "Direct Demonstration That Insulin and Insulin-Like Growth Factors Produce Their Metabolic and Growth Effects via Different Receptors," *Clinical Research* 27 (1979): 486A.

105. K. Megyesi et al., "Insulin and Non-Suppressible Insulin-Like Activity (NSILA-s): Evidence for Separate Plasma Membrane Receptor Sites," *Biochemical and Biophysical Research Communications* 57 (1974): 307-15.

106. ———, "Hypoglycemia in Association with Extrapancreatic Tumors: Demonstration of Elevated Plasma NSILA-s by a New Radioreceptor Assay," *Journal of Clinical Endocrinology and Metabolism* 38 (1974): 931-34.

107. ———, Radioreceptor Assay of Plasma NSILA-s in Man: Basal and Stimulated Levels in Normal and Pathologic States," in *Growth Hormone and Related Peptides*, Proceedings of the III International Symposium, Milan (Amsterdam: Excerpta Medica, 1975): 127-40.

108. Kahn et al., "Peptide Hormones" (See note 87).

109. Megyesi et al., "Separate Receptor Sites" (See note 105).

110. ———, "Plasma NSILA-s in Man" (See note 107).

111. ———, "New Radioreceptor Assay" (See note 106).

112. ———, "Plasma NSILA-s in Man" (See note 107).

113. M. A. Raben, "Human Growth Hormone," *Recent Progress in Hormone Research* 15 (1959): 71-114.

114. M. Cornblath and R. Schwartz, "Infant of the Diabetic Mother," in

Disorders of Carbohydrate Metabolism in Infancy, 2nd ed. (Philadelphia, W.B. Saunders Company, 1976): 115-54.

115. J. Roth, "Receptors for Peptide Hormones," in *Endocrinology,* vol. 3, ed. L. DeGroot (New York: Grune and Stratton, 1979): 2037-54.

116. J. Bottenstein et al., "The Growth of Cells in Serum-Free Hormone-Supplemented Media," in *Methods in Enzymology,* vol. 58, ed. W. B. Jakoby and I. H. Pastan (New York: Academic Press, 1979): 94-109.

117. J. Roth et al., "Hormone Receptors, Human Diseases, and Disorders in Receptor Design," in *Hormones and Cell Culture,* vol. 6, Proceedings of the Cold Spring Harbor Conference on Cell Proliferation (Cold Spring Harbor, New York: n.p., 1979): 167-86.

SIMILARITIES OF STRUCTURE AMONG STEROID RECEPTOR PROTEINS *

William T. Schrader and
Bert W. O'Malley †

SUMMARY

Considerable similarity exists among receptors for steroid hormones from many different organisms. In the present study receptors for progesterone (chick), estrogen (rat), androgen (hamster), and glucocorticoid (rat) were compared with respect to their chromatographic behavior. Receptor-hormone complexes in crude extracts were analyzed by DEAE-cellulose, DNA-cellulose, and phosphocellulose chromatography. The study concludes that all steroid receptors tested can be resolved into multiple components.

INTRODUCTION

Receptors for steroid hormones have been studied in many tissues and organisms. Perhaps the best studied from a structural and functional standpoint has been the chicken oviduct progesterone receptor.[1, 2] This receptor has been purified to apparent homogeneity and found to consist of two nonidentical progesterone-binding components. One, progestophilin A, has a molecular weight of 79,000 [3] and binds avidly to DNA.[4] The other, progestophilin B, has a molecular weight of 115,000 [5, 6] and binds weakly to DNA but strongly to chromatin from target cells.[7] Isolation of higher molecular weight aggregates of these proteins,[8] cross-

* The work reported here was supported by NIH grants HD-07857 and HD-07495 to the Baylor Center for Population Research and Studies in Reproductive Biology.

† Dr. O'Malley, professor and chairman of the Department of Cell Biology, Baylor College of Medicine, did not attend the conference.

linking studies,[9] and functional tests of chromatin transcription [10] have led to the belief that A and B are subunits of a larger AB complex.

Although A and B are not identical nor interconvertible, they have considerable similarity with respect to their hormone-binding site,[11] and with respect to liberation of fragments using specific proteases.[12] Due to the heterogeneity of other receptors under various experimental conditions, it was of interest to ask whether these others might also have subunit structure.

To test this idea receptor-hormone complexes from several sources were chromatographed, as described for progesterone receptor analysis.[13] The results show the heterogeneity of each receptor tested, and point to the similarities of structure among these interesting molecules.

METHODS OF PROCEDURE

Animals

White Leghorn immature female chicks were implanted with pellets of diethylstilbestrol in carbowax for two weeks before use; oviducts weighed 1 to 2 g. at the time of removal. Adult female rats castrated two weeks prior to use received one priming dose of estradiol subcutaneously twenty-four hours before use; uteri weighed 200 to 400 mg. Adult female rat livers for rat glucocorticoid receptor studies were removed after perifusion with Krebs Ringer bicarbonate containing 0.25 M sucrose. A guinea pig ductus deferens DDT-1 cell line was used as the source of androgen receptors.[14]

Preparation of Tissue Extracts

High-speed soluble fractions were prepared by homogenization of tissue or cells in 4 ml./g. tissue of Buffer A (10 mM Tris-HCl, 1 mM Na_2 EDTA, 12 mM 1-thioglycerol pH 7.4). The homogenates were centrifuged at 105,000×g. in a Beckman SW-50.1 rotor for sixty minutes to prepare cytosol.

Labeling with Radioactive Hormones

Tritiated steroids were obtained from New England Nuclear, dissolved in ethanol, and added to Buffer A before addition to the appropriate cytosol at 2×10^{-8} M. Steroids were [1,2-$^3H_2$]progesterone (50 Ci/mmole), [1,2,6,7-$^3H_4$]estradiol-17 (105 Ci/mmole), [1,2,6,7-$^3H_4$]dihydrotestosterone (100 Ci/mmole), and [1,2-$^3H_2$] triamcinolone acetonide (50 Ci/mmole). Aqueous samples containing these hormones were counted for ^3H, using a Beckman LS-250 scintillation spectrometer at a counting efficiency of 25 percent.

Ammonium Sulfate Precipitation

After complexing cytosol receptors with the appropriate labeled steroid for at least sixty minutes, receptors were precipitated by drop-wise addition of saturated ammonium sulfate in Buffer A to a final solution concentration of 40 percent saturation. After twenty minutes at 0°C, the precipitates were collected by centrifugation and redissolved in the original cytosol volume of Buffer A. The redissolved material was clarified by a second, brief centrifugation to remove insoluble denatured protein.

Column Chromatography

Analytical columns of DEAE-cellulose (Whatman DE-52), hydroxylapatite (Clarkson Hypatite), phosphocellulose (Whatman P-11), and DNA-cellulose (Bio-Rad Cellex N-1 containing 200 μg./ml. lyophilized calf thymus DNA) were equilibrated and washed as described previously.[15] The columns were nominally 3 ml. packed volumes, and were eluted using linear KCl gradients in Buffer A (DEAE, phosphocellulose and DNA-cellulose) or a linear gradient of potassium phosphate pH 7.5 (hydroxylapatite).

Glucocorticoid Receptor Purification

This was carried out exactly as described by W. A. Coty and coworkers,[16] except that rat liver glucocorticoid receptors labeled with [^3H]triamcinolone acetonide were used. After preparation of liver cytosol, the solution (200 ml.) was passed through a phosphocellulose column, and then a DNA-cellulose column equilibrated in Buffer A (both columns had packed volumes of 250 ml.). Aporeceptors, that is, those having no bound steroid, do not adsorb to these columns. The drop-through fraction was then brought to 40 percent saturation in ammonium sulfate, and the pellet was redissolved in Buffer A. This crude glucocorticoid receptor fraction was then labeled with [^{3}H]triamcinolone acetonide in excess and allowed to stand overnight. Subsequent chromatographic behavior of the labeled receptors is described in the text.

RESULTS

Chromatographic Analysis of Chick Progesterone Receptors

Our earlier studies of chick receptors for progesterone had shown that the preparation could be resolved into two equal amounts of hormone-binding activity following ammonium sulfate precipitation.[17] Figure 1 shows typical elution profiles of [^{3}H]progesterone-receptor complexes from DEAE-cellulose, phosphocellulose, and hydroxylapatite. In these tests ammonium sulfate-precipitated labeled receptor complexes were separated into three

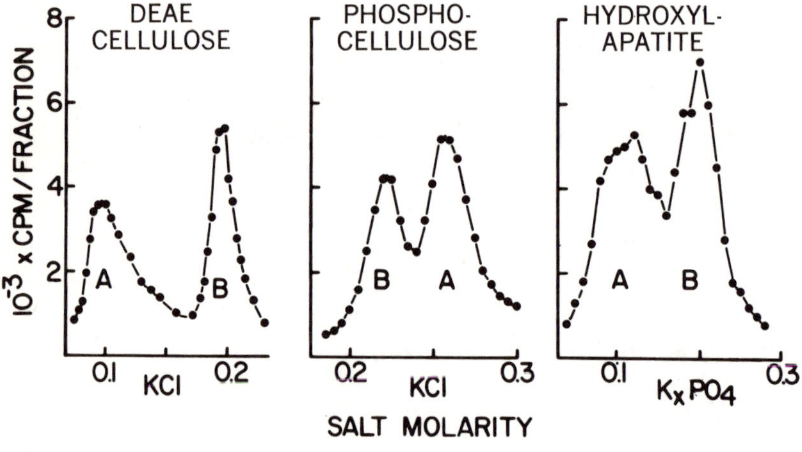

Figure 1. Column chromatography of chick progesterone-receptor complexes. Labeled progesterone receptors, precipitated by ammonium sulfate, were chromatographed on DEAE-cellulose, phosphocellulose, or hydroxylapatite. The columns were eluted by linear KCl gradients (DEAE and phosphocellulose), as shown on the abscissa. Hydroxylapatite was eluted by linear potassium phosphate gradient, as shown on the abscissa. Aliquots of each fraction were counted for ^{3}H using Amersham-Searle ACS counting fluid (• —— •).

fractions and chromatographed individually on these three columns, as already described. Generally, about 70 to 90 percent of the ^{3}H adsorbed to the columns. The elution profiles show two peaks of ^{3}H in each case; these have been analyzed more rigorously elsewhere and shown to be progestophilins A and B, as indicated in the figure. Since B is more acidic than A, their order of elution is reversed on phosphocellulose.

In all three cases equal peaks of ^{3}H are obtained from A and B; this pattern is not altered by inclusion of numerous protease inhibitors, nor does it vary with time of labeling. Thus it does not appear that the smaller A subunit is a product of B proteolysis. Since both are single polypeptide chains,[18] they are apparently closely related proteins.

DEAE-Cellulose Analysis of Other Steroid Receptors

In view of the results of Figure 1, it was of interest to compare the progesterone receptor's behavior with that of other receptors. For this purpose DEAE-cellulose chromatography was chosen. The results of analysis of four different receptors by this technique is shown in Figure 2. Labeled receptors for the four hormones were precipitated by 40 percent saturation of ammonium sulfate and were then applied and washed through 3 ml. DEAE-cellulose columns equilibrated in Buffer A as described under Methods. Two results can be seen in the figure: first, all four receptors adsorb to and elute from DEAE-cellulose as two equal peaks of radioactivity; second, their elution molarities with KCl are not identical from one hormone to the next, indicative of differences among the proteins themselves. The overall patterns, however, are remarkably similar. It should be noted here that, by ammonium sulfate precipitation, the receptors would be converted to their "activated" state, that is, their DNA-binding capacity would be maximal. It is clear that heterogeneity of some sort must exist for each receptor; by analogy to the progesterone case it is an attractive hypothesis to invoke subunits to explain these findings.

Finally, the lower two panels of Figure 2 show labeling of a cell culture extract with two different hormones. Both androgen

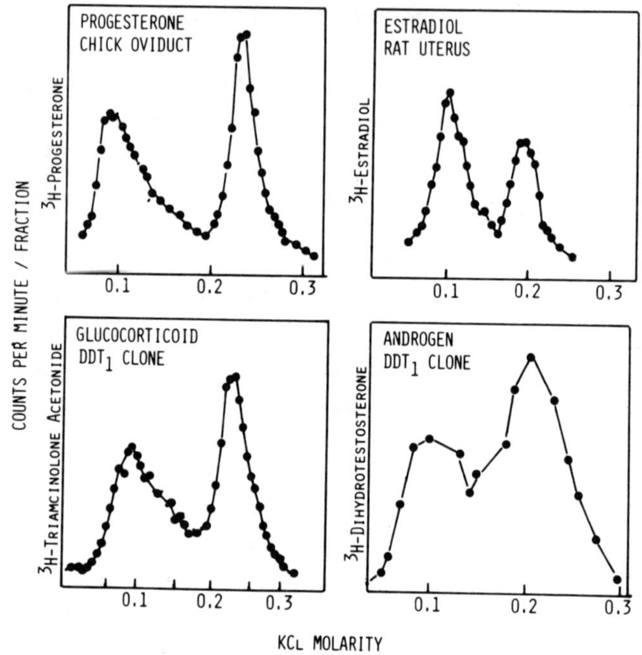

Figure 2. DEAE-cellulose column chromatography of labeled receptors for progesterone, estradiol, androgen (dihydrotestosterone), and glucocorticoid (triamcinolone acetonide). Tissue sources were as described in text. Labeled receptors prepared as in Figure 1 were chromatographed by KCl gradient elution as shown on the abscissas. Aliquots were counted for ^3H (● —— ●).

and glucocorticoid receptors are present in these cells,[19] but they do not behave identically on DEAE-cellulose. Thus even within the same cell the receptors may be structurally homologous but not identical.

Receptor Analysis by DNA-Cellulose Chromatography

The most intriguing aspect of our earlier findings with progesterone receptors concerned the different affinities of subunits A and B for nuclear constituents.[20] This difference has been central to our hypothesis that the intact complex requires both B and A as a functional dimer for biological activity.[21]

The simplest initial test for such functional heterogeneity of receptors forms would be to perform DNA-cellulose chromatography. Results of these assays done for progesterone, estrogen, and glucocorticoid receptors are shown in Figure 3. Androgen receptors were not used. On the right panel the figure shows resolution of progesterone receptor subunits B and A on this column. In this particular experiment B and A were first separated from each other by DEAE-cellulose (see Figure 1) in order to confirm which one eluted at higher KCl molarity. Receptor B elutes at about 0.04 M KCl, while receptor A elutes at 0.18 M KCl. If an ammonium sulfate precipitate is chromatographed directly (data not shown), exactly the same profile is obtained.[22]

Glucocorticoid and estrogen receptors were chromatographed in the other two panels. Again, following ammonium sulfate

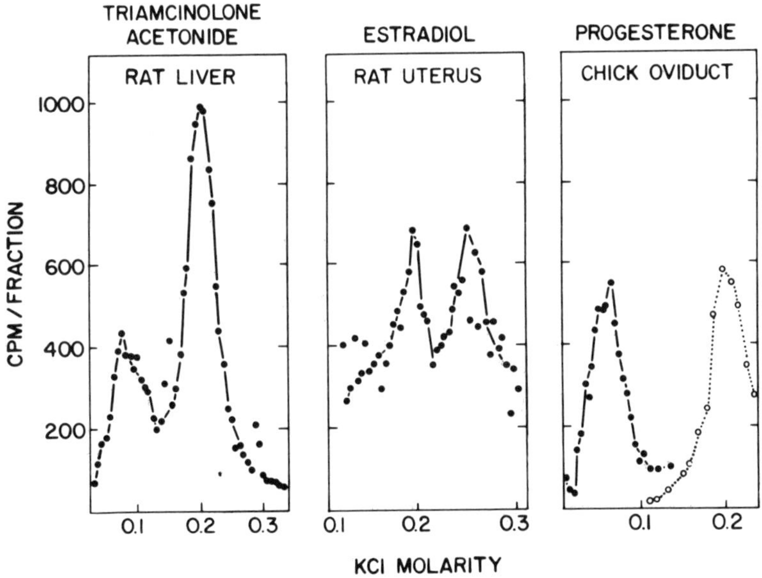

Figure 3. DNA-cellulose chromatography. *Left panel* and *middle panel,* triamcinolone acetonide and estradiol, respectively, were used to label receptors that were treated as in Figure 1. Aliquots were counted for ^3H (● —— ●). *Right panel,* progesterone receptor B (● —— ●) and A (○ — — ○) subunits were isolated by DEAE-cellulose chromatography as in Figure 2. The DNA cellulose columns were eluted by KCl gradients as indicated on the abscissas.

precipitation, two peaks of radioactive ligand-protein complexes were seen. The yield of glucocorticoid receptor adsorbed to the DNA was low; whether this is an experimental artifact or an unforseen aspect of its activity remains to be established. As in the progesterone case, the two peaks eluted at considerably different KCl molarities. No functional tests have yet been available to equate this behavior to that seen for progesterone receptors.

Purification of Glucocorticoid Receptors of Rat Liver

In view of the similarities seen among receptors in Figures 1-3, we speculated that a purification protocol suitable for progesterone receptor A subunit might be applicable to others as well. We used a method developed by Coty and colleagues,[23] which yields receptor A protein that is apparently homogeneous as assayed by a variety of gel electrophoresis systems.

Rat liver glucocorticoid receptor was chosen, and the initial stages of the purification were carried out exactly as described for progesterone receptor and outlined under Methods. After the ammonium sulfate precipitation, labeled [^{3}H]triamcinolone-receptor complexes were subjected to a series of column chromatograms as shown in Figure 4.

The top panel of the figure shows the DEAE-cellulose elution profile. As had been seen for hamster DDT-1 cell glucocorticoid receptor (Figure 1), two peaks of ^{3}H eluted. In analogy to the progesterone work, the earlier peak eluting at 0.1 M KCl was pooled and chromatographed on a 20 ml. DNA-cellulose column. The middle panel shows this profile. The large second peak elution at 0.3 M KCl was pooled and subjected to the final phosphocellulose column. A sharp peak eluted, as shown in the bottom panel.

When this entire pool was concentrated and subjected to gel electrophoresis under denaturing conditions in 0.1 percent Na dodecyl SO_4 by the method of U. K. Laemmli,[24] a single faint band of protein was observed. The molecular weight of this band was 70,000, a value consistent with molecular weight estimates for the 4S glucocorticoid receptor as determined by the method of L. M. Siegal and K. J. Monty.[25]

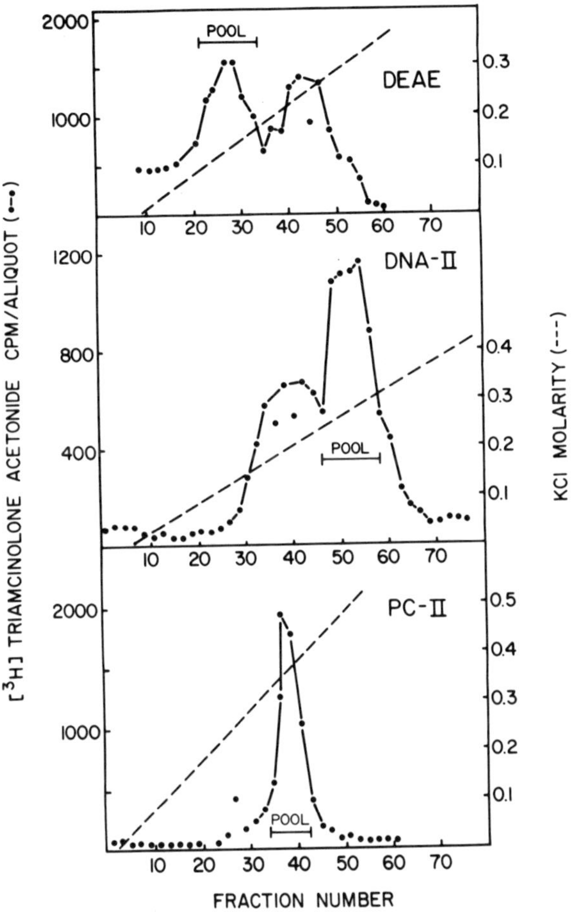

Figure 4. Purification protocol for glucocorticoid receptors of rat liver. Liver receptors were treated as described in the text for initial steps of purification. Labeled triamcinolone acetonide-receptor complexes were chromatographed successively by KCl gradient elution (- - -) on DEAE-cellulose (100 ml. packed volume) DNA-cellulose (20 ml. packed volume), and phosphocellulose (10 ml. packed volume). Aliquots were counted for ^3H (● —— ●), as in Figure 1. At each step the fractions indicated by the bar were pooled for the subsequent steps.

Although only one protein band was observed, it is not possible at this time to be certain it is the receptor and not a contaminant. The specific radioactivity of the final pool of sample was very high—as evidenced by a lack of many protein bands on the gel. The theoretical specific radioactivity (cpm bound/mg. protein), however, was not obtained. Additional studies on the protein will be needed to confirm this result.

The present information does point to the overall similarity among the various steroid receptors, as shown by electrostatic, adsorption, and purification tests. It is thus tempting to speculate that the mechanism by which these proteins act may also be common, with the differences being primarily a reflection of which genes are affected.

A Proposed Model for the Molecular Action of Steroid Receptors

Our current working hypothesis regarding the action of these interesting molecules is shown in Figure 5. The concept is de-

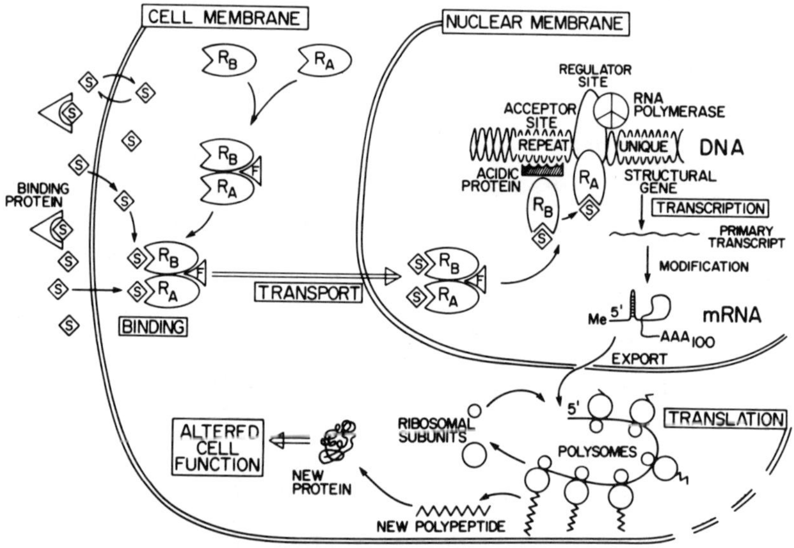

Figure 5. A proposed model for steroid receptor function. A two-subunit receptor is envisioned to function by sequential interaction at acceptor sites via the chromatin-binding capacity of the B subunit, followed by DNA helix destabilization caused by binding of the liberated A subunit. This DNA unwinding may then promote RNA chain initiation. See text for explanatory comments.

veloped entirely from the studies of steroid action in chick oviduct. But, as the foregoing figures document, there is reason to believe the hypothesis may be a useful one for studies of other steroid receptors.

The functional form of the receptor is envisioned to be a dimer of two dissimilar steroid-binding subunits. The sedimentation coefficient of the dimer is about 6S, whereas the subunits are about 4S each. The dimer has nuclei-binding activity following association with its ligand, thus causing a net accumulation of the complexes in the nucleus.[26] We have found that the dimer does not bind avidly to DNA, but rather exhibits the weak DNA attraction of the B subunit.[27]

Its initial interactions with the chromatin thus may be limited to the B subunit's chromatin-binding activity. This activity has been shown in earlier studies [28] to be target-tissue chromatin specific, and to involve adsorption to certain "acceptor sites," regions of the chromatin defined by the presence of a certain subclass of nonhistone chromosomal proteins.[29]

We speculate, but have no supporting evidence, that these acceptor sites are localized proximal to genes destined to be regulated by the hormone in question. There follows a dissociation of the A-B receptor complex to yield the highly DNA-reactive A subunit. This dissociation has been found to be facilitated by the binding of the hormone.[30]

The highly reactive A protein, now localized near a gene it is to regulate, has a functional DNA-binding site. In studies to be published elsewhere we have found that the protein has no detectable nucleotide-sequence specificity, but rather has high affinity for single-stranded DNA.[31] Such an activity might then be able to stabilize the DNA locally in an "unwound" configuration, and thereby present a more favorable RNA chain initiation site for RNA polymerase. Thus the receptor's presence in the chromatin would tend to stimulate initiation of RNA synthesis at the 5′ end of a gene. This concept is consistent with our earlier studies of RNA synthesis using purified receptors in vitro,[32] and also with the more recent studies of ovalbumin gene induction at the initiation step.[33]

The consequences of such a receptor dimeric structure are twofold. First, the model shown here would propose that hormone action would obey two-hit kinetics, since the dimer binds two moles of ligand per mole of receptor. Second, the fact that the dimer itself has occluded the A subunit's DNA-binding site would tend to limit or prevent association to the receptor with the wrong DNA. The active subunit might only be able to express DNA regulatory activity after localization at the genes of interest is accomplished via the B "specifier" subunit.[34] This mechanism would account for the fact that specific gene regulation can exist in the target cell for a given hormone, in the face of many thousands of active genes not affected by the signal.

DISCUSSION

The chromatographic analyses presented here are indicative of considerable structural homology among steroid receptors. They thereby suggest the existence of a common functional pathway. None of these findings, however, is of itself conclusive of this idea. It remains for receptor purification, coupled with in vitro functional tests, to establish and test these concepts more fully.

Not considered here are the many studies from other laboratories offering alternatives to this concept. Receptor "activation" and "transformation," for example, are descriptive terms at the present time for in vitro changes in receptor characteristics. To date there is no molecular explanation for these terms in any system. All of the agents so far tested that promote "activation" also promote receptor dissociation in vitro. These include ammonium sulfate, dialysis, warming, KCl, dilution, and pyridoxal phosphate. Conversely, those agents, which in our hands promote "inactivation" or "deactivation," also cause dimer stabilization. These include low ionic strength, lack of hormone, fluoride ion, and molybdate ion. Thus we also speculate that receptor "activation" and "transformation" are in fact in vitro manifestations of a simpler problem, namely, subunit-subunit assembly.

NOTES

1. W. T. Schrader and B. W. O'Malley, "Molecular Structure and Analysis of Progesterone Receptors," in *Receptors and Hormone Action,* vol. 2, ed. B. W. O'Malley and L. Birnbaumer (New York: Academic Press, 1978): 189-224.

2. W. V. Vedeckis, W. T. Schrader, and B. W. O'Malley, "The Chick Oviduct Progesterone Receptor," in *Biochemical Actions of Hormones,* vol. 5, ed. G. Litwack (New York: Academic Press, 1978): 321-72.

3. W. A. Coty, W. T. Schrader, and B. W. O'Malley, "Purification and Characterization of the Chick Oviduct Progesterone Receptor A Subunit," *Journal of Steroid Biochemistry* 10 (1979): 1-12.

4. W. T. Schrader, D. O. Toft, and B. W. O'Malley, "Progesterone-Binding Proteins of Chick Oviduct: Interaction with Nuclear Constituents," *Journal of Biological Chemistry* 247 (1972): 2401-07.

5. W. T. Schrader, R. W. Kuhn, and B. W. O'Malley, "Progesterone-Binding Components of Chick Oviduct: Receptor B Subunit Protein Purified to Apparent Homogeneity from Laying Hen Oviducts," *Journal of Biological Chemistry* 252 (1977): 299-307.

6. R. W. Kuhn et al., "Progesterone-Binding Components of Chick Oviduct: Biochemical Characterization of Purified Oviduct Progesterone Receptor B Subunit," *Journal of Biological Chemistry* 252 (1977): 308-17.

7. Schrader, Toft, and O'Malley, "Progesterone-Binding Proteins" (See note 4).

8. W. T. Schrader, S. S. Heuer, and B. W. O'Malley, "Progesterone Receptors of Chick Oviduct: Identification of 6S Receptor Dimers," *Biology of Reproduction* 12 (1975): 134-42.

9. M. E. Birnbaumer, W. T. Schrader, and B. W. O'Malley, "Progesterone Binding Components of Chick Oviduct: Chemical Crosslinking of Receptor Subunits Using a Reversible Bifunctional Crosslinking Agent," *Biochemistry Journal* 181 (1979): 201-13.

10. R. E. Buller et al., "Progesterone-Binding Components of Chick Oviduct: *In Vitro* Effect of Receptor Subunits on Gene Transcription," *Journal of Biological Chemistry* 251 (1976): 5178-86.

11. P. E. Hansen, W. T. Schrader, and B. W. O'Malley, "Kinetics of Progesterone Binding to the Chick Oviduct Receptor Protein," *Journal of Steroid Biochemistry* 7 (1976): 723-32.

12. W. V. Vedeckis, W. T. Schrader, and B. W. O'Malley, "Structural Relationships between the Chick Oviduct Progesterone Receptor A and B Proteins," in *Steroid Hormone Receptor Systems,* ed. W. W. Leavitt and J. H. Clark (New York: Plenum Publishing Corp., 1979): 309-27.

13. W. T. Schrader, "Methods for Extraction and Quantification of Receptors," *Methods in Enzymology* 36 (1975): 187-211.

14. J. S. Norris and P. O. Kohler, "Characterization of the Androgen Receptor from a Syrian Hamster Ductus Deferens Tumor Cell Line (DDT_1)," *Science* 192 (1976): 898-900.

15. Schrader, "Extraction and Quantification" (See note 13).

16. Coty, Schrader, and O'Malley, "Purification and Characterization" (See note 3).

17. Schrader, "Extraction and Quantification" (See note 13).

18. Schrader and O'Malley, "Progesterone Receptors" (See note 1).

19. Norris and Kohler, "Androgen Receptor" (See note 14).

20. Schrader, Toft, and O'Malley, "Progesterone-Binding Proteins" (See note 4).

21. Schrader and O'Malley, "Progesterone Receptors" (See note 1).

22. W. V. Vedeckis et al., manuscript submitted for publication.

23. Coty, Schrader, and O'Malley, "Purification and Characterization" (See note 3).

24. U. K. Laemmli, "Cleavage of Structural Proteins during the Assembly of the Head of Bacteriophage T4," *Nature* 227 (1970): 680-85.

25. L. M. Siegal and K. J. Monty, "Determination of Molecular Weights and Frictional Ratios of Proteins in Impure Systems by Use of Gel Filtration and Density Gradient Centrifugation, Application to Crude Preparations of Sulfite and Hydroxylamine Reductases," *Biochimica et Biophysica Acta* 112 (1966): 346-62.

26. R. E. Buller et al., "Progesterone-Binding Components of Chick Oviduct: Receptor Activation and Hormone-Dependent Binding to Purified Nuclei," *Journal of Biological Chemistry* 250 (1975): 801-08.

27. Schrader and O'Malley, "Progesterone Receptors" (See note 1).

28. Schrader, Toft, and O'Malley, "Progesterone-Binding Proteins" (See note 4).

29. T. C. Spelsberg et al., "Nuclear Binding Sites (Acceptors) for Progesterone in Avian Oviduct: Characterization of the Highest-Affinity Sites," *Annals of the New York Academy of Sciences* 286 (1977): 43-63.

30. Schrader and O'Malley, "Progesterone Receptors" (See note 1).

31. M. R. Hughes et al., manuscript in preparation.

32. Buller et al., "Progesterone-Binding Components" (See note 10).

33. D. R. Roop et al., "Transcription of Structural and Intervening Sequences in the Ovalbumin Gene and Identification of Potential Ovalbumin mRNA Precursors," *Cell* 15 (1978): 671-85.

34. B. W. O'Malley et al., "Mechanisms of Interaction of a Hormone Receptor Complex with the Genome of a Eucaryotic Target Cell," *Nature* 235 (1972): 141-44.

MYASTHENIA GRAVIS AND ACETYLCHOLINE RECEPTORS *

Daniel B. Drachman †

Myasthenia gravis (MG) is a neuromuscular disorder of man manifested clinically by weakness and fatigability of skeletal muscles. The neuromuscular junction has long been recognized as the site of the disorder, because of: 1) the similarities between MG and poisoning with curare, which blocks the neuromuscular junction; and 2) the favorable response to anticholinesterase agents, which enhance neuromuscular transmission.[1] The exact site and nature of the process remained elusive, however, until the availability of snake neurotoxins[2] permitted the specific identification of the acetylcholine receptor (AChR) abnormality in MG. It has now become clear that the basic defect in the disease is a reduction of available AChRs at neuromuscular junctions[3] brought about by an autoimmune attack.[4] This chapter will summarize some of the studies that have contributed to the current understanding of the AChR abnormality in MG.

NERVE-MUSCLE TRANSMISSION

To understand the pathophysiology of the defect in MG, some knowledge of the basic events of neuromuscular trans-

* The original work carried out in the author's laboratory was supported in part by National Institutes of Health grants 5 RO1 HD04817 and 5 PO1 NS10920.

† I am grateful to Ms. C. F. Barlow for expert assistance in the preparation of the manuscript. I am happy to acknowledge the participation of my colleagues in various phases of the studies carried out in our laboratory, including: R. N. Adams, C. W. Angus, D. M. Fambrough, K. H. Fischbeck, M. D. E. Griffin, G. J. Hoffman, I. Kao, J. D. Michelson, A. Murphy, A. Pestronk, S. Satyamurti, F. Slone, E. F. Stanley, and K. V. Toyka.

mission is necessary (Figure 1).[5] Acetylcholine (ACh), the natural transmitter, is synthesized chiefly in the motor nerve terminals and is stored in vesicles, which contain approximately 10^4 ACh molecules. Neurotransmission begins with release of the contents of the vesicles by a process of exocytosis that takes place at specialized release sites on the motor nerve terminal. The release sites are situated directly opposite the areas of highest concentration of AChRs on the postsynaptic membrane. When ACh combines with its receptor, a transient increase of permeability to sodium and potassium ions takes place, resulting in postsynaptic electrical depolarization. The contents of single vesicles may be released spontaneously, giving rise to miniature end-plate potentials (mepps). When a nerve impulse invades the motor nerve terminal a much larger amount of ACh (150-200 quanta) is re-

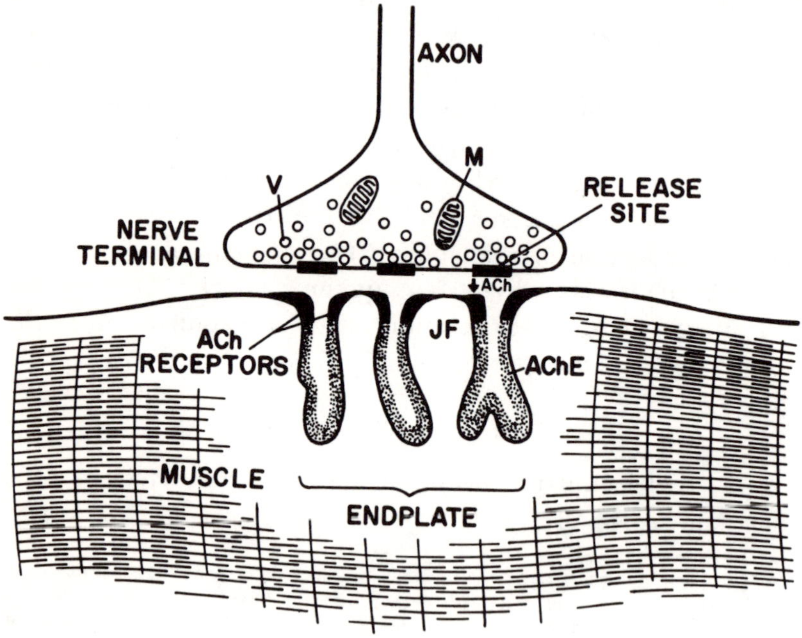

Figure 1. Diagram of neuromuscular junction. V = vesicle; M = mitochondrian; AChE = acetylcholinesterase; JF = junctional fold.

SOURCE: Reprinted, by permission, from D.B. Drachman, "Myasthenia Gravis," *New England Journal of Medicine* 298 (1978): 136.

leased, resulting in a larger postsynaptic depolarization, which in turn triggers muscle contraction.

ACh RECEPTOR DEFICIT IN MYASTHENIA GRAVIS

In the early 1960s D. Elmqvist and his associates reported a decrease in the amplitude of the mepps at neuromuscular junctions of patients with MG.[6] They originally postulated that this might be due to a deficiency of the amount of ACh released per quantum. We wondered, however, whether a postsynaptic abnormality in AChRs could account for the observed decrease of mepp amplitudes. In order to test this hypothesis, we made use of α-Bungarotoxin (α-BuTx) as a probe to measure the number of AChRs at neuromuscular junctions of patients with MG.[7] α-BuTx is a purified toxin from the venom of a Taiwanese snake (*Bungarus multicinctus*), which binds specifically and essentially irreversibly to AChRs.[8] In our initial study, muscle biopsies from patients with MG were incubated to saturation with ^{125}I-labeled α-BuTx, and the excess removed by washing. The number of AChRs per neuromuscular junction was computed from the amount of bound ^{125}I-α-BuTx, and the number of neuromuscular junctions present in the sample was determined histochemically. Our findings showed a 70 to 89 percent reduction in the number of AChRs at the junctions of myasthenic patients as compared with control individuals. This finding of a postsynaptic AChR deficit has now been amply confirmed by physiological, ultrastructural, and radiometric studies.[9-11]

ROLE OF ACh RECEPTOR DEFICIT IN MYASTHENIA GRAVIS

Our observations raised the question of whether the decrease of available ACh receptors per se could account for the physiological abnormalities in MG, or whether it merely represented a secondary phenomenon. In order to explore this question further we developed an experimental animal model with a reduced number of available AChRs.[12] For this purpose we used α-toxin from the cobra *Naja naja atra,* which also blocks AChRs in a highly specific fashion. The toxin was given to experimental rats in doses just sufficient to block a fraction of the junctional AChRs, and the resultant changes in neuromuscular transmission were compared with abnormalities occurring naturally in human MG. Evoked potentials were recorded from the calf muscles of the rats during direct stimulation of the exposed sciatic nerves with supramaximal pulses. The most effective method of administration of cobra toxin proved to be a single intravenous injection (12 to 20 μg.) followed by a long observation period of up to fourteen hours. Following equilibration of the toxin in the musculature, the "myasthenic" features were most readily seen. Rats treated in this way showed decremental responses to repetitive nerve stimulation, at 3 Hz, that reproduced the characteristic pattern in MG.[13] Furthermore, the decremental responses were exaggerated by minute doses of d-tubocurarine—one-thirtieth of the rat curarizing dose. This reaction is typical of neuromuscular junctions in human MG.[14] In all instances treatment of the rat with neostigmine, an anticholinesterase drug, produced marked improvement or recovery of the abnormal decremental responses. Finally, post-tetanic responses, thought to be particularly characteristic of MG,[15] were also reproduced in the cobra toxin model. Thus, from the point of view of the electrophysiological features, the cobra toxin model faithfully reproduced the abnormalities found in MG. Since the α-toxin used in this study has been shown to block AChRs specifically, our findings provided strong support for the hypothesis that a reduction of available AChRs per se could account for the defects of neuromuscular transmission in MG.

AUTOIMMUNE ABNORMALITY IN MG

The possibility that MG might be due to an autoimmune process was originally proposed on the basis of indirect evidence consisting of the frequent association of MG with other presumed autoimmune diseases,[16] a high rate of thymic abnormalities in myasthenic patients, and a favorable response to thymectomy in some cases.[17, 18] A further important clue that the autoimmune attack might be directed against AChRs came from the finding that animals immunized with purified AChR preparations from electric eels or rays in Freund's adjuvant developed a condition analogous to human MG.[19-21] The animals showed the typical reduction of mepp amplitudes,[22] decrease of available AChR sites,[23] and ultrastructural changes [24] that resemble the features of human MG in many, but not all, respects.

Based on the knowledge of a receptor deficit in MG and the analogy to the experimental model, the search for an autoimmune mechanism directed against AChRs in the human disease was soon begun. Antireceptor antibody was identified by several different methods, all of which depend on α-BuTx for their specificity.[25-28] By means of the most sensitive radioimmunoassay, based on the binding of antibody to AChRs, specific antibodies have been found in 80 to 90 percent of patients with MG. The antibody titer corresponds loosely but not exactly to the clinical state of the patients.

PASSIVE TRANSFER OF MG

The question of whether the circulating antibodies are instrumental in causing the receptor deficit, or merely represent a secondary response to receptor damage caused by some other agent, is of great importance in understanding the pathogenesis of MG. One well-documented clinical observation suggested that circulating antibodies might be pathogenic: approximately one out of six infants born to myasthenic mothers showed myasthenic weakness in the neonatal period, with complete recovery after several weeks.[29]

In the past, numerous attempts were made to test for a circulating humoral factor in MG by treating laboratory animals or nerve-muscle preparations with serum from affected patients.[30,31] The results of these studies were generally negative or subject to serious criticism.[32] The possibility of such a factor assumed new importance, however, in light of the discovery of antibodies to AChRs. In most previous studies of attempted transfer, animals or nerve-muscle preparations were exposed to myasthenic serum for a brief time.

Based on the idea that a relatively longer period of exposure might be required for the effect to take place, we designed an experiment in which physiological levels of human myasthenic IgG could be maintained in recipient mice for as long as fourteen days.[33] An immunoglobulin fraction was prepared from the serum of myasthenic patients or controls by precipitation with ammonium sulfate and by dialysis. Daily injections of the immunoglobulin preparations were given to mice in doses sufficient to maintain levels of human IgG equivalent to those in the patients. Some of the experimental mice became clinically weak within a few days, showing decremental responses to repetitive nerve stimulation. The most reliable tests of the effects of passive transfer were: 1) the measurements of mepp amplitudes; and 2) the binding of ^{125}I-α-BuTx to AChRs at neuromuscular junctions in the recipient mice. Immunoglobulins from fifteen of sixteen patients tested produced a decrease in mepp amplitudes, a reduction of AChRs, or both.[34] Injection of immunoglobulin from control individuals had no effect on AChRs.

By the use of this passive transfer model,[35] we found that the *IgG fraction* of patients' serum produced the features of myasthenia, while other serum immunoglobulin fractions had no effect. We also tested the influence of complement in this system by using mice depleted of C3 (by cobra venom factor) or genetically deficient in C5. These experiments showed that the early part of the complement system enhanced the effect of myasthenic IgG. Thus depletion of C3 diminished, but did not eliminate the effects of myasthenic patients' IgG or AChRs and mepps. The absence of C5 had no effect in this model. IgG therefore appears to play an important role in the pathogenesis of MG, enhanced by the early part of the complement cascade. Morphological studies, showing binding of IgG and C3 at human neuromuscular junctions, further support these findings.[36]

MECHANISMS OF ACh RECEPTOR REDUCTION BY MYASTHENIC IgG

Theoretically, circulating antibody against AChR might reduce the number of available AChRs at neuromuscular junctions by several different mechanisms:

• It might *block* the active site of the receptors, by binding at or near them.

• It might *alter the turnover of AChRs,* either by increasing the rate of degradation, or by decreasing the rate of synthesis, or both.

• It might *produce damage* to AChRs and AChR-containing postsynaptic membranes at neuromuscular junctions.

The evidence now available suggests that accelerated degradation of AChRs, receptor blockade, and postjunctional membrane damage may all be involved.

Studies of the effects of myasthenic IgG on receptor degradation and synthesis have been carried out in the tissue culture system, using primary cultures of rat skeletal muscle.[37] The cultured muscle fibers (myotubes) have AChRs distributed over their surfaces both diffusely and in higher density aggregates—"hot spots." The AChRs are specifically labeled with ^{125}I-α-BuTx. As the labeled receptors undergo degradation, the attached ^{125}I-α-BuTx is degraded to ^{125}I-tyrosine, which is released into the culture medium. The rate of receptor degradation is determined by measurement of the radioactivity released into the medium per unit time.[38] Normally approximately 4 percent of the surface AChRs are degraded per hour, in rat skeletal muscle. When IgG from myasthenic patients was added to the cultures the AChR degradation rate increased up to two- to threefold, as compared with cultures treated with IgG from control individuals.[39] Similar results have been reported from other laboratories, using immunoglobulin from myasthenic patients[40] or EAMG animals.[41] The antibody-induced acceleration is triggered by IgG alone, without requiring other humoral or cellular components of the immune system. Immunoglobulin from thirty-six of forty-seven patients tested produced a significant increase in the rate of AChR degradation in the tissue culture system.[42]

MECHANISM OF ACCELERATED ACETYLCHOLINE RECEPTOR DEGRADATION

Theoretically the myasthenic antibody might increase AChR degradation by: 1) accelerating the muscle cell's *overall receptor degrading mechanism*; or 2) altering a population of AChRs with bound antibody so that these receptors are *selectively degraded at a more rapid rate.*

To distinguish between these possibilities, we separately measured the degradation rates of AChRs *directly exposed to myasthenic* IgG and AChRs in identical cultures newly synthesized *just after* exposure of the cultures to myasthenic IgG.[43] Our findings showed that only the set of AChRs directly exposed to myasthenic IgG, and therefore having bound IgG, was degraded at an accelerated rate two to three times normal, while the second set of receptors without bound IgG was degraded at the control rate. This indicates that *direct exposure* of the AChRs to myasthenic antibody is necessary for acceleration of degradation. Further, it suggested that the binding of IgG alters the receptors in some way that causes them to be differentially selected for degradation.

We wondered whether this selective degradation effect might be due to the ability of antibodies to cross-link AChRs. IgG molecules are well known to be Y-shaped, with two arms capable of binding to two identical antigen molecules and therefore linking them. For this study [44] we prepared purified IgG, divalent $F(ab')_2$, and monovalent Fab fragments from myasthenic serum by standard enzymatic methods.[45,46] When added to muscle cultures the IgG and the divalent fragments produced acceleration of AChR degradation as before. The monovalent fragments were shown to bind to the AChRs in the muscle cultures, but did not accelerate their degradation (Figure 2, A,B,C). This suggested that cross-linking of the AChRs might account for the effect of the whole IgG molecule and the divalent fragments. To test the cross-linking hypothesis further we carried out two additional experiments:

- Monovalent (Fab) fragments were first added to the culture; a second, "piggyback" antibody, directed against the antibody fragments, was then added. The effect of the second antibody was to cross-link the Fab fragments, thus indirectly cross-linking the AChRs. This maneuver had the same acceleration effect on AChR degradation as the original divalent antibodies (Figure 2, D).

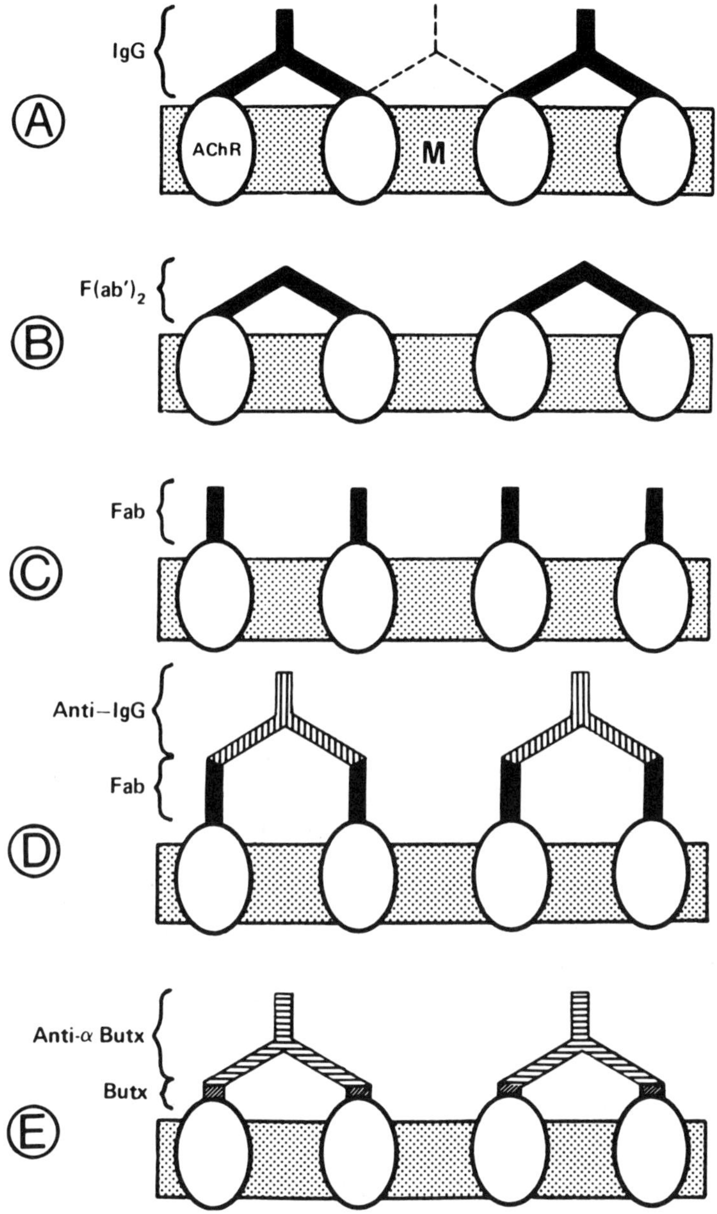

Figure 2. Diagrammatic representation of cross-linking of receptors by various antibody fragments. M = muscle membrane; AChR = acetylcholine receptor; α-BuTx = α-Bungarotoxin. Acceleration of degradation was observed in the experimental conditions shown in panels A, B, D, and E. Fab alone failed to accelerate AChR degradation.

SOURCE: Reprinted, by permission, from D. B. Drachman et al., "Myasthenic Antibodies Cross-Link Acetylcholine Receptors to Accelerate Degradation," *New England Journal of Medicine* 298 (1978): 1120.

• In this experiment we again used the "piggyback" technique, but here α-BuTx, rather than an antibody fragment, was attached to the AChRs. Antibodies against α-BuTx were prepared by immunizing rabbits with α-BuTx in Freund's adjuvant. When cultures labeled with ^{125}I-α-BuTx were treated with these antibodies, the degradation rate was again accelerated two- to threefold (Figure 2, E). This effect is attributable to cross-linking the AChRs via the interposed α-BuTx. The specificity of this effect was demonstrated by control experiments in which blockade of the antibody's binding sites with α-BuTx completely abolished its effect on degradation.

Taken together these findings strongly suggest that the ability of antibodies from myasthenic patients to cross-link AChRs accounts for their effect on receptor degradation. It remains to be shown how cross-linking accelerates receptor degradation. One possibility is that the cross-linked aggregates may be recognized by the muscle membrane, because of altered size or mobility within the membrane, and selected for degradation. Perhaps the cross-linking alters the relationship between the intramembranous AChR and the underlying cytoskeleton, thus favoring endocytosis.

In the culture system the accelerated degradation process accounts for part of the loss of AChR sites observed after exposure to myasthenic immunoglobulin. In addition, *blockade* of receptor sites by the antibody occurs in the skeletal muscle cultures [47,48] as well as in other assay systems.[49] We have tested the ability of myasthenic immunoglobulin to block AChR of cultured skeletal muscle by adding patients' or control immunoglobulin to cultures cooled to 4°C to eliminate degradation. The loss of α-BuTx binding sites in the cultures treated with myasthenic Ig was attributable to blocking, and occurred after treatment with immunoglobulin from forty-two of forty-nine patients. The blocking effect may be due to attachment of the antibody near the active site of the receptor, producing allosteric interference with the active site.[50]

Synthesis of AChRs does not appear to be affected by IgG from myasthenic patients. Using the appearance of new α-BuTx binding sites as a measure of synthesis and incorporation of AChRs in the rat muscle culture system, we found no change after treatment with IgG from myasthenic patients.

EFFECT OF MYASTHENIC Ig AT INTACT NEUROMUSCULAR JUNCTIONS

There are important differences between the *extrajunctional* AChRs of cultured muscle and AChRs at *intact neuromuscular junctions,* with respect to physical properties, pharmacology, and kinetics of turnover. These differences have raised questions about the applicability of the results of tissue culture experiments to the situation at intact neuromuscular junctions.[51] We have therefore studied the effects of myasthenic immunoglobulin on the turnover of AChRs at intact mammalian neuromuscular junctions, both in vivo and in vitro.[52]

In these experiments, AChRs of intact mouse diaphragms were first labeled by intrathoracic injection of ^{125}I-α-BuTx, with restriction of the mice to a vertical position to allow the toxin to gravitate to the diaphragms. The mice were then given daily intraperitoneal injections of immunoglobulin from individual myasthenic patients or from pooled control human sera. At intervals of zero to four days the mice were killed and the diaphragms removed and washed thoroughly. The remaining bound radioactivity was measured in a gamma counter. The diaphragms of mice treated with myasthenic immunoglobulin lost radioactivity approximately two to three times as rapidly as the diaphragms of immunoglobulin-treated control mice, indicating more rapid loss of junctional AChRs.

In order to determine whether the loss of receptors was due to degradation, diaphragms from mice treated in the same way were maintained in organ culture for more than twenty-four hours. The radioactive material released into the culture medium was examined by column chromatography. It consisted of a low molecular weight substance that comigrated with ^{125}I-tyrosine, thus confirming the degradative nature of the process. These and similar observations from other laboratories[53, 54] indicate that immunoglobulin from myasthenic patients increases the rate of degradation of AChRs at intact neuromuscular junctions of mammalian muscles. This process may represent an important antibody-mediated mechanism in human MG.

In addition to accelerated degradation and blocking of AChRs there is evidence of structural damage to neuromuscular junctions in MG. Morphologically the junctions show alterations, such as the apparent loss of bits of debris from the postsynaptic surface, and simplification of the normally complex pattern of junctional folds in more advanced cases.[55,56] A complement-mediated mechanism appears to play a role in the pathogenesis of MG, as demonstrated by the passive transfer model,[57] and by morphological studies of neuromuscular junctions of myasthenic patients.[58] The relative roles of accelerated degradation, blocking, and complement-mediated destruction at neuromuscular junctions of human MG patients have yet to be worked out, and probably differ from patient to patient.

DIFFERENT IMMUNE MECHANISMS IN PATIENTS WITH MG

Although the AChRs are probably always the target of the autoimmune process in MG, it is likely that the specific immune mechanisms differ in individual cases. The role of cell mediated mechanisms in myasthenia has not yet been well studied. It is known that lymphocytes from myasthenic patients are stimulated by the presence of purified AChR to undergo blast transformation.[59-61] It is not yet known, however, whether the sensitized lymphocytes participate only in the production of antibody, or, in addition, play a direct role involving cell-mediated damage to neuromuscular junctions.

ORIGIN OF THE AUTOIMMUNE RESPONE IN MG

One of the unsolved problems in MG concerns the origin of the autoimmune response. The thymus gland has long been suspected of involvement in this process, but the nature of the relationship is still poorly understood. Approximately 75 percent of myasthenic patients have thymic abnormalities, consisting of hyperplasia in most (germinal center formation), or thymic tumors in a small proportion.[62] Furthermore, alterations in the populations of B and T lymphocytes in the thymus, and possible changes in their antigenic properties, have been noted.[63] Thymic lymphocytes from myasthenic patients have also been shown to produce anti-AChR antibodies in tissue culture.[64]

In addition to lymphocytes, the thymus gland contains cells of other types, including muscle-like or "myoid" cells.[65] Typical skeletal muscle cells have recently been cultured from thymus glands of rats and of patients with MG.[66,67] These cells behave in all respects like normal mammalian skeletal muscle in tissue culture. It is of particular interest that the cultured cells have surface AChRs, as demonstrated by autoradiography and electrophysiological methods.[68] Because of their strategic location within the thymus the receptor-bearing muscle cells may be especially vulnerable to immune attack. Some alteration of either the muscle cells or the lymphocytes of the thymus may serve to break tolerance and thereby initiate an autoimmune response directed against AChRs. The possibility that a viral infection of the thymus could trigger this process has been suggested,[69] but there is as yet no evidence to support this concept.

CONCLUSIONS

The fundamental abnormality affecting the neuromuscular junction of myasthenic patients is a reduction of AChRs, due to an autoimmune attack. Antibodies to AChRs are present in most patients, and there is evidence that, aided by complement, they have a pathogenic role in the disease. The mechanism of antibody action involves acceleration of the rate of degradation of AChRs attributable to cross-linking of the receptors by the antibodies. In addition, antibodies block the AChRs and may participate in producing destructive changes at the neuromuscular junctions. The possibility that cell-mediated mechanisms may play a role in the immune responses of some myasthenic patients remains to be explored.

Perhaps the most important gap in the present understanding of MG concerns the origin of the autoimmune process, although there is indirect evidence that the thymus gland may be involved. Based on the growing knowledge of the pathogenesis of MG, future research will be directed toward developing specific therapeutic methods to deal with the immunological abnormalities of individual myasthenic patients.

NOTES

1. M. B. Walker, "Treatment of Myasthenia Gravis with Physostigmine," *Lancet* 1 (1934): 1200-01.

2. C. Y. Lee, "Chemistry and Pharmacology of Polypeptide Toxins in Snake Venoms," *Annual Review of Pharmacology* 12 (1972): 265-86.

3. D. M. Fambrough, D. B. Drachman, and S. Satyamurti, "Neuromuscular Junction in Myasthenia Gravis: Decreased Acetylcholine Receptors," *Science* 182 (1973): 293-95.

4. D. B. Drachman, "Myasthenia Gravis," *New England Journal of Medicine* 298 (1978): 136-42; 186-93.

5. J. I. Hubbard, "Microphysiology of Vertebrate Neuromuscular Transmission," *Physiological Reivews* 53 (1973): 674-723.

6. D. Elmqvist et al., "An Electrophysiological Investigation of Neuromuscular Transmission in Myasthenia Gravis," *Journal of Physiology* 174 (1964): 417-34.

7. Fambrough, Drachman, and Satyamurti, "Neuromuscular Junction" (See note 3).

8. Lee, "Snake Venoms" (See note 2).

9. E. X. Albuquerque et al., "An Electrophysiological and Morphological Study of the Neuromuscular Junctions in Patients with Myasthenia Gravis," *Experimental Neurology* 51 (1976): 536-63.

10. A. G. Engel et al., "Ultrastructural Localization of the Acetylcholine Receptor in Myasthenia Gravis and Its Experimental Autoimmune Models," *Neurology* 27 (1977): 307-15.

11. Y. Ito et al., "Acetylcholine Receptors and End-Plate Electrophysiology in Myasthenia Gravis," *Brain* 101 (1978): 345-68.

12. S. Satyamurti, D. B. Drachman, and F. Slone, "Blockade of Acetycholine Receptors: A Model of Myasthenia Gravis," *Science* 187 (1975): 955-57.

13. C. Ozdemir and R. R. Young, "The Results to Be Expected from Electrical Testing in the Diagnosis of Myasthenia Gravis," *Annals of the New York Academy of Sciences* 274 (1976): 203-22.

14. L. P. Rowland, H. Aranow, Jr., and P. F. A. Hoefer, "Observations on the Curare Test in the Differential Diagnosis of Myasthenia Gravis," in *Myasthenia Gravis*, ed. H. R. Viets (Springfield, Illinois: Charles C Thomas, 1961): Ch. 5, 411-34.

15. J. E. Desmedt, "The Neuromuscular Disorder in Myasthenia Gravis," in *New Developments in EMG and Clinical Neurophysiology*, ed. J. E. Desmedt (Basel: S. Karger, 1973): 241-304.

16. J. A. Simpson, "Myasthenia Gravis: A New Hypothesis," *Scottish Medical Journal* 4 (1960): 419-36.

17. B. Castleman, "The Pathology of the Thymus Gland in Myasthenia Gravis," *Annals of the New York Academy of Sciences* 135 (1966): 496-503.

18. A. Blalock et al., "Myasthenia Gravis and Tumors of the Thymic Region," *Annals of Surgery* 110 (1939): 544-61.

19. J. Patrick and J. Lindstrom, "Autoimmune Response to Acetylcholine Receptor," *Science* 180 (1973): 871-72.

20. H. Sugiyama et al., "Immunological Characterization of the Cholinergic Receptor Protein from Electrophorus Electricus," *Federation of European Biochemical Society Letters* 35 (1973): 124-28.

21. R. Tarrab-Hazdai et al., "Experimental Autoimmune Myasthenia Induced in Monkeys by Purified Acetylcholine Receptor," *Nature* 256 (1975): 128-30.

22. E. H. Lambert, J. M. Lindstrom, and V. A. Lennon, "End-Plate Potentials in Experimental Autoimmune Myasthenia Gravis in Rats," *Annals of the New York Academy of Sciences* 274 (1976): 300-18.

23. D. P. L. Green, R. Miledi, and A. Vincent, "Neuromuscular Transmission after Immunization against Acetylcholine Receptors," *Proceedings of the Royal Society of London B* 189 (1975): 57-68.

24. A. G. Engel et al., "Experimental Autoimmune Myasthenia Gravis: A Sequential and Quantitative Study of the Neuromuscular Junction Ultrastructure and Electrophysiologic Correlation," *Journal of Neuropathology and Experimental Neurology* 35 (1976): 563-87.

25. R. R. Almon, C. G. Andrew, and S. H. Appel, "Serum Globulin in Myasthenia Gravis: Inhibition of α-Bungarotoxin Binding to Acetylcholine Receptors," *Science* 186 (1974): 55-57.

26. A. N. Bender et al., "Myasthenia Gravis: A Serum Factor Blocking Acetylcholine Receptors of the Human Neuromuscular Junction," *Lancet* 1 (1957): 607-08.

27. S. H. Appel, R. R. Almon, and N. Levy, "Acetylcholine Receptor Antibodies in Myasthenia Gravis," *New England Journal of Medicine* 293 (1975): 760-61.

28. J. M. Lindstrom et al., "Antibody to Acetylcholine Receptor in Myasthenia Gravis: Prevalence, Clinical Correlates, and Diagnostic Value," *Neurology* 26 (1976): 1054-59.

29. T. Namba, S. B. Brown, and D. Grob," Neonatal Myasthenia Gravis: Report of Two Cases and Review of the Literature," *Pediatrics* 45 (1970): 488-504.

30. N. P. Bergh, "Biologic Assays in Myasthenia Gravis for Any Agents Causing a Neuromuscular Block," *Scandinavia Journal of Clinical Investigation* 5, Suppl. 5 (1953): 5-47.

31. W. L. Nastuk, A. J. L. Strauss, and K. E. Osserman, "Search for a Neuromuscular Blocking Agent in the Blood of Patients with Myasthenia Gravis," *American Journal of Medicine* 26 (1959): 394-409.

32. W. L. Nastuk and A. J. L. Strauss, "Further Developments in the Search for a Neuromuscular Blocking Agent in the Blood of Patients with Myasthenia Gravis," in *Myasthenia Gravis,* ed. H. R. Viets (Springfield, Illinois: Charles C Thomas, 1961): Ch. 3, 229-37.

33. K. V. Toyka et al., "Myasthenia Gravis: Passive Transfer from Man to Mouse," *Science* 190 (1975): 397-99.

34. K. V. Toyka et al., "Myasthenia Gravis: Study of Humoral Immune

Mechanisms by Passive Transfer to Mice," *New England Journal of Medicine* 296 (1977): 125-31.

35. Ibid.

36. A. G. Engel, E. H. Lambert, and F. H. Howard, "Immune Complexes (IgG and C3) at the Motor End-Plate in Myasthenia Gravis," *Mayo Clinic Proceedings* 52 (1977): 267-80.

37. I. Kao and D. B. Drachman, "Myasthenic Immunoglobulin Accelerates Acetylcholine Receptor Degradation," *Science* 196 (1977): 527-29.

38. P. N. Devreotes and D. M. Fambrough, "Acetylcholine Receptor Turnover in Membranes of Developing Muscle Fibers," *Journal of Cell Biology* 65 (1975): 335-58.

39. Ibid.

40. R. Anwyl, S. M. Appel, and T. Narahasi, "Myasthenia Gravis Serum Reduces Acetylcholine Sensitivity in Cultured Rat Myotubes," *Nature* 267 (1977): 262-63.

41. S. Bevan, R. W. Kullberg, and S. F. Heinemann, "Human Myasthenia Sera Reduce Actylcholine Sensitivity in Human Muscle Cells in Tissue Culture," *Nature* 267 (1977): 263-65.

42. D. B. Drachman, R. N. Adams, and L. F. Josifek, manuscript in preparation.

43. D. B. Drachman et al., "Effect of Myasthenic Patients' Immunoglobulin on Acetylcholine Receptor Turnover: Selectivity of Degradation Process," *Proceedings of the National Academy of Sciences* 75 (1978): 3422-26.

44. D. B. Drachman et al., "Myasthenic Antibodies Cross-Link Acetylcholine Receptors to Accelerate Degradation," *New England Journal of Medicine* 298 (1978): 1116-22.

45. C. A. Williams and M. W. Chase, eds., *Methods in Immunology and Immunochemistry* (New York: Academic Press, 1967).

46. F. W. Putnam et al., "The Cleavage of Rabbit γ Globulin by Papain," *Journal of Biological Chemistry* 237 (1962): 717-26.

47. Drachman, Adams, and Josifek (See note 42).

48. D. B. Drachman et al., "Effect of Myasthenic Immunoglobulin on Acetylcholine Receptors of Cultured Muscle," *Annals of Neurology* 1 (1977): 504.

49. Almon, Andrew, and Appel, "Serum Globulin" (See note 25).

50. Drachman, Adams, and Josifek (See note 42).

51. A. G. Engel, E. H. Lambert, and F. M. F. Howard, "Immune Complexes (IgG and C3) at the Motor End-Plate in Myasthenia Gravis," *Neurology* 27 (1977): 307-15.

52. E. F. Stanley and D. B. Drachman, "Effect of Myasthenic Immunoglobulin on Acetylcholine Receptor of Intact Mammalian Neuromuscular Junctions," *Science* 200 (1978): 1285-87.

53. S. Heinemann, J. Merlie, and J. M. Lindstrom, "Modulation of Acetylcholine Receptor in Rat Diaphragm by Anti-Receptor Sera," *Nature* 274 (1978): 65-67.

54. C. G. Reiness and C. V. Weinberg, "Antibody Acetylcholine Receptor Increases Degradation of Junctional and Extrajunctional Receptors in Adult Muscle," *Nature* 274 (1978): 68-69.

55. A. G. Engel and T. Santa, "Histometric Analysis of the Ultrastructure

of the Neuromuscular Junction in Myasthenia Gravis and in the Myasthenic Syndrome," *Annals of the New York Academy of Sciences* 183 (1971): 46-63.

56. J. E. Rash et al., "Studies on Human Myasthenia Gravis. Electrophysiological and Ultrastructural Evidence Compatible with Antibody Labeling of the Acetylcholine Receptor Complex," *Proceedings of the National Academy of Sciences* 73 (1976): 4584-88.

57. Toyka et al., "Humoral Immune Mechanisms" (See note 34).

58. Engel, Lambert, and Howard, "Immune Complexes" (See note 36).

59. O. Abramsky et al., "Cellular Immune Response to Acetylcholine Receptor-Rich Fraction in Patients with Myasthenia Gravis," *Clinical Experimental Immunology* 19 (1975): 11-16.

60. D. P. Richman, J. Patrick, and B. G. W. Arnason, "Cellular Immunity in Myasthenia Gravis," *New England Journal of Medicine* 294 (1976): 694-98.

61. B. Conti-Tronconi et al., "Cellular Immune Response against Acetylcholine Receptor in Myasthenia Gravis," *Neurology* 29 (1979): 496-501.

62. B. Castleman, "The Pathology of the Thymus Gland in Myasthenia Gravis," *Annals of the New York Academy of Sciences* 135 (1966): 496-503.

63. N. I. Abdou et al., "The Thymus in Myasthenia Gravis. Evidence for Altered Cell Populations," *New England Journal of Medicine* 291 (1974): 1271-75.

64. A. Vincent et al., "Antiacetylcholine Receptor Antibody Synthesis in Culture," in *Plasmapheresis and the Immunobiology of Myasthenia Gravis,* ed. P. Dau (Boston: Houghton Mifflin, 1979): 59-71.

65. R. L. Van de Velde and N. B. Friedman, "Thymic Myoid Cells and Myasthenia Gravis," *American Journal of Pathology* 59 (1970): 347-68.

66. I. Kao and D. B. Drachman, "Thymic Muscle Cells Bear Acetylcholine Receptors: Possible Relation to Myasthenia Gravis," *Science* 195 (1977): 74-75.

67. H. Wekerle et al., "Striated Muscle Fibres Differentiate in Monolayer Cultures of Adult Thymus Reticulum," *Nature* 256 (1975): 493-94.

68. Kao and Drachman, "Thymic Muscle Cells" (See note 66).

69. S. K. Datta and R. S. Schwartz, "Infectious (?) Myasthenia," *New England Journal of Medicine* 291 (1974): 1304-05.

GENETICS AND BIOCHEMISTRY OF THE LIPOPROTEIN RECEPTOR DEFECT IN FAMILIAL HYPERCHOLESTEROLEMIA

Joseph L. Goldstein
and Michael S. Brown *

INTRODUCTION

Of all of the diseases producing hyperlipidemia in man, familial hypercholesterolemia (FH) is the best defined clinically, genetically, and biochemically; it remains the most clearcut example of a simply inherited defect that leads to coronary heart disease. The disorder is characterized *chemically* by an elevated plasma concentration of low-density lipoprotein (LDL), the major cholesterol-transport protein in human plasma; *clinically* by deposition of cholesterol in specific sites to form xanthomas and premature coronary heart disease; *genetically* by autosomal dominant inheritance; and *biochemically* by a deficiency in a cell-surface receptor that regulates the degradation of plasma LDL. Although the FH gene is expressed clinically as an autosomal trait, the disorder is more severe in individuals inheriting two doses of the gene (homozygotes) than in those inheriting one dose (heterozygotes).[1-3]

GENETICS

Heterozygotes for this autosomal dominant trait manifest hypercholesterolemia from the time of birth, but usually remain asymptomatic until adult life when tendon xanthomas and pre-

* Dr. Brown, a member of the Departments of Molecular Genetics and Internal Medicine, University of Texas Health Science Center at Dallas, did not attend the conference.

mature coronary heart disease of varying severity develop. The total plasma cholesterol level in heterozygotes is generally in the range of 270 to 550 mg./dl and rarely exceeds 600 mg./dl. The wide variability in cholesterol levels among heterozygotes is probably a reflection of two factors: 1) the "normal" allele at the FH locus may exist in variant forms among affected individuals; and 2) the mutant FH gene exerts its effect within a system of interacting exogenous (environmental) and endogenous (polygenic) factors that are known to modify cholesterol levels.

In contrast to FH heterozygotes, FH homozygotes, who inherit a double dose of the mutant gene and thus have no normal allele, show unique clinical characteristics that are much more uniform and severe, and are easily distinguished from normal and other hypercholesterolemic states. The syndrome in homozygotes is characterized by a profound hypercholesterolemia (650 to 1,000 mg./dl), a distinct form of custaneous xanthoma developing in the first four years of life, and signs of rapidly progressive coronary, cerebral, and peripheral vascular disease beginning in childhood and eventuating in death from myocardial infarction, often before age twenty.[4-7]

From analysis of more than 325 reported families, the following conclusions regarding the biology of the FH gene seem warranted: [8-10]

- When either the plasma total cholesterol or LDL-cholesterol level is used as a genetic marker for analyzing known affected families, the gene is highly penetrant at all ages. The calculated penetrance value has ranged from 90 to 100 percent, that is 90 to 100 percent of persons carrying the gene will have a plasma cholesterol value greater than the 95th percentile value of the general population.
- Although a single dose of the gene has a significant impact on longevity in heterozygotes, it does not appear to cause a diminution of their reproductive fitness.
- The expression of the FH gene does not appear to be influenced by obesity, diabetes mellitus, or hypertension.
- Evidence from family studies suggests that the gene re-

sponsible for FH may be located on the same chromosome as the gene specifying the third component of complement.

Biochemical studies in fibroblasts cultured from the skin of patients with the clinical phenotype of homozygous FH have disclosed the existence of at least three different mutations affecting the activity of the cell surface LDL receptor. Thus FH, like most inherited disorders of metabolism, is genetically heterogeneous; a similar clinical picture may result from distinct but related mutations at a single genetic locus.[11]

The frequency of FH heterozygotes among unselected patients with coronary heart disease has been determined by family analysis of consecutively studied survivors of myocardial infarction. Collected by investigators in London,[12] Seattle,[13] and Helsinki,[14] these data show that 3 to 6 percent of survivors are FH heterozygotes. While no such direct determination has yet been made for the general population, a minimal frequency for heterozygous FH has been estimated at one in 500 Caucasian individuals.[15-17] If the heterozygote frequency is assumed to be one in 500, then the homozygote frequency should theoretically be one in 1 million ($1{:}500 \times 1{:}500 \times 1{:}4$), and one in 250,000 couples in the general population would be at genetic risk of producing a homozygote.

CLINICAL ASPECTS

The elevated total cholesterol content in the plasma of heterozygotes and homozygotes is entirely attributable to an elevation in the concentration of LDL. The mean total cholesterol and LDL-cholesterol concentrations in heterozygotes at all ages are roughly twice the mean values of normal subjects of similar age; the mean total cholesterol and LDL-cholesterol concentrations in homozygotes are about two to three times those of heterozygotes, and about six times those of normal subjects.

Associated with the increase of LDL-cholesterol concentration in plasma, there is a deposition of cholesterol in several tissues of the body, especially in tendons (xanthomas) and in arterial plaques (atheromas). The occurrence of xanthomas in

FH is a function of the severity and duration of the elevation in plasma LDL. Local factors, however, including motion and local trauma, also predispose to the deposition of LDL-cholesterol. Both homozygotes and heterozygotes may have tendon xanthomas, especially in the Achilles tendons and in the extensor tendons of the hand, and tuberous xanthomas, especially over the elbows. Elevated orange-yellow planar xanthomas lying superficially in the skin over the extremities, buttocks, and hands—especially in the interdigital web between the first and second fingers—are unique to homozygotes and do not occur in heterozygotes. Xanthomas appear in most homozygotes during the first four years of life, but they do not usually become manifest in heterozygotes until after age twenty.[18-21]

FH is the outstanding example of a single gene mutation that regularly produces atherosclerosis. The cardiac disease in homozygotes is rapidly progressive, and the clinical findings in these patients are remarkably uniform. Angina pectoris, myocardial infarction, or sudden death—three of the cardinal features of adult ischemic heart disease—usually occur in homozygotes between the ages of five and twenty. One homozygote is recorded as having had an acute myocardial infarction as early as eighteen months, and very few survive past age thirty.[22]

The pattern of cardiac disease in heterozygotes with FH is much more variable than in homozygotes. For heterozygous men, the chance of having a myocardial infarction before age sixty is about 85 percent.[23,24] This is to be contrasted with a 15 percent risk of coronary artery disease among unaffected men by age sixty. In female heterozygotes the risk of developing coronary artery disease by age sixty is about 50 percent, as compared with 10 percent in unaffected females.[25,26] Thus, despite the presence of the same genetic abnormality and similarly elevated plasma LDL levels, heterozygous women manifest coronary heart disease less often and at a later age than do heterozygous men. This sex difference, which is also a characteristic feature of the usual form of atherosclerotic heart disease in normolipidemic individuals, does not seem to be present in homozygotes.

DEFECTIVE LDL RECEPTOR AS PRIMARY BIOCHEMICAL DEFECT IN FH

The LDL Pathway in Normal Cells

Cholesterol is required for the normal function of the plasma membrane of all mammalian cells. Studies over the past twenty years have demonstrated that most of the cholesterol found in the membranes of nonhepatic cells is not synthesized locally, but is derived from cholesterol synthesized in the liver or absorbed from the intestine.[27,28] The liver distributes endogenously synthesized or intestinally derived cholesterol to extrahepatic tissues by means of lipoproteins secreted in the plasma. LDL is the main cholesterol-carrying lipoprotein in human plasma. LDL consists of a core of neutral lipid composed predominantly of cholesteryl esters. Surrounding this neutral lipid core is a polar coat composed of phospholipid, free cholesterol, and a protein called apoprotein B. Body cells require free cholesterol for their membranes, not cholesteryl esters. Thus cells require a mechanism by which they can extract the cholesteryl esters from LDL and hydrolyze those esters so as to generate free cholesterol for metabolic use.

The pathway by which cells are able to utilize the cholesteryl esters of LDL has been elucidated through studies of human fibroblasts in tissue culture.[29-31] Figure 1 summarizes this pathway as it has emerged from studies conducted in Dallas over the past several years. When cultured human fibroblasts are in need of cholesterol they synthesize a receptor molecule that becomes localized on the plasma membrane. This receptor specifically binds the protein component of the lipoprotein. Once LDL binds to the receptor on the surface of these cells, the lipoprotein is internalized by adsorptive endocytosis and delivered to cellular lysosomes. Within the lysosomes the protein component of LDL is hydrolyzed to amino acids and the cholesteryl esters of LDL are exposed to an acid lipase that hydrolyzes the cholesteryl esters so as to generate free cholesterol. The free cholestrol is able to cross the lysosomal membrane to enter the cellular compartment, where it is used by the cell for membrane synthesis.

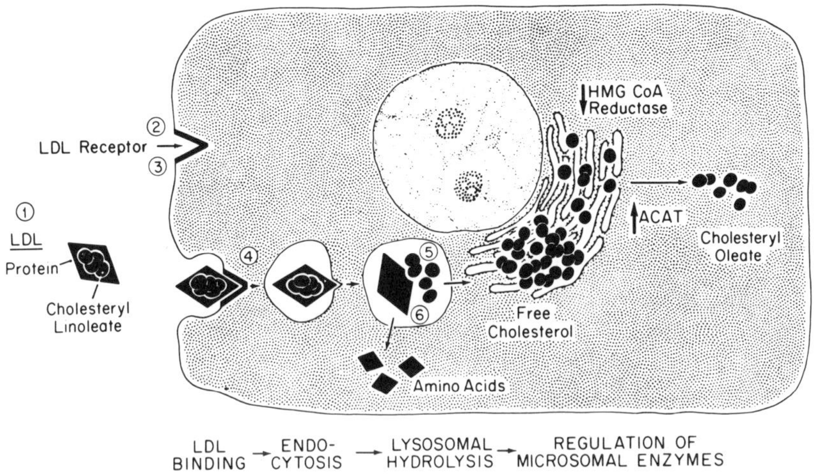

Figure 1. Sequential steps in the LDL pathway in cultured human fibroblasts. HMGCoA reductase denotes 3-hydroxy-3-methylglutaryl coenzyme A reductase; ACAT denotes acyl-coenzyme A:cholesterol acyltransferase.

Within the cell the cholesterol derived from the catabolism of LDL regulates three metabolic events: 1) it suppresses the activity of 3-hydroxy-3-methylglutaryl coenzyme A reductase (HMG CoA reductase), which is the rate-controlling enzyme in cholesterol biosynthesis; 2) the cholesterol activates an acyl-coenzyme A:cholesterol acyltransferase (ACAT), which reesterifies the incoming cholesterol. The newly formed cholesteryl esters are stored within the cell as cytoplasmic lipid droplets; 3) the incoming cholesterol turns off the synthesis of the LDL receptor itself. This feedback regulation of LDL receptor synthesis stops the uptake of LDL and prevents cells from becoming overloaded with cholesterol when exposed to LDL.[32, 33]

The LDL pathway was originally elucidated through biochemical studies in which the metabolism of various types of radiolabeled LDL was studied in human fibroblasts. To study this pathway at the morphological level, we have recently worked in collaboration with R. G. W. Anderson to attach LDL covalently to an electron-dense probe, ferritin. When this ferritin-labeled LDL was incubated with fibroblasts, we observed that the lipoprotein bound to specialized regions of the plasma membrane, where the membrane is indented and where it appears to be coated on its cytoplasmic side by a fuzzy material.[34,35] These so-called coated pits have been observed in many other cell types, but their precise function has not been known.

Even though the coated pits account for only about 2 percent of the surface of the human fibroblasts, they contain 50-80 percent of the LDL receptors. The importance of the binding of LDL to the coated pits lies in the observation that these coated regions are specialized for the rapid endocytosis of receptor-bound molecules. For example, if one binds LDL-ferritin to the cell at 4°C and then warms up the cell for only one minute, the majority of the ferritin-labeled coated pits have invaginated into the cell; within two minutes they pinch off to form coated endocytic vesicles that carry the bound LDL-ferritin into the cell; and within five minutes the LDL-ferritin particles that were bound to the coated pits have reached cellular lysosomes.[36]

The LDL Pathway in FH Cells

Fibroblasts cultured from forty-nine patients with the classic clinical phenotype of homozygous FH have all displayed evidence of a primary abnormality in the function of the LDL receptor.[37] To date, three different mutations in the gene that specifies the LDL receptor have been identified.[38-43] One of these mutant alleles, R^{b°, specifies a protein that has no detectable binding activity (<1 percent of normal). The second allele, R^{b-}, specifies a protein that has a reduced binding activity (1 to 10 percent of normal). The third mutant allele, R^{b+,i°, is an extremely rare allele that specifies a protein that can bind LDL normally, but cannot catalyze the internalization of the receptor-bound lipoprotein. The distribution of these three mutant alleles, the presumed genotypes, and the biochemical phenotypes as they occur in fibroblast strains derived from forty-nine patients with the clinical diagnosis of homozygous FH is tabulated in Table I.

In the receptor-negative class of homozygotes, which consists of twenty-eight of the forty-nine fibroblast strains, the fibroblasts exhibit no functional LDL receptors, that is, these mutant cells exhibit less than 2 percent of the amount of high-affinity

TABLE I. GENETIC ANALYSIS OF FIBROBLAST STRAINS FROM 49 SUBJECTS WITH THE CLINICAL DIAGNOSIS OF HOMOZYGOUS FAMILIAL HYPERCHOLESTEROLEMIA

Number of Subjects	Genotype at LDL Receptor Locus	Biochemical Phenotype
28	$R^{b°} / R^{b°}$	Receptor-negative
20	$R^{b-}/R^{b°}$ or R^{b-}/R^{b-}	Receptor-defective
1	$R^{b°}/R^{b+, i°}$	Internalization defect

LDL binding observed in normal cells studied under identical conditions. As a result, LDL is not taken up and degraded at a normal rate, its cholesterol is not transferred into the cell, HMG CoA reductase activity and cholesterol synthesis are not suppressed, and cholesteryl ester formation is not activated. Genetic and biochemical studies suggest that receptor-negative patients are homozygous for the $R^{b°}$ allele. Fibroblasts from the parents of these subjects, who are obligate heterozygotes and whose geneotype is $+/R^{b°}$, express about one-half of the normal number of LDL receptors.

The second class of FH homozygotes, designated receptor-defective, has been observed to consist of twenty of the forty-nine fibroblast strains. Cells from this group of individuals are able to bind from 2 to 20 percent of the normal number of LDL particles. In each case the relative amount of receptor binding correlates with the relative ability of LDL to partially suppress HMG CoA reductase activity and to produce a limited activation of cholesteryl ester formation. We have suggested that the receptor-defective variant represents a genetic compound state in which two different mutant alleles are present at the FH locus, in a manner analogous to other genetic compounds, such as SC hemoglobinopathy.[44] By this hypothesis one mutant allele is the same as the functionless allele present in receptor-negative individuals ($R^{b°}$), and the other mutant allele specifies an LDL receptor that is functional but manifests reduced affinity and capacity (R^{b-}). Equally likely is the possibility that some of the subjects with the receptor-defective phenotype are homozygous for the R^{b-} allele. Thus far, no consistent clinical differences have been noted between receptor-negative and receptor-defective homozygotes.

A third type of mutation has also been identified in fibroblasts from one patient with the typical clinical and pedigree features of homozygous FH.[45,46] The LDL receptor in this unique cell strain shows a normal ability to bind LDL, as judged by both its affinity and capacity for binding the lipoprotein. Once bound to the receptor, however, the LDL cannot be taken up into the cell. As a result, cells from this patient, designated internalization-defective, are unable to degrade LDL or utilize its cholesterol, and hence they display abnormalities in the regulation of cholesterol metabolism that are the same as those in cells from receptor-negative homozygotes.

Ultrastructural studies using ferritin-labeled LDL indicate that the failure of these cells to internalize receptor-bound LDL results from an abnormality in the location of the LDL receptor on the cell membrane. Whereas in normal cells the receptor is concentrated in coated regions that invaginate into the cell every five minutes, in the internalization-defective cells the receptor is located at random along the membrane and is not found in coated regions, although a normal number of such regions is present.[47] Fibroblasts from homozygotes with the receptor-negative mutation also show a normal number of coated regions; in contrast to the internalization-defective cells, however, these mutant cells show no LDL-ferritin binding by electron microscopy.[48-50]

Both parents of the patient with the internalization defect exhibit hypercholesterolemia and the other typical clinical features of heterozygous FH. Study of their fibroblasts has revealed that this patient is a genetic compound and carries two different mutant alleles at the LDL receptor locus. The father of this patient is a heterozygote for the R^{b+,i° allele (genotype, $+/R^{b+,i^\circ}$). Approximately half of his LDL receptors, that is, those specified by the normal + allele, are able to carry their bound LDL into the cell; the other half, specified by the R^{b+,i° allele, are unable to reach the coated pits and so do not carry their bound LDL into the cells. These observations have been made biochemically, using ^{125}I-LDL,[51,52] and have been confirmed at the ultrastructural level using ferritin-labeled LDL.[53]

The defect in metabolic regulation in fibroblasts from all three classes of FH homozygotes can be overcome if cholesterol is delivered to the cells in an artificial nonlipoprotein form that permits the sterol to enter the cell without a requirement for the LDL receptor.[54,55] For example, when the cholesterol content of a homozygote's fibroblasts is raised by the addition to the culture medium of cholesterol dissolved in ethanol, HMG CoA reductase activity is suppressed and cholesteryl ester formation is activated in the same manner as in normal cells. These findings indicate that the homozygote cells possess all the intracellular factors necessary to respond to cholesterol, provided the sterol can be delivered to the proper intracellular site. Of considerable interest is the observation that certain oxygenated derivatives of cholesterol, such as 7-ketocholesterol and 25-hydroxycholesterol, are up to 100 times more potent on a molar basis than is cholesterol itself in suppressing HMG CoA reductase and in activating ACAT in cultured cells. These oxygenated sterols, like cholesterol, are equally effective in cells from homozygotes and normals.[56,57]

In addition to cultured fibroblasts, the LDL receptor defect in homozygous FH has been demonstrated in long-term cultured lymphoid cells;[58,59] cultured aortic smooth muscle cells;[60] lymphocytes freshly isolated from the bloodstream;[61,62] and fetal amniotic fluid cells.[63] Measurement of the binding of ^{125}I-LDL to monolayers of cultured fibroblasts has shown that the heterozygous parents of homozygotes with the receptor-negative and receptor-defective phenotypes, as well as other patients with the phenotype of heterozygous FH, possess about one-half the normal number of LDL receptors.[64] As expected from genetic considerations, these functional receptors in heterozygotes show a normal affinity for LDL, and thus appear to represent the product of the normal allele at the FH locus. Under assay conditions in which the concentration of LDL is saturating, the heterozygote cells bind half as many LDL particles as do normal cells, and thus they degrade the lipoprotein and form cholesteryl esters at half the rate of normal cells. Moreover, when high levels of LDL are added to these fibroblasts, the rate at which they suppress HMG CoA reductase is reduced.[65]

A more interesting aspect of the heterozygous state emerges when the LDL binding curve is studied at LDL levels below saturation. Whenever the concentration of LDL surrounding the heterozygote cells is twice as high as that surrounding normal cells, the cells from both genotypes bind and degrade the lipoprotein at the same rate.[66,67] Since the extent of suppression of HMG CoA reductase activity and the degree of stimulation of cholesteryl ester formation in fibroblasts are both proportional to the amount of receptor-bound LDL, each of these processes is similar in cells from both genotypes whenever the level of LDL in the fluid surrounding the heterozygote is twice that in the normal. The ability of the heterozygote cells to compensate for their receptor deficiency when the LDL level is twice normal has important implications for the pathogenesis of the clinical syndrome.[68]

Since FH is transmitted by a single gene mechanism, all of the metabolic abnormalities observed in vivo must ultimately be explained in some direct or indirect way by the receptor deficiency. Speculations as to how the cellular receptor abnormality is expressed in vivo in living FH patients are presented elsewhere.[69]

USE OF THE $R^{b+,i^{\circ}}$ MUTATION TO FORMULATE A MODEL FOR INCORPORATION OF LDL RECEPTORS INTO COATED PITS

From analysis of the properties of the mutant $R^{b+,i^{\circ}}$ allele that occurs in the unique FH patient with the LDL internalization defect, we have formulated a working model for the way in which LDL receptors are incorporated into coated pits on the cell surface. This model is based on our genetic studies of the LDL receptor,[70] as well as on studies by others on the assembly of lipid envelope viruses in mammalian cells.[71] As shown in Figure 2, we envision the LDL receptor as a transmembrane protein made within the cell, where it is presumably glycosylated and inserted

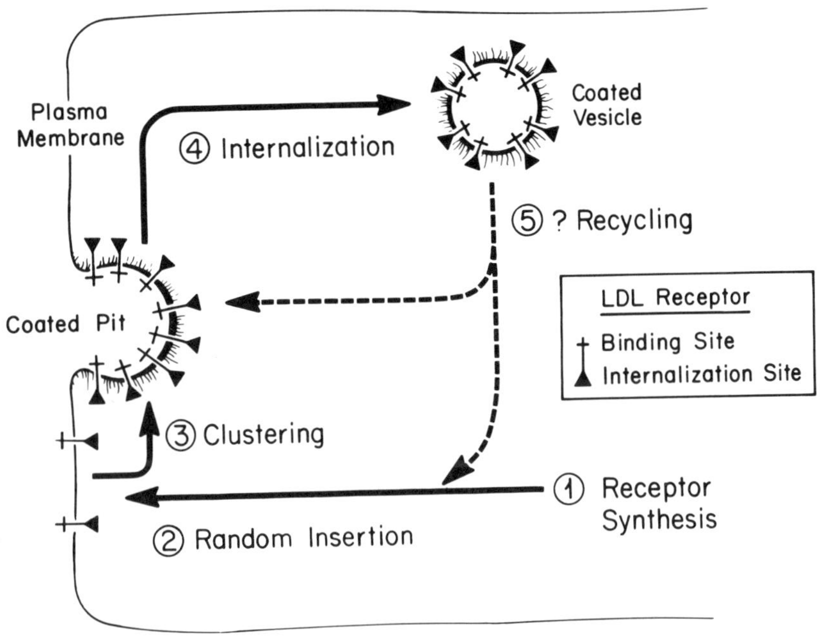

Figure 2. Schematic illustration of the proposed pathway by which LDL receptors become localized to coated pits on the plasma membrane of human fibroblasts. The sequential steps in this process are: 1) synthesis of LDL receptors on polyribosomes; 2) insertion of LDL receptors at random sites along noncoated segments of plasma membrane; 3) clustering together of LDL receptors in coated pits; 4) internalization of LDL receptors as coated pits invaginate to form coated endocytic vesicles; and 5) recycling of internalized LDL receptors back to the plasma membrane.

into the plasma membrane. The receptor is postulated to have two active sites: one of these, on the external surface of the membrane, is the binding site for LDL; a second, on the cytoplasmic side of the membrane, contains an amino-acid sequence that is necessary for the receptor to become incorporated into coated pits. We call this the internalization site; the receptor is initially inserted into the plasma membrane at random. If the LDL receptor has an active internalization site, however, it is clustered together with other receptors to form a coated pit. This clustering possibly occurs under the influence of a cytoplasmic protein not yet identified. Once the LDL receptors are clustered together and the coated pit becomes full, it invaginates, pinches off, and forms a coated vesicle. The vesicles then fuse with lysosomes and the LDL is degraded. Indirect kinetic evidence suggests that the receptors are recycled back to the cell surface after they discharge their LDL within the lysosomes.

When the LDL receptor carries a mutation in the internalization site, as in the patient with the internalization defect, the receptor is inserted into the membrane, but it is unable to become incorporated into coated pits. As a result the receptor accumulates at a site just proximal to the metabolic block, that is, at random locations on noncoated segments of the plasma membrane.

Further studies of the biochemistry and cell biology of the coated pit mechanism, as well as of the molecular genetics of the LDL receptor, should provide new insights into the fundamental question of how cell-surface receptors transmit their regulatory signals.

NOTES

1. M. S. Brown and J. L. Goldstein, "Familial Hypercholesterolemia: Genetic, Biochemical and Pathophysiologic Considerations, "*Advances in Internal Medicine* 20 (1975): 273-96.

2. D. S. Fredrickson, J. L. Goldstein, and M. S. Brown, "The Familial Hyperlipoproteinemias," in *The Metabolic Basis of Inherited Disease,* ed. J. B. Stanbury, J. B. Wyngaarden, and D. S. Fredrickson, 5th ed. (New York: McGraw-Hill Book Company, 1978): 604-55.

3. R. J. Havel, J. L. Goldstein, and M. S. Brown, "Lipoproteins and Lipid Transport," in *Diseases of Metabolism,"* ed. P. K. Bondy and L. E. Rosenberg, 8th ed. (Philadelphia: W. B. Saunders Company, 1979): forthcoming.

4. Brown and Goldstein, "Familial Hypercholesterolemia" (See note 1).

5. Fredrickson, Goldstein, and Brown, "Familial Hyperlipoproteinemias" (See note 2).

6. Havel, Goldstein, and Brown, "Lipoproteins" (See note 3).

7. A. K. Khachadurian, "The Inheritance of Essential Familial Hypercholesterolemia," *American Journal of Medicine* 37 (1964): 402-07.

8. Brown and Goldstein, "Familial Hypercholesterolemia" (See note 1).

9. Fredrickson, Goldstein, and Brown, "Familial Hyperlipoproteinemias" (See note 2).

10. Havel, Goldstein and Brown, "Lipoproteins" (See note 3).

11. M. S. Brown and J. L. Goldstein, "Familial Hypercholesterolemia: Model for Genetic Receptor Disease," *Harvey Lectures,* series 73 (New York: Academic Press, 1979): 163-201.

12. D. Patterson and J. Slack, "Lipid Abnormalities in Male and Female

Survivors of Myocardial Infarction and Their First-Degree Relatives," *Lancet* i (1972): 393-99.

13. J. L. Goldstein et al., "Hyperlipidemia in Coronary Heart Disease. II. Genetic Analysis of Lipid Levels in 176 Families and Delineation of a New Inherited Disorder, Combined Hyperlipidemia," *Journal of Clinical Investigation* 52 (1973): 1544-68.

14. E. A. Nikkilä and A. Aro, "Family Study of Serum Lipids and Lipoproteins in Coronary Heart Disease," *Lancet* i (1973): 954-58.

15. Brown and Goldstein, "Familial Hypercholesterolemia" (See note 1).

16. Fredrickson, Goldstein, and Brown, "Familial Hyperlipoproteinemias" (See note 2).

17. Goldstein et al., "Combined Hyperlipidemia" (See note 13).

18. Brown and Goldstein, "Familial Hypercholesterolemia" (See note 1).

19. Fredrickson, Goldstein, and Brown, "Familial Hyperlipoproteinemias" (See note 2).

20. Havel, Goldstein, and Brown, "Lipoproteins" (See note 3).

21. Khachadurian, "Inheritance of Hypercholesterolemia" (See note 7).

22. Fredrickson, Goldstein, and Brown, "Familial Hyperlipoproteinemias" (See note 2).

23. N. J. Stone et al., "Coronary Artery Disease in 166 Kindred with Familial Type II Hyperlipoproteinemia," *Circulation* 49 (1974): 476-88.

24. J. Slack and N. C. Nevin, "Hyperlipidaemic Xanthomatosis," *Journal of Medical Genetics* 5 (1968): 4-8.

25. Stone et al., "Coronary Artery Disease" (See note 23).

26. Slack and Nevin, "Xanthomatosis" (See note 24).

27. J. M. Dietschy and J. D. Wilson, "Regulation of Cholesterol Metabolism," *New England Journal of Medicine* 282 (1970): 1128-38; 1179-83; 1241-49.

28. M. S. Brown and J. L. Goldstein, "Receptor-Mediated Control of Cholesterol Metabolism," *Science* 191 (1976): 150-54.

29. __________, "Familial Hypercholesterolemia: Model" (See note 11).

30. __________, "Receptor-Mediated Control" (See note 28).

31. J. L. Goldstein and M. S. Brown, "The Low-Density Lipoprotein Pathway and Its Relation to Atherosclerosis," *Annual Review of Biochemistry* 46 (1977): 897-930.

32. Brown and Goldstein, "Receptor-Mediated Control (See note 28).

33. Goldstein and Brown, "Low-Density Lipoprotein" (See note 31).

34. R. G. W. Anderson, J. L. Goldstein, and M. S. Brown, "Localization of Low-Density Lipoprotein Receptors on Plasma Membrane of Normal Human Fibroblasts and Their Absence in Cells from a Familial Hypercholesterolemia Homozygote," *Proceedings of the National Academy of Sciences* 73 (1976): 2434-38.

35. R. G. W. Anderson, M. S. Brown, and J. L. Goldstein, "Role of the Coated Endocytic Vesicle in the Uptake of Receptor-Bound Low-Density Lipoprotein in Human Fibroblasts," *Cell* 10 (1977): 351-64.

36. Ibid.

37. Brown and Goldstein, "Familial Hypercholesterolemia: Model" (See note 11).

38. Ibid.

39. M. S. Brown and J. L. Goldstein, "Familial Hypercholesterolemia: Defective

Binding of Lipoproteins to Cultured Fibroblasts Associated with Impaired Regulation of 3-Hydroxy-3-Methylglutaryl Coenzyme A Reductase Activity," *Proceedings of the National Academy of Sciences* 71 (1974): 788-92.

40. ———, "Analysis of a Mutant Strain of Human Fibroblasts with a Defect in the Internalization of Receptor-Bound Low-Density Lipoprotein," *Cell* 9 (1976): 663-74.

41. J. L. Goldstein and M. S. Brown, "Genetics of the LDL Receptor: Evidence That the Mutations Affecting Binding and Internalization Are Allelic," *Cell* 12 (1977): 629-41.

42. R. G. W. Anderson, J. L. Goldstein, and M. S. Brown, "A Mutation That Impairs the Ability of Lipoprotein Receptors to Localise in Coated Pits on the Cell Surface of Human Fibroblasts," *Nature* 270 (1977): 695-99.

43. D. W. Bilheimer et al., "Genetics of the Low-Density Lipoprotein Receptor: Diminished Receptor Activity in Lymphocytes from Heterozygotes with Familial Hypercholesterolemia," *Journal of Clinical Investigation* 61 (1978): 678-96.

44. J. L. Goldstein et al., "Genetic Heterogeneity in Familial Hypercholesterolemia: Evidence for Two Different Mutations Affecting Functions of Low-Density Lipoprotein Receptor," *Proceedings of the National Academy of Sciences* 72 (1975): 1092-96.

45. Brown and Goldstein, "Analysis of Mutant Strain" (See note 40).

46. Goldstein and Brown, "Genetics of LDL Receptor" (See note 41).

47. Anderson, Goldstein, and Brown, "Ability of Lipoprotein Receptors" (See note 42).

48. ———, "Lipoprotein Receptors" (See note 34).

49. Anderson, Brown, and Goldstein, "Coated Endocytic Vesicle" (See note 35).

50. Anderson, Goldstein, and Brown, "Ability of Lipoprotein Receptors" (See note 42).

51. Goldstein and Brown, "Genetics of LDL Receptors" (See note 41).

52. Anderson, Goldstein, and Brown, "Ability of Lipoprotein Receptors" (See note 42).

53. Ibid.

54. Brown and Goldstein, "Receptor-Mediated Control" (See note 28).

55. Goldstein and Brown, "Low-Density Lipoprotein" (See note 31).

56. Brown and Goldstein, "Receptor-Mediated Control" (See note 28).

57. Goldstein and Brown, "Low-Density Lipoprotein" (See note 31).

58. H. J. Kayden, L. Hatam, and N. G. Beratis, "Regulation of 3-Hydroxy-3-Methylglutaryl Coenzyme A Reductase Activity and the Esterification of Cholesterol in Human Long-Term Lymphoid Cell Lines," *Biochemistry* 15 (1976): 521-28.

59. Y. K. Ho et al., "Binding, Internalization, and Hydrolysis of Low-Density Lipoprotein in Long-Term Lymphoid Cell Lines from a Normal Subject and a Patient with Homozygous Familial Hypercholesterolemia," *Journal of Experimental Medicine* 144 (1976): 444-55.

60. M. S. Brown, R. G. W. Anderson, and J. L. Goldstein, "Mutations Affecting the Binding, Internalization, and Lysosomal Hydrolysis of Low-Density Lipoprotein in Cultured Human Fibroblasts, Lymphocytes, and Aortic Smooth Muscle Cells," *Journal of Supramolecular Structure* 6 (1977): 85-94.

61. Bilheimer et al., "Genetics of LDL Receptor" (See note 43).

62. Y. K. Ho et al., "Regulation of Low-Density Lipoprotein Receptor Activity

in Freshly Isolated Human Lymphocytes," *Journal of Clinical Investigation* 58 (1976): 1465-74.

63. M. S. Brown et al., "Prenatal Diagnosis of Homozygous Familial Hypercholesterolaemia," *Lancet* i (1978): 526-29.

64. M. S. Brown and J. L. Goldstein, "New Directions in Human Biochemical Genetics: Understanding the Manifestations of Receptor-Deficiency States," in *Progress in Medical Genetics,* n.s., vol. 1, ed. A. G. Steinberg et al. (Philadelphia: W. B. Saunders Company, 1976): 103-19.

65. Ibid.

66. Ibid.

67. J. L. Goldstein et al., "Heterozygous Familial Hypercholesterolemia: Failure of Normal Allele to Compensate for Mutant Allele at a Regulated Genetic Locus," *Cell* 9 (1976): 195-203.

68. Brown and Goldstein, "Human Biochemical Genetics" (See note 64).

69. J. L. Goldstein and M. S. Brown, "Familial Hypercholesterolemia: Pathogenesis of a Receptor Disease," *Johns Hopkins Medical Journal* 143 (1978)): 8-16.

70. Brown and Goldstein, "Familial Hypercholesterolemia: Model" (See note 11).

71. D. M. Knipe, D. Baltimore, and H. F. Lodish, "Separate Pathways of Maturation of the Major Structural Proteins of Vesicular Stomatitis Virus," *Journal of Virology* 21 (1977): 1128-39.

IV. TOXINS, DRUGS, AND PARASITES

ROLE FOR ADENOSINE DIPHOSPHATE (ADP)-RIBOSYLATION IN THE REGULATION OF ADENYLATE CYCLASE BY BACTERIAL TOXINS AND AVIAN ENZYMES

Joel Moss
and Martha Vaughan *

SUMMARY

The effects of choleragen, an enterotoxin of *Vibrio cholerae,* on vertebrate cells are believed to result from an increase in intracellular adenosine 3′,5′-monophosphate (cyclic AMP), secondary to activation of adenylate cyclase. Studies by D. M. Gill established that the in vitro activation of adenylate cyclase required nicotinamide-adenine dinucleotide (NAD). The subsequent demonstration that choleragen catalyzes the hydrolysis of NAD to adenosine diphosphate-ribose (ADP-ribose) and nicotinamide, as well as the transfer of ADP-ribose from NAD to arginine and several proteins, provided a basis for the hypothesis that adenylate cyclase activation results from a choleragen-catalyzed ADP-ribosyltransferase reaction in which NAD and an unidentified cellular protein serve as substrates. Further support for this proposal was obtained from studies of the *E. coli* heat-labile enterotoxin, a protein structurally similar to but distinct from choleragen. This enterotoxin, which is thought to be responsible in part for the symptomatology of "traveler's diarrhea," also

* Dr. Vaughan, head of the Laboratory of Cellular Metabolism, National Heart, Lung, and Blood Institute, did not attend the conference. The authors wish to thank Miss Sally J. Stanley for expert technical assistance, and Mrs. Marie Sherwood for her assistance in preparing this manuscript.

appears to exert its effects on cells by activating adenylate cyclase. It, too, requires NAD for activation of adenylate cyclase in vitro, and it possesses NAD glycohydrolase and ADP-ribosyltransferase activities. Although it is unclear whether animal cells utilize a similar enzymatic mechanism to activate adenylate cyclase, a transferase possessing catalytic properties similar to those of choleragen and the *E. coli* heat-labile enterotoxin has been isolated from avian erythrocytes. It acts as a NAD glycohydrolase and an ADP-ribosyltransferase using both arginine and proteins as ADP-ribose acceptors. This protein also will activate adenylate cyclase in a NAD-dependent reaction.

INTRODUCTION

Cholera is characterized by the production of a profuse watery diarrhea following infection of the intestinal tract with the microorganism, *Vibrio cholerae.* The symptomatology results from the action of an enterotoxin secreted by *V. cholerae.* This enterotoxin, called choleragen or cholera toxin, apparently binds to receptors on intestinal cells and causes an increase in intracellular cAMP. The accumulation of cAMP leads to enhanced secretion of fluid and electrolytes. The mechanism of choleragen action has not been completely deciphered. It is known that in addition to acting on intestinal cells choleragen can increase cAMP in almost all vertebrate cells studied.[1-18]

Choleragen is an oligomeric protein consisting of three dissimilar polypeptide chains, A_1, A_2, and B. The holotoxin is composed of one A_1 and one A_2, linked through a single disulfide bond, and 4 to 6 B subunits in noncovalent association (Table I).[19-38] The toxin initially binds to the cell surface receptor through a specific interaction with the monosialoganglioside G_{M1}, which acts as a "receptor." [39-50] The B subunit is responsible for the interaction of the toxin with the ganglioside.[51-66] The manner in which the bound toxin is subsequently internalized and interacts with the adenylate cyclase system remains unclear. Toxin bound to the cell surface does undergo patching and capping, and lateral motion on the surface may be involved in choleragen action.[67-71]

TABLE 1. CHOLERAGEN: STRUCTURE AND FUNCTION

$(A_1\text{-s-s-}A_2)B_{4-6}$

Peptide	Molecular weight [1]	Function [2]
A_1	~ 23,500 (1) [3]	Activation of adenylate cyclase
A_2	~ 5,500 (1)	Unknown
B	11,600 (4-6)	Binding to the cell receptor, probably ganglioside G_{M1}

NOTES: 1) See notes 19-38. 2) See notes 3, 36, 40, 54-66, 74, 77, 78, 84. 3) Numbers in parenthesis represent peptides per oligomer.

The intact choleragen oligomer appears to be required for maximal activity of the toxin on intact cells; the isolated A subunit possesses only a fraction of the activity of the holotoxin.[72-78] In cell-free particulate, membrane, or detergent-solubilized preparations, only the A subunit or A_1 peptide of choleragen is necessary,[79-84] along with NAD, as first demonstrated by Gill [85,86] and subsequently confirmed by other investigators.[87-99] Specifically, β-NAD is required; α-NAD, nicotinamide, and ADP-ribose were inactive in the pigeon erythrocyte system.[100] NADH, NADP, and NADPH were inactive in the rat liver system.[101] Some NAD analogues were active in crude membrane preparations, but their effectiveness may result from their conversion to NAD.

In its dependence on NAD, choleragen is similar to two other bacterial toxins, diphtheria and Pseudomonas toxins, which also require NAD for activity.[102-130] These toxins inhibit protein synthesis in susceptible cells and in cell-free systems.[131-142] The toxins cause inactivation of elongation factor II by a NAD-dependent mechanism; NAD apparently serves as a substrate in an ADP-ribosylation reaction (Reaction 1), in which the toxin catalyzes the transfer

$$\text{NAD} + \text{elongation factor II} \rightleftharpoons \text{ADP-ribose-elongation factor II} + \text{nicotinamide} + H^+ \qquad (1)$$

of the ADP-ribose moiety of NAD to an amino acid in elongation factor II.[143-151] In the absence of elongation factor II, both toxins can catalyze the hydrolysis of NAD to ADP-ribose and nicotinamide (Reaction 2).[152,153] The occurrence of this reaction indicates

$$\text{NAD} + H_2O \rightarrow \text{ADP-ribose} + \text{nicotinamide} + H^+, \qquad (2)$$

that activation of the ribosyl-nicotinamide bond of NAD does not require the presence of an appropriate acceptor protein.

It was evident that if choleragen utilized NAD as a substrate in an ADP-ribosylation reaction, similar in mechanism to that catalyzed by diphtheria and Pseudomonas toxins, then it should catalyze reactions analogous to Reactions 1 and 2. Unfortunately, in the case of choleragen, a purified protein acceptor for ADP-ribose analogous to elongation factor II was not available. Investigation of the mechanism of choleragen action, therefore, required that a model for Reaction 1 be discovered that is independent of a purified protein acceptor. Reaction 2, the hydrolysis of NAD, is clearly independent of an acceptor protein. Initial experiments were therefore designed to determine whether choleragen possessed NAD glycohydrolase activity.

NAD GLYCOHYDROLASE AND ADP-RIBOSYLTRANSFERASE ACTIVITIES OF CHOLERAGEN

Highly purified choleragen preparations hydrolyzed NAD to ADP-ribose and nicotinamide (Reaction 1, Table 2).[154,155] The rate of the reaction was enhanced by thiols such as dithiothreitol and glutathione (Table 3).[156] Initiation of the reaction with toxin results in a delay prior to attaining a constant reaction rate; incubation of the toxin with thiol prior to assay abolishes this delay (Figure 1). It is presumed that thiol acts by releasing the A_1 peptide of choleragen from the holotoxin. It had earlier been

TABLE 2. REACTIONS CATALYZED BY CHOLERAGEN, *E. Coli* HEAT-LABILE ENTEROTOXIN AND AVIAN ERYTHROCYTE TRANSFERASE

		Notes
I. ADP-ribosyltransferase		
A. NAD + protein ⟶	ADP-ribose-protein + nicotinamide + H^+	171, 180-183
B. NAD + L-arginine ⟶	ADP-ribose-L-arginine + nicotinamide + H^+	168, 169, 171, 180
II. NAD glycohydrolase		
A. NAD + H_2O ⟶	ADP-ribose + nicotinamide + H^+	154, 155, 168, 169, 180

TABLE 3. EFFECT OF ASSAY COMPONENTS ON NAD GLYCOHYDROLASE AND ADP-RIBOSYLTRANSFERASE ACTIVITIES OF CHOLERAGEN

Addition	Activity (nmol/90 min.) — Arg	+ Arg
I. A. Dithiothreitol (mM)		
0.3	3	2
3.0	13	37
30.0	76	235
B. Glutathione		
0.3	5	3
3.0	3	16
30.0	34	106

NOTE: Assay mixes contained 400 mM potassium phosphate (pH 7.0), 8 mM [carbonyl-^{14}C]NAD (39,800 cpm), 75 mM arginine (where indicated), choleragen (50 μg.), and thiol. Reactions were initiated with toxin, and run for 90 min. at 30°C. Two 0.1 ml. samples were removed and chromatographed on AG 1-X2 to isolate the [carbonyl-^{14}C]nicotinamide, as described previously. All assays were performed in duplicate.

established that cleavage of the disulfide linkage joining the A_1 and A_2 peptides was necessary for activation of adenylate cyclase by choleragen.[157,158] Thus the requirements for demonstration of NAD glycohydrolase activity and for adenylate cyclase activation appear to be similar, suggesting that these are functions of the same protein. In the presence of thiol, the NAD glycohydrolase activity was also enhanced by relatively high concentrations of potassium phosphate, NaCl, or sodium acetate in the assay.[159]

R. M. Tait and S. van Heyningen have suggested that the NAD glycohydrolase activity found in purified choleragen preparations is due to a contaminating enzyme(s).[160] The bulk of the activity in their choleragen preparation, which appears to be impure, did not migrate with the toxin protein on electrophoresis. In contrast, Moss and his coworkers found that all of the NAD glycohydrolase activity of choleragen was recovered with the toxin protein after electrophoresis and cochromatographed with the A subunit of the toxin on Bio-Gel P-60 in acid urea (Figure 2).[161] In addition, it has been observed that all the NAD glycohydrolase activity of choleragen is present in preparations of the A_1 pep-

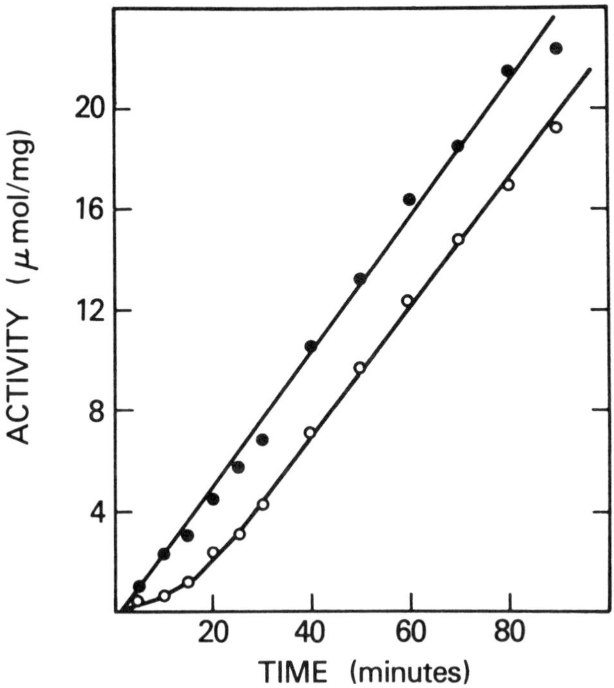

Figure 1. Effect of dithiothreitol on the rate of the choleragen-catalyzed ADP-ribosyltransferase reaction. Both assay mixes contained 1.2 mmol potassium phosphate (pH 7.0), 60 μmol dithiothreitol, 225 μmol arginine methyl ester, and choleragen (0.5 mg.) in a volume of 2.73 ml. Mix A (●——●) was incubated at 30°C for ten minutes prior to addition of 24 μmol of [carbonyl-^{14}C]NAD (4.6×10^5 cpm) in 0.27 ml.; Mix B (○——○) was not incubated at 30°C before addition of [carbonyl-^{14}C]NAD. At indicated times, two 0.1 ml. samples were withdrawn for isolation of [carbonyl-^{14}C]nicotinamide.

SOURCE: J. Moss, V. C. Manganiello, and M. Vaughan, "Hydrolysis of Nicotinamine Adenine Dinucleotide by Choleragen and Its A Protomer: Possible Role in the Activation of Adenylate Cyclase," *Proceedings of the National Academy of Sciences* 73 (1976): 4424-27.

tide, the turnover number of A_1 being identical to that of the holotoxin.[162] No activity was present in the A_2 [163] or B [164] peptides (Figure 2, Table 4). These results would suggest that, although crude choleragen preparations may contain NAD glycohydrolase activity that is not associated with the toxin, choleragen itself does possess an NAD glycohydrolase activity. This finding is not

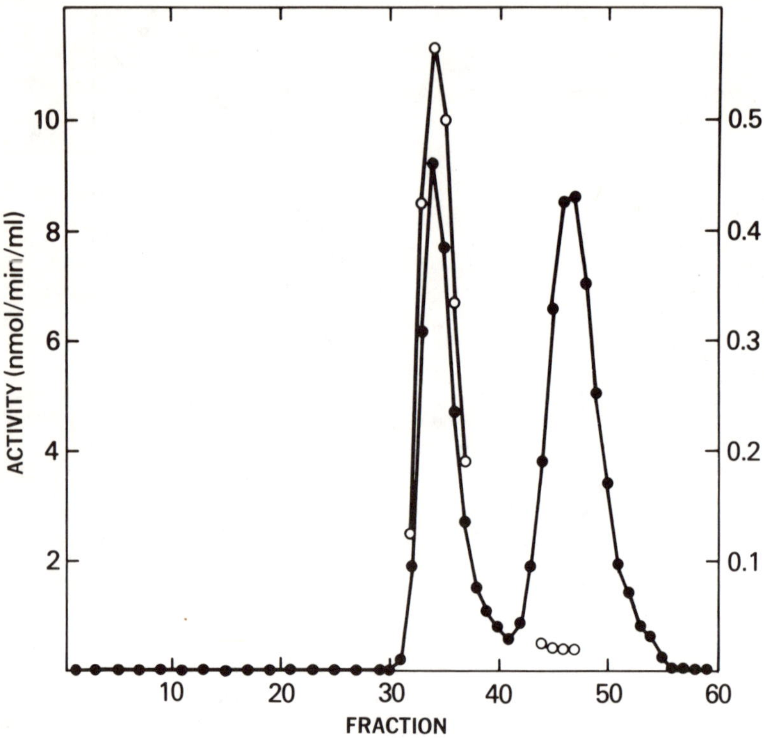

Figure 2. Cochromatography of the "A" subunit and the NAD glycohydrolase activity in choleragen preparations. Choleragen (5 mg. in 1 ml.) was dialyzed against 6.5 M urea, 0.1 M glycine, pH 3.2 according to Finkelstein and coworkers,[22] and chromatographed on a Bio-Gel P-60 column equilibrated with the same buffer. The optical density at 280 nanometer (O.D.$_{280\ nm}$) (●——●) was determined; the fractions containing the highest O.D.$_{280\ nm}$ were dialyzed against 50 mM Tris(Cl$^-$) (ph 7.4), 200 mM NaCl, 3 mM NaN_3, 1 mM EDTA, and assayed for NAD glycohydrolase activity (○——○) in the presence of 200 mM potassium phosphate (pH 7.0), 2 mM [carbonyl-^{14}C]NAD (33,800 cpm), and 20 mM dithiothreitol.

SOURCE: J. Moss, V. C. Manganiello, and M. Vaughan, "Hydrolysis of Nicotinamine Adenine Dinucleotide by Choleragen and Its A Protomer: Possible Role in the Activation of Adenylate Cyclase," *Proceedings of the National Academy of Sciences* 73 (1976): 4424-27.

unexpected in that all the other NAD-dependent ADP-ribosyltransferases, such as diphtheria toxin and Pseudomonas Exotoxin A, also catalyze NAD hydrolysis.[165,166] In addition, as will be noted later, a heat-labile enterotoxin from *E. coli,* which catalyzes the NAD-dependent activation of adenylate cyclase,[167] also

TABLE 4. ACTIVITY OF CHOLERAGEN PEPTIDES

Peptide	ADP-Ribosyltransferase Activity	Interaction with Surface Receptor	Notes
A_1	Yes	No	162
A_2	No	No	162
B	No	Yes	3, 36, 40, 54-66

possesses NAD glycohydrolase activity.[168] Thus, based on the analogy with other toxins, as well as the direct evidence obtained with purified choleragen preparations, it would appear that NAD glycohydrolase activity is intrinsic to choleragen and, in particular, to the A_1 peptide.

In addition to catalyzing the hydrolysis of NAD to ADP-ribose and nicotinamide, choleragen catalyzes the transfer of the ADP-ribose moiety of NAD to arginine (Reaction 1, Table 2) to yield ADP-ribose-L-arginine (Figure 3).[169] Several other guanidino compounds can also serve as ADP-ribose acceptors, as noted in Table 5.[170,171] Basic amino acids such as lysine and histidine, which do not have a guanidino group, are inactive.[172] Citrulline, in which the guanidino group of arginine has been replaced with a ureido group, is also inactive.[173] Thus the importance of the guanidino moiety in determining the ability of a compound to serve as an ADP-ribose acceptor is clear. In fact guanidine itself is active, suggesting that the remainder of the molecule is not critical for acceptor activity. Molecular structure in the vicinity of the guanidino moiety does, however, influence acceptor activity. Arginine methyl ester, in which the carboxyl moiety of arginine is blocked, and agmatine, in which the carboxyl is

TABLE 5. ACTIVITY OF GUANIDINO DERIVATIVES AND RELATED COMPOUNDS IN THE CHOLERAGEN-CATALYZED ADP-RIBOSYLTRANSFERASE REACTION

Active	Inactive	Notes
Agmatine $\geq$	Lysine	169, 171
Arginine methyl ester $>$	Histidine	
D-arginine $\cong$	Citrulline	
L-arginine $\geq$	Ornithine	
Guanidine		

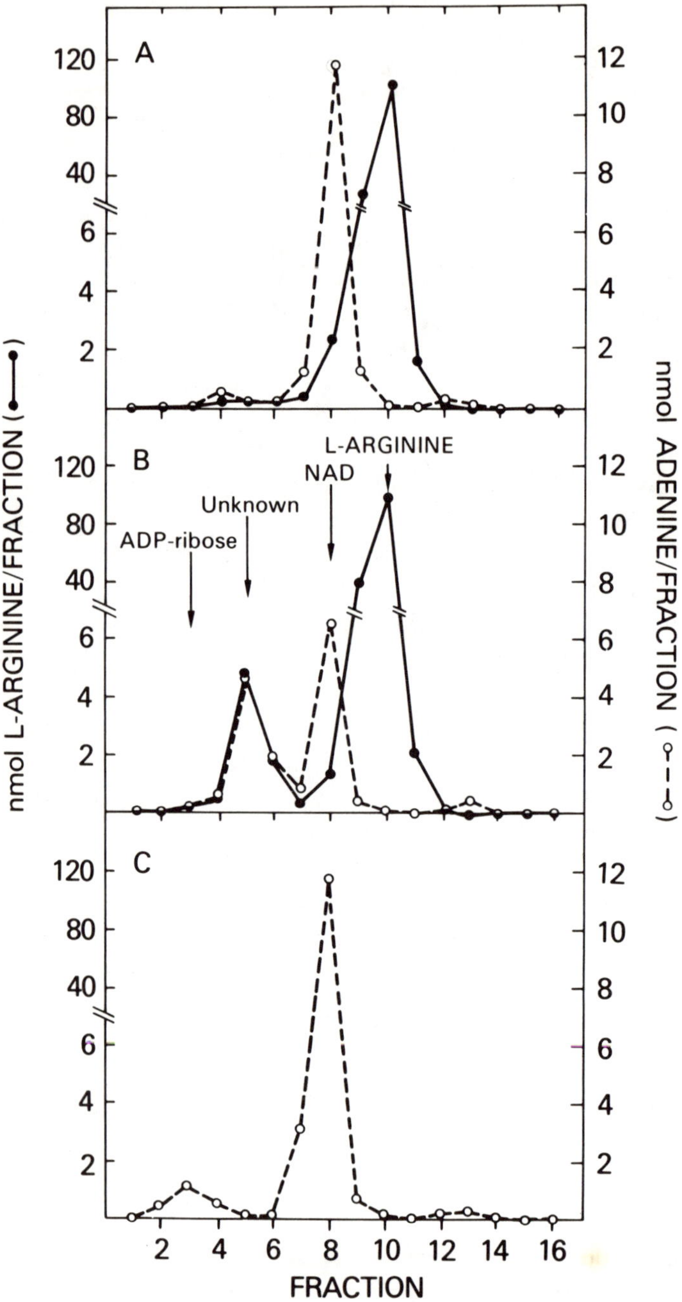

A
B
C
L-ARGININE
NAD
Unknown
ADP-ribose
nmol L-ARGININE/FRACTION
nmol ADENINE/FRACTION
FRACTION

absent, are both more effective acceptors than arginine. These results are consistent with the presumption that an arginine moiety in a protein is the cellular ADP-ribose acceptor. In that case both the carboxyl and amino groups would likely be involved in peptide linkages. Hence arginine methyl ester and agmatine might more effectively mimic the natural acceptor than does free arginine. Choleragen is unstable at low protein concentrations in the transferase assay; stability can be enhanced by addition of protein or nonionic detergents.[174]

N. J. Oppenheimer investigated the stereochemistry of the ADP-ribosylation of arginine and determined that the product is the α-anomer.[175] The choleragen-catalyzed reaction thus has a stereospecificity like that of the reaction catalyzed by poly(ADP-ribose) synthetases.[176] The α-ADP-ribose-L-arginine anomerizes relatively rapidly, however, and a mixed population of the α- and β-forms results.[177]

In addition to utilizing arginine as an ADP-ribose acceptor, choleragen catalyzes the ADP-ribosylation of numerous proteins, including lysozyme, polyarginine, histones, glycopeptide hormones, and the A subunit of the toxin itself.[178-180] Choleragen also catalyzes the ADP-ribosylation of several membrane proteins of different molecular weights.[181-184] It would appear that the reaction is an ADP-ribosylation rather than a poly(ADP-ribosylation) reaction, since digestion of the protein reaction product, labeled

---

Figure 3. Identification of the reaction product formed from [adenine-U-^{14}C]NAD and [^{3}H]-L-arginine in the presence of choleragen. Reaction mixes (final volume 0.295 ml.) were prepared containing 50.8 mM potassium phosphate (pH 7.0), 10 mM [adenine-U-^{14}C]NAD (4.8×10^5 cpm), 20.3 mM dithiothreitol, 76.3 mM [^{3}H]-L-arginine (2.9×10^6 cpm), and choleragen (250 μg.), where indicated. After 220 minutes at 37°C, 2 μl. were applied to DEAE-cellulose thin-layer plates (Brinkmann Instruments) and developed for three hours in isobutyric acid:NH_4OH:H_2O, 66:1:33. The plates were then sliced and counted. Frame A: [^{3}H]-L-arginine alone; frame B: [^{3}H]-L-arginine and choleragen; frame C: choleragen alone. Origin = fraction 1; Soluent Front = Fraction 16. A new compound was generated from [adenine-^{14}C]NAD and [^{3}H]-L-arginine by incubation with choleragen that contained adenine and arginine in a 1:1 ratio. Similar experiments with [^{32}P]NAD confirm the presence of ^{32}P in the compound and support its identification as ADP-ribose-L-arginine.

SOURCE: J. Moss and M. Vaughan, "Mechanism of Action of Choleragen. Evidence for ADP-Ribosyltransferase Activity with Arginine as an Acceptor," *Journal of Biological Chemistry* 252 (1977): 2455-57.

with an [adenine-U-^{14}C]NAD substrate, resulted in the release of only 5′-AMP rather than 5′-AMP and 2′-(5″-phosphoribosyl)-5′-AMP.[185] Although it has not been determined whether the acceptor amino acid in the protein is an arginine, in the cases of lysozyme, histones, and polyarginine it has been shown that arginine can inhibit the ADP-ribosylation, whereas other amino acids that are not ADP-ribose acceptors, and presumably do not gain access to the catalytic site, do not block the ADP-ribosylation.[186]

ROLE OF ADP-RIBOSYLATION IN CHOLERAGEN ACTIVATION OF ADENYLATE CYCLASE

In view of the identification of NAD glycohydrolase and ADP-ribosyltransferase activities as intrinsic properties of choleragen, it has been postulated that the toxin exerts its effects on cells by catalyzing the ADP-ribosylation of a protein that is critical to the activation of adenylate cyclase. Difficulty has been encountered in the identification of the natural ADP-ribose acceptor substrate for the choleragen-catalyzed reaction. In part the problem is related to the ability of choleragen to catalyze the NAD-dependent ADP-ribosylation of almost all proteins examined. It is important therefore to be able to assess the function of the ADP-ribosylated substrate in order to evaluate its role in the activation of adenylate cyclase.

D. Cassel and T. Pfeuffer have obtained convincing evidence that the protein acceptor in the ADP-ribosylation reaction may be the GTP-binding protein component of the adenylate cyclase system.[187] Working with pigeon erythrocytes, they observed that activation of adenylate cyclase by choleragen required the presencc of a guanine nucleotide. When a choleragen-treated, solubilized membrane preparation was treated with GTP-Sepharose, responsiveness of the adenylate cyclase to guanine nucleotides was diminished. Elution of the GTP-Sepharose with guanine nucleotide yielded a fraction that, when added to the cyclase, restored responsiveness to guanine nucleotides. This fraction contained a GTP-binding protein of 42,000 daltons that could be ADP-ribosylated by choleragen. They postulated that this protein is the substrate in the choleragen-catalyzed activation of adenylate cyclase. Gill and R. Meren have also demonstrated ADP-ribosylation of what is probably the same protein,[188] although they have not reported any studies on its functional role. H. R. Bourne and coworkers have examined the labeling of membrane proteins from S49 mouse lymphoma cells in the presence of choleragen and [^{32}P]NAD.[189] Several proteins appear to be labeled in the wild-type cells. In an adenylate cyclase-deficient variant, which is believed to lack the components necessary for activation by choleragen, guanine nucleotides, or NaF, minimal toxin-related radiolabeling was observed. These findings are consistent with a functional role for an ADP-ribosylated protein in the activation of adenylate cyclase by choleragen.

It is not clear whether the apparent lack of specificity of choleragen and its ability to ADP-ribosylate a number of cellular proteins reflect accurately what occurs in the intact cell or are artifacts of the model systems studied thus far. Similarly, the relatively rapid anomerization of the α-ADP-ribosyl-L-arginine produced in the model reaction catalyzed by choleragen [190] may or may not be representative of the stability of the structure in an ADP-ribosylated protein, where molecular architecture may interfere with anomerization. If anomerization does occur, however, enzymatic removal of the ADP-ribose may be hindered, since a cellular enzyme would presumably be specific for either the α- or β-anomer. Conceivably, two proteins with differing specificity may be present. Alternatively, anomerization may explain the persistence of choleragen activation of adenylate cyclase.

A possible mechanism for the effect of choleragen on adenylate cyclase has been proposed by D. Cassel and Z. Selinger.[191] It had been observed in several systems that guanylyl imidodiphosphate (Gpp(NH)p), a nonhydrolyzable derivative of GTP, enhanced adenylate cyclase activity, whereas GTP had little or no effect. Following treatment with choleragen, however, the effects of Gpp(NH)p and GTP were comparable.[192-195] Thus GTP appeared to behave as if its hydrolysis were diminished. Cassel and Selinger found that isoproterenol, a β-adrenergic agonist, increased both adenylate cyclase activity and the hydrolysis of

GTP in turkey erythrocyte membranes.[196] Choleragen further enhanced the activity of the isoproterenol-stimulated adenylate cyclase, but inhibited GTP hydrolysis. They proposed that a specific GTPase was associated with the GTP-binding site on adenylate cyclase; since GTP enhances adenylate cyclase activity in the presence of choleragen, they proposed that choleragen exerts its effects by inhibiting GTP hydrolysis at the specific site.[197]

SIMILARITIES BETWEEN CHOLERAGEN AND *E. COLI* HEAT-LABILE ENTEROTOXIN

Certain strains of *E. coli* that appear to be responsible for the symptoms of "traveler's diarrhea" result, at least in part, from the effects of two enterotoxins, one heat-labile and one heat-stable, secreted by the microorganism.[198-203] The heat-labile enterotoxin is a protein that has been isolated as species with molecular weights varying from 20,000 to several million.[204-209] Recent reports suggest that the toxin may be structurally similar to choleragen and possess subunits analogous to the A and B protomers of choleragen.[210] The effects of the heat-labile enterotoxin apparently result from the accumulation of intracellular cAMP secondary to activation of adenylate cyclase.[211-216] Thus the elevation in cAMP may be central to the effects of both choleragen and the heat-labile enterotoxin.

The mechanism of action of the heat-labile enterotoxin appears to be similar to that of choleragen; it, too, requires the presence of NAD for adenylate cyclase activation.[217] In addition it utilizes NAD as a substrate in the same three reactions catalyzed by choleragen (Table 2): 1) the hydrolysis of NAD to ADP-ribose and nicotinamide;[218] 2) the stereospecific β-NAD-dependent ADP-ribosylation of L-arginine to form α-ADP-ribose-L-arginine;[219,220] and 3) the ADP-ribosylation of proteins.[221] With both toxins there was a lag in reaction rate when the reaction was initiated with toxin, as well as a dependence on dithiothreitol or other thiol for maximal enzymatic activity (Table 6).[222,223]

There appear to be significant differences between the toxins in both the affinities for the reactants and the optimal assay con-

TABLE 6. COMPARISON OF NAD GLYCOHYDROLASE AND THE ADP-RIBOSYLTRANSFERASE ACTIVITIES OF CHOLERAGEN AND *E. Coli* HEAT-LABILE ENTEROTOXIN

Characteristic[1]	*E. coli* Heat-Labile Enterotoxin	Choleragen	Notes
Lag in activity with holotoxin	Yes	Yes	168
Dithiothreitol requirement	Yes	Yes	154, 168
Effect of high phosphate concentrations on activity	Slight inhibition	Stimulation	154, 168 169
Activity at pH 6.2 relative to 7.0	Lower	Same	154, 168
K_m for arginine methyl ester	240 mM	50 mM	168, unpublished data
K_m for NAD	8 mM	4 mM	154, 168

NOTE: 1) Assay conditions optimal for *E. coli* heat-labile enterotoxin or choleragen.

ditions.[224-227] The *E. coli* heat-labile enterotoxin is inhibited by high ionic strength, whereas choleragen is stimulated. *E. coli* enterotoxin is less active at pH 6.2 than at 7.0, whereas choleragen shows identical activity at both pHs (Table 6). Furthermore, the K_m for arginine methyl ester is approximately four times higher for the *E. coli* enterotoxin than it is for choleragen. Thus, although recent evidence would suggest that the catalytic unit of *E. coli* heat-labile enterotoxin is similar in size to the A subunit of choleragen, the kinetic data indicate that there are probably structural differences between the two toxins.[228-231] The fact that two structurally distinct toxins catalyze the NAD-dependent activation of adenylate cyclase, as well as possess NAD glycohydrolase and ADP-ribosyltransferase activity, suggests that the ADP-ribosylation of an arginine may indeed be a model for the reaction critical to the activation of adenylate cyclase.

ISOLATION OF AN ENZYME FROM AVIAN ERYTHROCYTE POSSESSING ADP-RIBOSYLTRANSFERASE ACTIVITY

Although the evidence presented previously would suggest that bacterial toxins activate adenylate cyclase by catalyzing the

ADP-ribosylation of a cellular protein critical to cyclase activity, it remains unclear whether animal cells utilize a similar mechanism to control cyclase activity. An avian erythrocyte enzyme has recently been discovered that catalyzes: 1) the hydrolysis of NAD to ADP-ribose and nicotinamide; and 2) the NAD-dependent ADP-ribosylation of arginine and proteins (Table 2).[232] The ADP-ribosylation of L-arginine is stereospecific and results in the formation of α-ADP-ribose-L-arginine.[233] Thus the avian ADP-ribosyltransferase appears to be catalytically equivalent to both choleragen and *E. coli* heat-labile enterotoxin.[234, 235] There are, however, significant differences between the bacterial toxins and the avian erythrocyte transferase. The transferase is an intracellular enzyme with an approximate molecular weight estimated by gel filtration chromatography of 25,000,[236] that is, slightly greater than those of the catalytic units of both toxins (Table 1).[237] The most striking functional difference between the transferases is in their affinities for NAD. The K_m for NAD, which is only 30 μM for the avian enzyme, is 4 and 8 mM for choleragen and *E. coli* heat-labile enterotoxin, respectively.[238-240] The avian erythrocyte transferase, like choleragen and the *E. coli* toxin, catalyzes the NAD-dependent activation of adenylate cyclase (Table 7).[241]

TABLE 7. EFFECT OF NAD ON THE ACTIVATION OF ADENYLATE CYCLASE BY THE AVIAN ERYTHROCYTE TRANSFERASE

	Adenylate Cyclase Activity	
Transferase Protein	No NAD	2 mM NAD
	(pmol/min./mg. protein)	
None	36	35
26 μg.	34	49

NOTE: ADP-ribosyltransferase was purified from turkey erythrocytes as described previously.[179] Samples of the adenylate cyclase preparation (particulate fraction from bovine brain, 225 μg. of protein) were incubated with 2 mM ATP, 10 mM $Mg.Cl_2$, 10 mM Tris(Cl^-) (pH 8.0), and/or ADP-ribosyltransferase, as indicated in a total volume of 60 μl. for 10 min. at 30°C. Adenylate cyclase assays were then initiated with the addition of 40 μl. of a solution, such that the final reaction system contained 2.2 mM [^{32}P]ATP (110,000 cpm), 11 mM $Mg.Cl_2$, 31 mM Tris(Cl) (pH 8.0), and 100 μm. GTP. After incubation for 10 min. at 30°C, cAMP was isolated by the method of Y. Salomon and coworkers for radioassay.[242]

Thus it would appear that the avian enzyme is similar to the bacterial toxins in that it can utilize the cellular protein, which is critical to cyclase activity, as an ADP-ribose acceptor. It remains to be determined whether that protein is a natural acceptor for the avian transferase in the intact erythrocyte, or whether it is ADP-ribosylated only under the special conditions used in vitro with the purified enzyme.

NOTES

1. R. A. Finkelstein, "Cholera," *CRC Critical Reviews in Microbiology* 2 (1973): 553-623.

2. J. Holmgren, "Immunologic and Receptor-Binding Properties of Cholera Toxin and Its Subunits," *Proceedings of the 9th U.S.-Japan Cholera Conference* (Washington: U. S. Department of State Publication 8762, 1973): 196-213.

3. W. E. van Heyningen, "Gangliosides as Membrane Receptors for Tetanus Toxin, Cholera Toxin and Serotonin," *Nature* 249 (1974): 415-17.

4. R. A. Finkelstein, "Immunology of Cholera," *Current Topics in Microbiology and Immunology* 69 (1975): 137-96.

5. V. Bennett et al., "Structure and Function of Cholera Toxin and Hormone Receptors," *Journal of Supramolecular Structure* 4 (1976): 99-120.

6. V. Bennett and P. Cuatrecasas, "Cholera Toxin Receptors," in *Methods in Receptor Research* pt. 1, ed. M. Blecher (New York: Marcel Dekker, Inc., 1976): 73-98.

7. ———, "Cholera Toxin: Membrane Gangliosides and Activation of Adenylate Cyclase," in *The Specificity and Action of Animal, Bacterial and Plant Toxins. Receptors and Recognition* series B, vol. 1, ed. P. Cuatrecasas (London: Chapman and Hall, 1976): 1-66.

8. R. A. Finkelstein, "Possibilities of Immunization Against Cholera and Related Enterotoxic Enteropathies," in *Proceedings of the 14th Congress of International Association of Biological Standardization, Douglas, Isle of Man 1975. Development of Biological Standardization* vol. 33, acting eds. R. H. Regamey et al. (Basel: Karger, 1976): 102-07.

9. ———, "Progress in the Study of Cholera and Related Enterotoxins," in *Mechanisms in Bacterial Toxinology*, ed. A. W. Bernheimer (New York: John Wiley and Sons, Inc., 1976): 54-84.

10. J. Holmgren and L. Lindholm, "Cholera Toxin, Ganglioside Receptors and the Immune Response," *Immunological Communications* 5 (1976): 737-56.

11. I. Lönnroth, "Cholera Toxin. Structure-Function Relationship," (Thesis, Institute of Medical Microbiology, University of Göteborg, Sweden, 1976).

12. W. E. van Heyningen and C. A. King, "The Role of Gangliosides in the Action of Cholera Toxin," in *Ganglioside Function: Biochemical and Pharmacological Implications*, ed. G. Porcellati, B. Ceccarelli, and G. Tettamanti (New York: Plenum Publishing Corporation, 1976): 205-14.

13. W. E. van Heyningen, S. van Heyningen, and C. A. King, "The Nature and Action of Cholera Toxin," in *Acute Diarrhoea in Childhood, Ciba Foundation*

Symposium vol. 42, ed. K. Elliott and J. Knight (Amsterdam: Elsevier Publishing Co., 1976): 73-88.

14. D. M. Gill, "Mechanism of Action of Cholera Toxin," in *Advances in Cyclic Nucleotide Research* vol. 8, ed. P. Greengard and G. A. Robison (New York: Raven Press, 1977): 85-118.

15. S. van Heyningen, "Cholera Toxin," *Biological Reviews* 52 (1977): 509-49.

16. J. Holmgren, "Cholera Toxin and the Cell Membrane," in *Bacterial Toxins and Cell Membranes,* ed. J. Jeljaszewicz and T. Wadström (New York: Academic Press, 1978): 333-66.

17. J. Holmgren and A.-M. Svennerholm, "Mechanisms of Disease and Immunity in Cholera: A Review," *Journal of Infectious Diseases* 136, suppl. (1977): S105-12.

18. M. Vaughan and J. Moss, "Mechanisms of Action of Choleragen," *Journal of Supramolecular Structure* 8 (1978): 473-88.

19. R. A. Finkelstein, M. K. LaRue, and J. J. LoSpalluto, "Properties of the Cholera Exo-Enterotoxin: Effects of Dispersing Agents and Reducing Agents in Gel Filtration and Electrophoresis," *Infection and Immunity* 6 (1972): 934-44.

20. J. J. LoSpalluto and R. A. Finkelstein, "Chemical and Physical Properties of Cholera Exo-Enterotoxin (Choleragen) and its Spontaneously Formed Toxoid (Choleragenoid)," *Biochimica et Biophysica Acta* 257 (1972): 158-66.

21. I. Lönnroth and J. Holmgren, "Subunit Structure of Cholera Toxin," *Journal of General Microbiology* 76 (1973): 417-27.

22. R. A. Finkelstein et al., "Dissociation and Recombination of the Subunits of the Cholera Enterotoxin (Choleragen)," *Journal of Immunology* 113 (1974): 145-50.

23. J. W. Jacobs, H. D. Niall, and G. W. G. Sharp, "The Amino Terminal Sequence of Cholera Toxin Subunits," *Biochemical and Biophysical Research Communications* 61 (1974): 391-95.

24. S. van Heyningen, "Cholera Toxin: Interaction of Subunits with Ganglioside G_{M1}," *Science* 183 (1974): 656-57.

25. J. Holmgren and I. Lönnroth, "Oligomeric Structure of Cholera Toxin: Characteristics of H and L Subunits," *Journal of General Microbiology* 86 (1975): 49-65.

26. E. Mendez, C. Y. Lai, and A. Wodnar-Filipowicz, "Location and the Primary Structure Around the Disulfide Bonds in Cholera Toxin," *Biochemical and Biophysical Research Communications* 67 (1975): 1435-43.

27. J. Sattler et al., "Studies of the Subunit Structure of Choleragen," *European Journal of Biochemistry* 57 (1975): 309-16.

28. D. M. Gill, "The Arrangement of Subunits in Cholera Toxin," *Biochemistry* 15 (1976): 1242-48.

29. D. G. Klapper, R. A. Finkelstein, and J. D. Capra, "Subunit Structure and N-Terminal Amino Acid Sequence of the Three Chains of Cholera Enterotoxin," *Immunochemistry* 13 (1976): 605-11.

30. A. Kurosky et al., "Chemical Characterization of the Structure of Cholera Toxin and Its Natural Toxoid," *Journal of Infectious Diseases* 133, suppl. (1976): S14-S22.

31. C. Y. Lai, E. Mendez, and D. Chang, "Chemistry of Cholera Toxin: The Subunit Structure," *Journal of Infectious Diseases* 133, suppl. (1976): S23-S30.

32. N. Ohtomo et al., "Size and Structure of the Cholera Toxin Molecule and Its Subunits," *Journal of Infectious Diseases* 133, suppl. (1976): S31-S40.

33. S. van Heyningen, "The Subunits of Cholera Toxin: Structure, Stoichiometry, and Function," *Journal of Infectious Diseases* 133, suppl. (1976): S5-S13.

34. A. Kurosky, D. E. Markel, and J. W. Peterson, "Covalent Structure of the β Chain of Cholera Enterotoxin," *Journal of Biological Chemistry* 252 (1977): 7257-64.

35. C.-Y. Lai, "Determination of the Primary Structure of Cholera Toxin B Subunit," *Journal of Biological Chemistry* 252 (1977): 7249-56.

36. J. J. Mekalanos, R. J. Collier, and W. R. Romig, "Simple Method for Purifying Choleragenoid, the Natural Toxoid of *Vibrio Cholerae*," *Infection and Immunity* 16 (1977): 789-95.

37. C. Y. Lai et al., "Primary Structure of Cholera Toxin B-Subunit," *Biochemical and Biophysical Research Communications* 74 (1977): 215-22.

38. P. B. Sigler et al., "Cholera Toxin Crystals Suitable for X-Ray Diffraction," *Science* 197 (1977): 1277-79.

39. P. Cuatrecasas, "Gangliosides and Membrane Receptors for Cholera Toxin," *Biochemistry* 12 (1973): 3558-66.

40. D. M. Gill and C. A. King, "The Mechanism of Action of Cholera Toxin in Pigeon Erythrocyte Lysates," *Journal of Biological Chemistry* 250 (1975): 6424-32.

41. J. Holmgren et al., "Interaction of Cholera Toxin and Membrane G_{M1} Ganglioside of Small Intestine," *Proceedings of the National Academy of Sciences* 72 (1975): 2520-24.

42. T. Révész and M. Greaves, "Ligand-Induced Redistribution of Lymphocyte Membrane Ganglioside GM1," *Nature* 257 (1975): 103-06.

43. M. Basu et al., "Lectin and Cholera Toxin Binding to Guinea Pig Tumor (104C1) Cell Surfaces before and after Glycosphingolipid Incorporation," *Biochemical and Biophysical Research Communications* 71 (1976): 385-92.

44. J. N. Kanfer, T. P. Carter, and H. M. Katzen, "Lipolytic Action of Cholera Toxin on Fat Cells. Re-examination of the Concept Implicating G_{M1} Ganglioside as the Native Membrane Receptor," *Journal of Biological Chemistry* 251 (1976): 7610-19.

45. C. A. King, W. E. van Heyningen, and N. Gascoyne, "Aspects of the Interaction of *Vibrio Cholerae* Toxin with the Pigeon Red Cell Membrane," *Journal of Infectious Diseases* 133, suppl. (1976): S75-S81.

46. L. Manuelidis and E. E. Manuelidis, "Cholera Toxin-Peroxidase: Changes in Surface Labeling of Glioblastoma Cells with Increased Time in Tissue Culture," *Science* 193 (1976): 588-90.

47. J. Moss et al., "Functional Incorporation of Ganglioside into Intact Cells: Induction of Choleragen Responsiveness," *Proceedings of the National Academy of Sciences* 73 (1976): 1034-37.

48. T. Révész et al., "Differential Expression of Cell Surface Binding Sites for Cholera Toxin in Acute and Chronic Leukaemias," *British Journal of Pharmacology* 34 (1976): 623-30.

49. H.-A. Hansson, J. Holmgren, and L. Svennerholm, "Ultrastructural Localization of Cell Membrane G_{M1} Ganglioside by Cholera Toxin," *Proceedings of the National Academy of Sciences* 74 (1977): 3782-86.

50. S. van Heyningen, "Activity of Covalently Cross-Linked Cholera Toxin with

the Adenylate Cyclase of Intact and Lysed Pigeon Erythrocytes," *Biochemical Journal* 168 (1977): 457-63.

51. Van Heyningen, "Ganglisides as Receptors" (See note 3).

52. Mekalanos, Collier, and Romig, "Method for Purifying Choleragenoid" (See note 36).

53. Gill and King, "Action of Cholera Toxin" (See note 40).

54. P. Cuatrecasas, "*Vibrio Cholerae* Choleragenoid. Mechanism of Inhibition of Cholera Toxin Action," *Biochemistry* 12 (1973): 3577-81.

55. J. Holmgren, "Comparison of the Tissue Receptors for *Vibrio Cholerae* and *Escherichia Coli* Enterotoxins by Means of Gangliosides and Natural Cholera Toxoid," *Infection and Immunity* 8 (1973): 851-59.

56. J. Holmgren, I. Lönnroth, and L. Svennerholm, "Tissue Receptor for Cholera Exotoxin: Postulated Structure from Studies with G_{M1} Ganglioside and Related Glycolipids," *Infection and Immunity* 8 (1973): 208-14.

57. L. M. Lichtenstein et al., "Effects of Cholera Toxin on in Vitro Models of Immediate and Delayed Hypersensitivity. Further Evidence for the Role of Cyclic Adenosine 3',5'-Monophosphate," *Journal of Clinical Investigation* 52 (1973): 691-97.

58. N. F. Pierce, "Differential Inhibitory Effects of Cholera Toxoids and Ganglioside on the Enterotoxins of *Vibrio Cholerae* and *Escherichia Coli,*" *Journal of Experimental Medicine* 137 (1973): 1009-23.

59. M. Field, "Mode of Action of Cholera Toxin: Stabilization of Catecholamine-Sensitive Adenylate Cyclase in Turkey Erythrocytes," *Proceedings of the National Academy of Sciences* 71 (1974): 3299-3303.

60. W. A. Walker, M. Field, and K. J. Isselbacher, "Specific Binding of Cholera Toxin to Isolated Intestinal Microvillous Membranes," *Proceedings of the National Academy of Sciences* 71 (1974): 320-24.

61. V. Bennett and P. Cuatrecasas, "Mechanism of Action of *Vibrio Cholerae* Enterotoxin," *Journal of Membrane Biology* 22 (1975): 1-28.

62. V. Bennett, E. O'Keefe, and P. Cuatrecasas, "Mechanism of Action of Cholera Toxin and the Mobile Receptor Theory of Hormone Receptor-Adenylate Cyclase Interactions," *Proceedings of the National Academy of Sciences* 72 (1975): 33-37.

63. A. Haksar, D. V. Maudsley, and F. G. Péron, "Stimulation of Cyclic Adenosine 3':5'-Monophosphate and Corticosterone Formation in Isolated Rat Adrenal Cells by Cholera Enterotoxin Comparison with the Effects of ACTH," *Biochimica et Biophysica Acta* 381 (1975): 308-23.

64. C. A. King and W. E. van Heyningen, "Evidence for the Complex Nature of the Ganglioside Receptor for Cholera Toxin," *Journal of Infectious Diseases* 131 (1975): 643-48.

65. I. Lönnroth and C. Lönnroth, "Interaction of Cholera Toxin and Its Subunits with Lymphocytes. The Effects on Intracellular cAMP," *Experimental Cell Research* 104 (1977): 15-24.

66. J. Sattler et al., "Studies of the Ligand Binding to Cholera Toxin. II. The Hydrophilic Moiety of Sialoglycolipids," *Hoppe-Seyler's Zeitschrift fur Physiologische Chemie* 358 (1977): 159-63.

67. Révész and Greaves, "Redistribution of Ganglioside GMI" (See note 42).

68. Hansson, Holmgren, and Svernnerholm, "Localization of Cell Membrane" (See note 49).

69. V. Bennett and P. Cuatrecasas, "Mechanism of Activation of Adenylate Cyclase by *Vibrio Cholerae* Enterotoxin," *Journal of Membrane Biology* 22 (1975): 29-52.

70. S. W. Craig and P. Cuatrecasas, "Mobility of Cholera Toxin Receptors on Rat Lymphocyte Membranes," *Proceedings of the National Academy of Sciences* 72 (1975): 3844-48.

71. H. H. Sedlacek et al., "Cholera Toxin Induced Redistribution of Sialoglycolipid Receptor at the Lymphocyte Membrane," *Federation of the European Biochemical Societies Letters* 61 (1976): 272-76.

72. van Heyningen, "Subunits of Cholera Toxin" (See note 33).

73. Gill and King, "Action of Cholera Toxin (See note 40).

74. N. Sahyoun and P. Cuatrecasas, "Mechanism of Activation of Adenylate Cyclase by Cholera Toxin," *Proceedings of the National Academy of Sciences* 72 (1975): 3438-42.

75. F. Berkenbile and R. Delaney, "Stimulation of Adenylate Cyclase by *Vibrio Cholerae* Toxin and Its Active Subunit," *Journal of Infectious Diseases* 133, suppl. (1976): S82-S88.

76. S. T. Donta, S. R. Kreiter, and G. Wendelschafer-Crabb, "Morphological and Steroidogenic Changes in Cultured Adrenal Tumor Cells Induced by a Subunit of Cholera Enterotoxin," *Infection and Immunity* 13 (1976): 1479-82.

77. S. van Heyningen, "Activation by Cholera Toxin of Adenylate Cyclase Solubilized from Rat Liver," *Biochemical Journal* 157 (1976): 785-87.

78. A. Wodnar-Filipowicz and C. Y. Lai, "Stimulation of Adenylate Cyclase in Washed Pigeon Erythrocyte Membrane with Cholera Toxin and Its Subunits," *Archives of Biochemistry and Biophysics* 176 (1976): 465-71.

79. Gill and King, "Action of Cholera Toxin (See note 40).

80. King and van Heyningen, "Nature of Ganglioside Receptor" (See note 64).

81. Sahyoun and Cuatrecasas, "Activation of Adenylate Cyclase" (See note 74).

82. van Heyningen, "Activation of Cholera Toxin" (See note 77).

83. Wodnar-Filipowicz and Lai, "Stimulation of Adenylate Cyclase" (See note 78).

84. S. van Heyningen and C. A. King, "Subunit A from Cholera Toxin Is an Activator of Adenylate Cyclase in Pigeon Erythrocytes," *Biochemical Journal* 146 (1975): 269-71.

85. D. M. Gill, "Involvement of Nicotinamide Adenine Dinucleotide in the Action of Cholera Toxin *In Vitro*," *Proceedings of the National Academy of Sciences* 72 (1975): 2064-68.

86. ———, "Multiple Roles of Erythrocyte Supernatant in the Activation of Adenylate Cyclase by *Vibrio Cholerae* Toxin *In Vitro*," *Journal of Infectious Diseases* 133, suppl. (1976): S55-S63.

87. Moss et al., "Induction of Choleragen Responsiveness" (See note 47).

88. van Heyningen, "Activation of Cholera Toxin" (See note 77).

89. Wodnar-Filipowicz and Lai, "Stimulation of Adenylate Cyclase" (See note 78).

90. M. W. Bitensky et al., "Cholera Toxin Activation of Adenylate Cyclase in Cancer Cell Membrane Fragments," *Proceedings of the National Academy of Sciences* 72 (1975): 2572-76.

91. J. Flores, P. Witkum, and G. W. G. Sharp, "Activation of Adenylate

Cyclase by Cholera Toxin in Rat Liver Homogenates," *Journal of Clinical Investigation* 57 (1976): 450-58.

92. Y. Matuo, M. A. Wheeler, and M. W. Bitensky, "Small Fragments from the A Subunit of Cholera Toxin Capable of Activating Adenylate Cyclase," *Proceedings of the National Academy of Sciences* 73 (1976): 2654-58.

93. B. R. Mullin et al., "Cholera Toxin Interactions with Thyrotropin Receptors on Thyroid Plasma Membranes," *Proceedings of the National Academy of Sciences* 73 (1976): 1679-83.

94. M. A. Wheeler et al., "Interaction of *Vibrio Cholerae* Toxin with Sarcoma 180 Cell Membranes," *Journal of Infectious Diseases* 133, suppl. (1976): S89-S96.

95. B. R. Martin, M. D. Houslay, and E. L. Kennedy, "Cholera Toxin Requires Oxidized Nicotinamide-Adenine Dinucleotide to Activate Adenylate Cyclase in Purified Rat Liver Plasma Membranes," *Biochemical Journal* 161 (1977): 639-42.

96. G. L. Johnson and H. R. Bourne, "Influence of Cholera Toxin on the Regulation of Adenylate Cyclase by GTP," *Biochemical and Biophysical Research Communications* 78 (1977): 792-98.

97. S. L. Levinson and A. J. Blume, "Altered Guanine Nucleotide Hydrolysis as Basis for Increased Adenylate Cyclase Activity after Cholera Toxin Treatment," *Journal of Biological Chemistry* 252 (1977): 3766-74.

98. J. Moss and M. Vaughan, "Choleragen Activation of Solubilized Adenylate Cyclase: Requirement for GTP and Protein Activator for Demonstration of Enzymatic Activity," *Proceedings of the National Academy of Sciences* 74 (1977): 4396-4400.

99. J. Fischer et al., "Studies on the Time Course and Rate-Limiting Step in the Activation of Adenylate Cyclase in Rat Liver by Cholera Toxin," *Biochemical Journal* 173 (1978): 59-64.

100. Gill, "Action of Cholera Toxin" (See note 85).

101. Martin, Houslay, and Kennedy, "Cholera Toxin Requires NAD" (See note 95).

102. N. Strauss and E. D. Hendee, "The Effect of Diphtheria Toxin on the Metabolism of HeLa Cells," *Journal of Experimental Medicine* 109 (1959): 145-63.

103. N. Strauss, "The Effect of Diphtheria Toxin on the Metabolism of HeLa Cells. II. Effect of Nucleic Acid Metabolism," *Journal of Experimental Medicine* 112 (1960): 351-59.

104. I. Kato and A. M. Pappenheimer, Jr., "An Early Effect of Diphtheria Toxin on the Metabolism of Mammalian Cells Growing in Culture," *Journal of Experimental Medicine* 112 (1960): 329-49.

105. I. Kato, "Mode of Action of Diphtheria Toxin on Protein Synthesis. I. Effect of Diphtheria Toxin on C^{14}-Amino Acids Incorporation into Microsomes and Mitochondria *in Vitro*," *Japanese Journal of Experimental Medicine* 32 (1962): 335-43.

106. R. J. Collier and A. M. Pappenheimer, Jr., "Studies on the Mode of Action of Diphtheria Toxin. II. Effect of Toxin on Amino Acid Incorporation in Cell-Free Systems," *Journal of Experimental Medicine* 120 (1964): 1019-39.

107. R. J. Collier, "Effect of Diphtheria Toxin on Protein Synthesis: Inactivation of One of the Transfer Factors," *Journal of Molecular Biology* 25 (1967): 83-98.

108. R. S. Goor and A. M. Pappenheimer, Jr., "Studies on the Mode of Action

of Diphtheria Toxin. III. Site of Toxin Action in Cell-Free Extracts," *Journal of Experimental Medicine* 126 (1967): 899-912.

109. __________, "Studies on the Mode of Action of Diphtheria Toxin. IV. Specificity of the Cofactor (NAD) Requirement for Toxin Action in Cell-Free Systems," *Journal of Experimental Medicine* 126 (1967): 913-21.

110. T. Honjo et al., "Diphtheria Toxin-Dependent Adenosine Diphosphate Ribosylation of Aminoacyl Transferase II and Inhibition of Protein Synthesis," *Journal of Biological Chemistry* 243 (1968): 3553-55.

111. R. J. Collier and H. A. Cole, "Diphtheria Toxin Subunit Active In Vitro," *Science* 164 (1969): 1179-82.

112. D. M. Gill et al., "Studies on the Mode of Action of Diphtheria Toxin. VII. Toxin-Stimulated Hydrolysis of Nicotinamide Adenine Dinucleotide in Mammalian Cell Extracts," *Journal of Experimental Medicine* 129 (1969): 1-21.

113. R. S. Goor and E. S. Maxwell, "The Diphtheria Toxin-Dependent Adenosine Diphosphate Ribosylation of Rat Liver Aminoacyl Transferase II," *Journal of Biological Chemistry* 245 (1970): 616-23.

114. R. J. Collier and J. Kandel, "Structure and Activity of Diphtheria Toxin. I. Thiol-Dependent Dissociation of a Fraction of Toxin into Enzymically Active and Inactive Fragments," *Journal of Biological Chemistry* 246 (1971): 1496-1503.

115. R. Drazin, J. Kandel, and R. J. Collier, "Structure and Activity of Diphtheria Toxin. II. Attack by Trypsin at a Specific Site within the Intact Toxin Molecule," *Journal of Biological Chemistry* 246 (1971): 1504-10.

116. D. M. Gill and L. L. Dinius, "Observations on the Structure of Diphtheria Toxin," *Journal of Biological Chemistry* 246 (1971): 1485-91.

117. D. M. Gill and A. M. Pappenheimer, Jr., "Structure-Activity Relationships in Diphtheria Toxin," *Journal of Biological Chemistry* 246 (1971): 1492-95.

118. T. Honjo et al., "Adenosine Diphosphate Ribosylation of Aminoacyl Transferase II and Inhibition of Protein Synthesis by Diphtheria Toxin," *Journal of Biological Chemistry* 246 (1971): 4251-60.

119. O. R. Pavlovskis and F. B. Gordon, *"Pseudomonas Aeruginosa* Exotoxin: Effect on Cell Culture," *Journal of Infectious Diseases* 125 (1972): 631-36.

120. J. Kandel, R. J. Collier, and D. W. Chung, "Interaction of Fragment A from Diphtheria Toxin with Nicotinamide Adenine Dinucleotide," *Journal of Biological Chemistry* 249 (1974): 2088-97.

121. O. R. Pavlovskis and A. H. Shackelford, *"Pseudomonas Aeruginosa* Exotoxin in Mice. Localization and Effects on Protein Synthesis," *Infection and Immunity* 9 (1974): 540-46.

122. E. A. Robinson, O. Henriksen, and E. S. Maxwell, "Elongation Factor 2. Amino Acid Sequence at the Site of Adenosine Diphosphate Ribosylation," *Journal of Biological Chemistry* 249 (1974): 5088-93.

123. R. J. Collier, "Diphtheria Toxin. Model of Action and Structure," *Bacteriological Reviews* 39 (1975): 54-85.

124. B. H. Iglewski and D. Kabat, "NAD-Dependent Inhibition of Protein Synthesis by *Pseudomonas Aeruginosa* Toxin," *Proceedings of the National Academy of Sciences* 72 (1975): 2284-88.

125. E. Bermek, "Interactions of Adenosine Diphosphate-Ribosylated Elongation Factor 2 with Ribosomes," *Journal of Biological Chemistry* 251 (1976): 6544-49.

126. S. H. Leppla, "Large-Scale Purification and Characterization of the Exotoxin of *Pseudomonas Aeruginosa,*" *Infection and Immunity* 14 (1976): 1077-86.

127. D. W. Chung and R. J. Collier, "Enzymatically Active Peptide from the Adenosine Diphosphate-Ribosylating Toxin of *Pseudomonas Aeruginosa,*" *Infection and Immunity* 16 (1977): 832-41.

128. B. H. Iglewski, P. V. Liu, and D. Kabat, "Mechanism of Action of *Pseudomonas Aeruginosa* Exotoxin A: Adenosine Diphosphate-Ribosylation of Mammalian Elongation Factor 2 in Vitro and in Vivo," *Infection and Immunity* 15 (1977): 138-44.

129. A. M. Pappenheimer, Jr., "Diphtheria Toxin," *Annual Review of Biochemistry* 46 (1977): 69-94.

130. M. L. Vasil, D. Kabat, and B. H. Iglewski. "Structure-Activity Relationships of an Exotoxin of *Pseudomonas Aeruginosa,*" *Infection and Immunity* 16 (1977): 353-61.

131. Strauss and Hendee, "Effect of Diphtheria Toxin" (See note 102).

132. Strauss, "Effect on Metabolism" (See note 103).

133. Kato and Pappenheimer, "Early Effect of Diphtheria Toxin" (See note 104).

134. Kato, "Action of Diphtheria Toxin" (See note 105).

135. Collier and Pappenheimer, "Effect in Cell-Free Systems" (See note 106).

136. Collier, "Inactivation of Transfer Factors" (See note 107).

137. Goor and Pappenheimer, "Site of Toxin Action" (See note 108).

138. Honjo et al., "Inhibition of Protein Synthesis" (See note 118).

139. Pavlovskis and Gordon, "Effect on Cell Culture" (See note 119).

140. Pavlovskis and Shackelford, "Effects on Protein Synthesis" (See note 121).

141. Iglewski and Kabat, "NAD-Dependent Inhibition" (See Note 124).

142. Leppla, "Exotoxin of *Pseudomonas Aeruginosa*" (See note 126).

143. Honjo et al., "Diphtheria Toxin-Dependent ADP-Ribosylation" (See note 110).

144. Collier and Cole, "Diphtheria Toxin Subunit" (See note 111).

145. Goor and Maxwell, "ADP-Ribosylation of Rat Liver" (See note 113).

146. Honjo et al., "Inhibition of Protein Synthesis" (See note 118).

147. Robinson, Henriksen, and Maxwell, "Amino Acid Sequence" (See note 122).

148. Collier, "Diphtheria Toxin" (See note 123).

149. Leppla, "Exotoxin of *Pseudomonas Aeruginosa*" (See note 126).

150. Iglewski, Liu, and Kabat, "Action of Exotoxin A" (See note 128).

151. Pappenheimer, "Diphtheria Toxin" (See note 129).

152. Kandel, Collier, and Chung, "Interaction of Fragment A" (See note 120).

153. Chung and Collier, "Enzymatically Active Peptide" (See note 127).

154. J. Moss, V. C. Manganiello, and M. Vaughan, "Hydrolysis of Nicotinamine Adenine Dinucleotide by Choleragen and Its A Protomer: Possible Role in the Activation of Adenylate Cyclase," *Proceedings of the National Academy of Sciences* 73 (1976): 4424-27.

155. J. Moss et al., "Effect of Gangliosides and Substrate Analogues on the Hydrolysis of Nicotinamine Adenine Dinucleotide by Choleragen," *Proceedings of the National Academy of Sciences* 74 (1977): 74-78.

156. Moss, Manganiello, and Vaughan, "Hydrolysis of NAD" (See note 154).

157. Wodnar-Filipowicz and Lai, "Stimulation of Adenylate Cyclase" (See note 78).

158. Gill, "Roles of Erythrocyte Supernatant" (See note 86).

159. Moss, Manganiello, and Vaughan, "Hydrolysis of NAD" (See note 154).

160. R. M. Tait and S. van Heyningen, "The Adenylate Cyclase-Activating Activity of Cholera Toxin Is Not Associated with a Nicotinamide-Adenine Dinucleotide Glycohydrolase Activity," *Biochemical Journal* 175 (1978): 1059-62.

161. Moss, Manganiello, and Vaughan, "Hydrolysis of NAD" (See note 154).

162. J. Moss, S. J. Stanley, and M. C. Lin, "NAD Glycohydrolase and ADP-Ribosyltransferase Activities Are Intrinsic to the A_1 Peptide of Choleragen," manuscript submitted for publication.

163. Ibid.

164. Moss, Manganiello, and Vaughan, "Hydrolysis of NAD" (See note 154).

165. Kandell, Collier, and Chung, "Interaction of Fragment A" (See note 120).

166. Chung and Collier, "Enzymatically Active Peptide" (See note 127).

167. D. M. Gill, D. J. Evans, Jr., and D. G. Evans, "Mechanism of Activation of Adenylate Cyclase In Vitro by Polymyxin-Released Heat-Labile Enterotoxin of *Escherichia Coli*," *Journal of Infectious Diseases* 133, suppl. (1976): S103-07.

168. J. Moss and S. H. Richardson, "Activation of Adenylate Cyclase by Heat-Labile *E. Coli* Enterotoxin: Evidence for ADP-Ribosyltransferase Activity Similar to that of Choleragen," *Journal of Clinical Investigation* 62 (1978): 281-85.

169. J. Moss and M. Vaughan, "Mechanism of Action of Choleragen. Evidence for ADP-Ribosyltransferase Activity with Arginine as an Acceptor," *Journal of Biological Chemistry* 252 (1977): 2455-57.

170. Ibid.

171. J. Moss et al., "NAD-Dependent ADP-Ribosylation of Arginine and Proteins by *Escherichia coli* Heat-Labile Enterotoxin," *Journal of Biological Chemistry*, in press.

172. Moss and Vaughan, "Action of Choleragen" (See note 169).

173. Ibid.

174. J. Moss, P. S. Ross, and M. Vaughan, "The NAD Glycohydrolase and ADP-Ribosyltransferase Activities of Choleragen: Effect of Membranes, Detergents, and Thiol on Catalytic Activity," *Proceedings of the 13th Joint U.S.-Japanese Conference on Cholera* (Washington: U.S. Department of Health, Education, and Welfare No. (NIH) 78-1590, 1978): 382-95.

175. N. J. Oppenheimer, "Structural Determination and Stereospecificity of the Choleragen-Catalyzed Reaction of NAD^+ with Guanidines," *Journal of Biological Chemistry* 253 (1978): 4907-10.

176. A. M. Ferro and N. J. Oppenheimer, "Structure of a Poly(Adenosine Diphosphoribose) Monomer: 2'-(5''-Phosphoribosyl)-5'-Adenosine Monophosphate," *Proceedings of the National Academy of Sciences* 75 (1978): 809-13.

177. Oppenheimer, "Structural Determination" (See note 175).

178. J. B. Trepel, D.-M. Chuang, and N. H. Neff. "Transfer of ADP-Ribose from NAD to Choleragen: A Subunit Acts as Catalyst and Acceptor Protein," *Proceedings of the National Academy of Sciences* 74 (1977): 5440-42.

179. ———, "ADP-Ribosyltransferase Activity of Choleragen: Choleragen A Subunit and Some Peptide Hormones Act as Acceptor Proteins," *Federation Proceedings* 37 (1978): 798.

180. J. Moss and M. Vaughan, "Isolation of an Avian Erythrocyte Protein

Possessing ADP-Ribosyltransferase Activity and Capable of Activating Adenylate Cyclase," *Proceedings of the National Academy of Sciences* 75 (1978): 3621-24.

181. G. L. Johnson, H. R. Kaslow, and H. R. Bourne, "Genetic Evidence That Cholera Toxin Substrates Are Regulatory Components of Adenylate Cyclase," *Journal of Biological Chemistry* 253 (1978): 7120-23.

182. D. Cassel and T. Pfeuffer, "Mechanism of Cholera Toxin Action: Covalent Modification of the Guanyl Nucleotide-Binding Protein of the Adenylate Cyclase System," *Proceedings of the National Academy of Sciences* 75 (1978): 2669-73.

183. D. M. Gill and R. Meren, "ADP-Ribosylation of Membrane Proteins Catalyzed by Cholera Toxin: Basis of the Activation of Adenylate Cyclase," *Proceedings of the National Academy of Sciences* 75 (1978): 3050-54.

184. D. M. Gill, "Cholera Toxin-Catalyzed ADP-Ribosylation of Erythrocyte Proteins: General Properties," *Journal of Supramolecular Structure,* in press.

185. Moss and Vaughan, "Isolation of Avian Erythrocyte Protein" (See note 180).

186. Ibid.

187. Cassel and Pfeuffer, "Cholera Toxin Action" (See note 182).

188. Gill and Meren, "ADP-Ribosylation of Membrane Proteins" (See note 183).

189. Johnson, Kaslow, and Bourne, "Genetic Evidence" (See note 181).

190. Oppenheimer, "Structural Determination" (See note 175).

191. D. Cassel and Z. Selinger, "Mechanism of Adenylate Cyclase Activation by Cholera Toxin: Inhibition of GTP Hydrolysis at the Regulatory Site," *Proceedings of the National Academy of Sciences* 74 (1977): 3307-11.

192. Johnson and Bourne, "Influence of Cholera Toxin" (See note 96).

193. Levinson and Blume, "Cholera Toxin Treatment" (See note 97).

194. Moss and Vaughan, "Choleragen Activation" (See note 98).

195. Cassel and Selinger, "Activation by Cholera Toxin" (See note 191).

196. ———, "Catecholamine-Stimulated GTPase Activity in Turkey Erythrocyte Membranes," *Biochimica et Biophysica Acta* 452 (1976): 538-51.

197. ———, "Activation by Cholera Toxin" (See note 191).

198. T. M. Jacks and B. J. Wu, "Biochemical Properties of *Escherichia Coli* Low-Molecular-Weight, Heat-Stable Enterotoxin," *Infection and Immunity* 9 (1974): 342-47.

199. H. S. Kantor, P. Tao, and S. L. Gorbach, "Stimulation of Intestinal Adenyl Cyclase by *Escherichia Coli* Enterotoxin: Comparison of Strains from an Infant and an Adult with Diarrhea," *Journal of Infectious Diseases* 129 (1974): 1-9.

200. E. G. Shore et al., "Enterotoxin-Producing *Escherichia Coli* and Diarrheal Disease in Adult Travelers: A Prospective Study." *Journal of Infectious Diseases* 129 (1974): 577-82.

201. S. L. Gorbach et al., "Travelers' Diarrhea and Toxigenic *Escherichia Coli,*" *New England Journal of Medicine* 292 (1975): 933-36.

202. R. A. Finkelstein et al., "Clinical Cholera Caused by Enterotoxigenic *Escherichia Coli,*" *Journal of Clinical Microbiology* 3 (1976): 382-84.

203. M. H. Merson et al., "Travelers' Diarrhea in Mexico. A Prospective Study of Physicians and Family Members Attending a Congress." *New England Journal of Medicine* 294 (1976): 1299-1305.

204. Holmgren, "Comparison of Tissue Receptors" (See note 55).

205. S. Larivière, C. L. Gyles, and D. A. Barnum, "Preliminary Characterization

of the Heat-Labile Enterotoxin of *Escherichia Coli* F11(P155)," *Journal of Infectious Diseases* 128 (1973): 312-20.

206. D. J. Evans, Jr., D. G. Evans, and S. L. Gorbach, "Polymyxin B-Induced Release of Low-Molecular-Weight, Heat-Labile Enterotoxin from *Escherichia Coli*," *Infection and Immunity* 10 (1974): 1010-17.

207. O. Söderlind, R. Möllby, and T. Wadström, "Purification and Some Properties of a Heat-Labile Enterotoxin from *Escherichia Coli*," *Zentralblatt für Bakteriologie, Parasitenkunde, Infektionskrankheiten und Hygiene; Erste Abteilung: Originale, Reihe A* 229 (1974): 190-204.

208. F. Dorner, H. Jaksche, and W. Stöckl, "*Escherichia Coli* Enterotoxin: Purification, Partial Characterization, and Immunological Observations," *Journal of Infectious Diseases* 133, suppl. (1976): S142-56.

209. R. A. Finkelstein et al., "Isolation and Properties of Heat-Labile Enterotoxin(s) from Enterotoxigenic *Escherichia Coli*," *Journal of Infectious Diseases* 133, suppl. (1976): S120-37.

210. D. C. Robertson, S. L. Kunkel, and P. H. Gilligan, "Purification and Characterization of Heat-Labile Enterotoxin Produced by ENT$^+$ *E. Coli:* Application of Hydrophobic Chromatography and Use of Defined Media," *Proceedings of the 14th Joint Conference U. S.-Japan Cooperative Medical Science Program Cholera Panel, Karatsu, Japan, 1978* (Karatsu: Japanese Cholera Panel, 1978): 75.

211. Gill, Evans, and Evans, "Activation of Adenylate Cyclase" (See note 167).

212. Kantor, Tao, and Gorbach, "Stimulation of Intestinal Adenyl Cyclase" (See note 199).

213. Dorner, Jaksche, and Stöckl, "*E. Coli* Enterotoxin" (See note 208).

214. D. J. Evans, Jr. et al., "Stimulation of Adenyl Cyclase by *Escherichia Coli* Enterotoxin," *Nature New Biology* 236 (1972): 137-38.

215. E. L. Hewlett et al., "Toxins of *Vibrio Cholerae* and *Escherichia Coli* Stimulate Adenyl Cyclase in Rat Fat Cells," *Nature* 249 (1974): 371-73.

216. S. Hynie et al., "Stimulation of Intestinal and Liver Adenyl Cyclase by Enterotoxin from Strains of *Eschericia Coli* Enteropathogenic for Calves," *Toxicon* 12 (1974): 173-79.

217. Gill, Evans, and Evans, "Activation of Adenylate Cyclase" (See note 167).

218. Moss and Richardson, "Evidence for ADP-Ribosyltransferase Activity" (See note 168).

219. Ibid.

220. Moss et al., "ADP-Ribosylation of Arginine" (See note 171).

221. Ibid.

222. Moss, Manganiello, and Vaughan, "Hydrolysis of NAD" (See note 154).

223. Moss and Richardson, "Evidence for ADP-Ribosyltransferase Activity" (See note 168).

224. Moss, Manganiello, and Vaughan, "Hydrolysis of NAD" (See note 154).

225. Moss and Richardson, "Evidence for ADP-Ribosyltransferase Activity" (See note 168).

226. Moss and Vaughan, "Action of Choleragen" (See note 169).

227. Moss et al., "ADP-Ribosylation of Arginine" (See note 171).

228. Moss, Manganiello, and Vaughan, "Hydrolysis of NAD" (See note 154).

229. Moss and Richardson, "Evidence for ADP-Ribosyltransferase Activity" (See note 168).

230. Moss and Vaughan, "Action of Choleragen" (See note 169).

231. Moss et al., "ADP-Ribosylation of Arginine" (See note 171).

232. Moss and Vaughan, "Isolation of Avian Erythrocyte Protein" (See note 180).

233. J. Moss, S. J. Stanley, and N. J. Oppenheimer, "Substrate Specificity and Purification of a Stereospecific NAD- and Guanidine-Dependent ADP-Ribosyltransferase from Avian Erythrocytes," *Journal of Biological Chemistry* 254 (1979): 8891-94.

234. Moss et al., "ADP-Ribosylation of Arginine" (See note 171).

235. Oppenheimer, "Structural Determination" (See note 175).

236. Moss, Stanley, and Oppenheimer (See note 233).

237. Robertson, Kunkel, and Gilligan, "Hydrophobic Chromatography" (See note 210).

238. Moss, Manganiello, and Vaughan, "Hydrolysis of NAD" (See note 154).

239. Moss and Richardson, "Evidence for ADP-Ribosyltransferase Activity" (See note 168).

240. Moss, Stanley, and Oppenheimer (See note 233).

241. Moss and Vaughan, "Isolation of Avian Erythrocyte Protein" (See note 180).

242. Y. Salomon, C. Londos, and M. Rodbell, "A Highly Sensitive Adenylate Cyclase Assay," *Analytical Biochemistry* 58 (1974): 541-48.

PROTEINS THAT CROSS CELL MEMBRANES: PARALLELS BETWEEN DIPHTHERIA AND CHOLERA TOXINS

D. Michael Gill

INTRODUCTION

Ten years ago the proposal was made that diphtheria toxin had the specific ability to cross the cell membrane and catalytically perform some intracellular mayhem. It was suggested that the cell membrane was crossed in a particular and specific manner, and that an enzymically active fragment of the toxin came to lie within the cytosol, and not simply inside a lysosome. Now it seems that diphtheria toxin is just one of the group of ten or more toxins, listed in Table 1, that act inside cells. Generalizing from these ten it seems that toxins—and perhaps other proteinaceous effectors—that operate within the cytosol of a living cell, and that must therefore cross the plasma membrane to the cystosol, normally come in two parts: one traverses the cell membrane, and the other is in some way responsible for the passage.[1]

Each of the toxins listed in the table acts inside cells, modifying some essential cellular target by a reaction that can be duplicated to one degree or another with purer components in the test tube. The transfer of ADP-ribose from NAD to elongation factor (EF) 2 is catalyzed both by fragments of diphtheria toxin and of Pseudomonas exotoxin A.[2,3] Although they catalyze the same reaction they are unrelated proteins. Cholera toxin catalyzes an analogous reaction between NAD and the GTP regulatory component of adenylate cyclase,[4,5] as does the heat-labile toxin of *Escherichia coli* carrying an ent^{LT} plasmid,[6] and presumably of other organisms carrying a like plasmid. These are the four

TABLE 1. PROTEIN TOXINS THAT ACT INSIDE CELLS

Diphtheria toxin
Pseudomonas aeruginosa exotoxin A
NAD + EF 2 ⟶ ADP-ribose EF 2 + Nico + H^+
Cholera toxin
Heat-labile enterotoxin of *E. coli*
NAD + Guanyl nucleotide regulator ⟶ ADP-ribose GNR + Nico + H^+
Abrin
Ricin
Modeccin
60S ribosome function impaired
Colicin E3
Cloacin DF13
16S RNA split
Colicin E2
DNase

proven cases of bacterial toxins acting within eukaryotic cells. The other six examples in Table 1 are either plant toxins acting on eukaryotic cells or bacterial toxins (bacterocins) that act on bacterial cells and have little, if any, effect on eukaryotes. Abrin, ricin, and modeccin catalytically alter ribosomes in a manner that is not completely understood, but which modifies their interaction with EF 1, and to a lesser extent EF 2.[7-9] A number of other products catalytically modify the 60S ribosome in much the same way, for example, Pokeweed antiviral protein, Crotin II, Curcin II, Enomycin, Phenomycin, and Alpha Sarcin. These are not included in our list because they display little ability to enter cells; they are perhaps simply equivalent to the enzymically active A chain of ricin or abrin, rather than to the whole molecule.[10] Colicin E3 [11] and Cloacin DF13 [12] catalyze the specific cleavage of 16S ribosomal RNA, and thus prevent proper recognition of messenger RNA; Colicin E2 appears to be a DNase.[13] All of these toxins come in two pieces. In some cases the pieces are linked by peptide bonds, and we recognize specific domains involved in the enzymic and entry functions; in other cases the peptide chains are linked solely by disulfide bonds or by weak interactions between subunits. Just as they all catalyze different reactions inside the cells, they may have a variety of ways of getting there. In this paper I shall concentrate on diphtheria and cholera toxins, which certainly are the best understood; which probably have very different modes of transit across the plasma membrane; and which may serve as models for the other toxins.

DIPHTHERIA TOXIN

The basic facts of diphtheria toxin are well known. If one injects the toxin into a guinea pig, the animal dies. For a given dose of toxin and the weight of the animal, the time of death can be predicted quite accurately, and if, shortly before it would have died, the guinea pig is killed and the various organs removed, sliced, and incubated in an appropriate medium, all of them are found to have an impaired ability to synthesize protein. In each case the amount of EF 2 that has not been ADP-ribosylated is reduced (Table 2). EF 2 is a soluble protein that binds to a ribosomal site between the formation of each peptide bond. It ferries GTP to the ribosome. The GTP is split, the nascent chain-tRNA-mRNA complex translocates from the so-called "A site" to the "P-site," and the EF 2 is released with GDP bound. If the EF 2 is ADP-ribosylated it binds ribosomes in the presence of GTP, but it does not promote translocation. Thus the reduction in EF 2 explains the reduction in protein synthesis and, through a series of steps we do not understand clearly, ultimately causes death. The proximal cause of death no doubt varies with toxin

TABLE 2. REDUCTIONS IN RATE OF PROTEIN SYNTHESIS AND IN EF 2 CONTENT

	Protein Synthesis	Unmodified EF 2
	(Percent of Control Values)	
Heart	47	62
Liver	99	86
Kidney	81	74
Pancreas	74	67
Spleen	45	33
Lung	48	47

NOTE: Measured twenty-four hours after the injection of 0.2 μg. diphtheria toxin intravenously into a guinea pig.

dose. EF 2 is found to be modified at a single amino acid residue, which appears to be a basic amino acid that has been posttranslationally modified, but so far has resisted precise identification. It seems likely from the work of T. J. and J. M. Moehring that EF 2 not so modified may not be a substrate for the toxin.[14]

In vivo or in cultured cells the reaction of NAD with EF 2 is virtually irreversible, but we can drive it backwards in vitro if we add nicotinamide, lower the pH, and remove the NAD that is regenerated by the reverse reaction. If we take homogenates of the tissues of the dying guinea pig and incubate them under such conditions, we can regenerate active EF 2, thus proving that the missing EF 2 is indeed ADP-ribosylated. The reaction identified in the test tube really does occur in the whole animal. In the test tube it is of course convenient to use radioactive NAD and to measure the radioactive acid-insoluble macromolecular product. If, for example, we take the total soluble proteins of the rabbit reticulocyte, react them with toxin and NAD, and then separate the products on a SDS gel, we find 100 percent of the EF 2 to be labeled, while there is almost no labeling of any other protein. A typical result is shown in Figure 1,a. Sometimes we see labeled breakdown products of EF 2, such as a 43,000 dalton band,[15] and if we expose the autoradiogram for a longer period we can detect other proteins that are specifically ADP-ribosylated in the presence of diphtheria toxin. We can, for example, detect ADP-ribosylated diphtheria toxin itself.[16] Even if it is not absolute, it is clear that the discrimination in favor of EF 2 is very high.

Since the modification of EF 2 can occur both in intact cells and in cell extracts, it is convenient to distinguish the "toxicity" of diphtheria toxin on the intact cells from its enzymic "activity" in the test tube. Figure 2 shows some of the species that are toxic or active. The initially secreted intact diphtheria toxin is often nicked before purification in a specific small stretch of peptide between the A and B regions.[17] Both intact and nicked toxins are very toxic, but neither is active against EF 2 in the text tube. The nicked toxin is, however, active when S-S bonds are reduced, specifically the S-S bonds connecting fragments A and B, by which

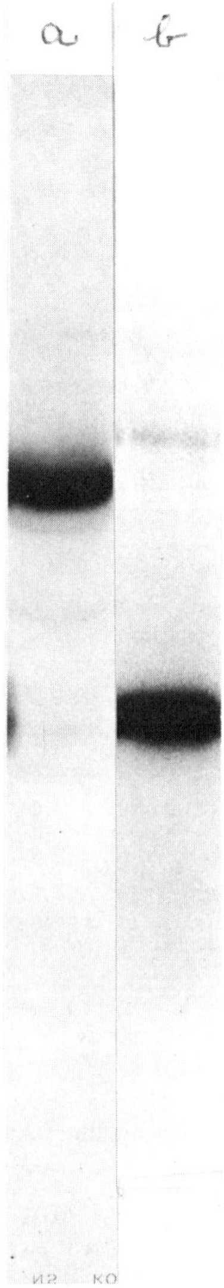

Figure 1. Autoradiograms of the products formed from radioactive NAD: a) EF 2 in rabbit reticulocyte cytosol incubated with diphtheria toxin. b) 42,000 dalton protein in plasma membranes isolated from pigeon erythrocyte lysate incubated with cholera toxin.

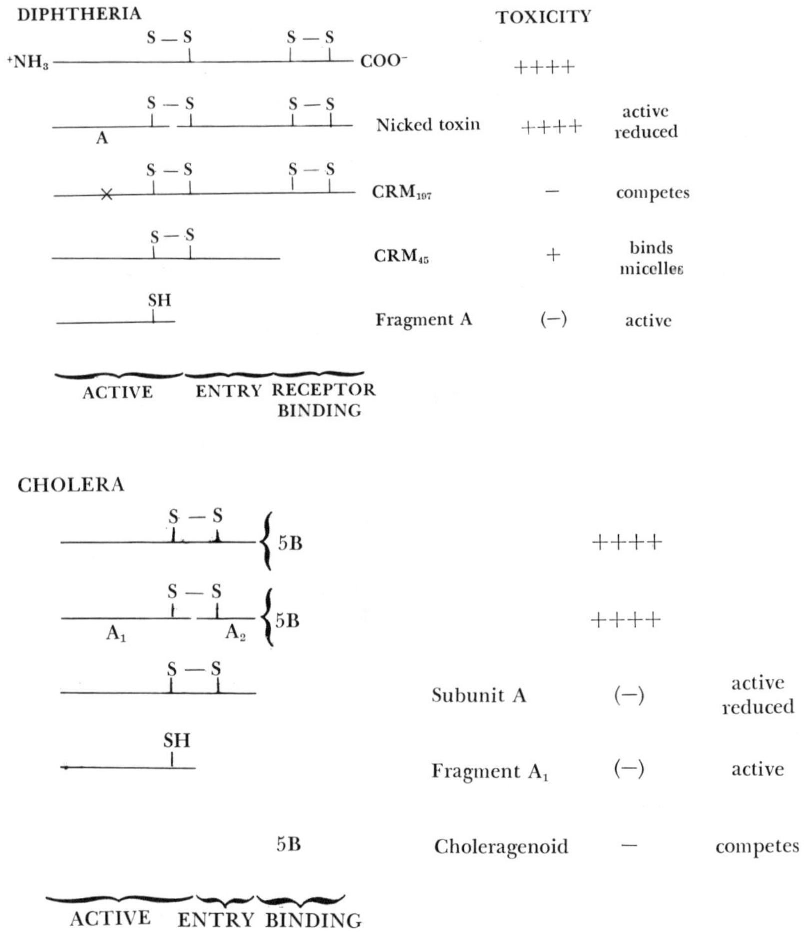

Figure 2. Diphtheria and cholera toxins and related proteins available for study.

means some conformational constraint on the activity of A is lifted. Intact toxin is activated by treatment with trypsin and dithiothreitol. It is commonly believed that when intact diphtheria toxin intoxicates a whole cell some mechanism must exist to break the peptide bond and the disulfide bond. Most preparations of diphtheria toxin that we deal with have been partially proteolysed, so they consist of a mixture of the intact and the nicked form. If we spread them out on a SDS gel without reduction they run together at about 62,000 daltons, and the enzymic activity—the ability to ADP-ribosylate EF 2—travels with the toxin. (Notice that the material can be taken directly from a SDS gel and assayed without any particular efforts to renature it, except for dilution of the SDS.) If we prereduce the toxin, however, the intact protein runs at 62,000 daltons and is inactive, while the liberated fragment A runs at 21,500 daltons and is fully active.

Although the released fragment A has a very high enzymic activity, it has, when separated from B, lost all but 0.001 percent of its toxicity. Thus the B region or fragment must be involved in enabling fragment A to reach its intracellular target. The major unsolved problem of diphtheria toxin is exactly how the B and the membrane collaborate to get A across. What we do know comes largely from the study of products of a number of mutant toxin genes, which enables us to investigate the function of the B region a little more closely. CRM_{197} is not a nontoxic protein of the full size, that is, 62,000 daltons, more or less, which must have an amino acid replacement or small deletion in the A region, for it is neither toxic nor active when reduced and nicked. Yet it has the important biological property of competing with wild-type toxin and, to some extent, preventing the toxin from acting on whole cells.

It appears likely that CRM_{197} binds to some compound or other on the surface of the intact cell involved with the entry of diphtheria toxin fragment A, which we would like to call the toxin receptor. We know the B region of CRM_{197} is fully functional because it can be recombined in vitro with fragment A to form toxin. A second mutant gene, presumably carrying a nonsense mutation, codes for the incomplete protein CRM_{45}, which is dramatically less toxic than toxin, but which has a functional A region. Unlike both fragment A and native toxin, CRM_{45} binds detergent micelles. In other words it displays a hydrophobic region that is absent from fragment A and that is presumably concealed in the toxin until the protein is denatured.[18] Recent work of P. Falmagne located this hydrophobic region probably toward the N terminus of the B region of the toxin [19] where there is at least one tryptic peptide soluble in butanol.

It is most interesting that CRM_{45}, which can bind these micelles, and therefore presumably bind the phospholipid bilayer of the cell membrane, although much less toxic than diphtheria toxin is nevertheless much more toxic than isolated fragment A.[20] Thus we can recognize in the B region of diphtheria toxin several distinct functions. As we might have predicted, binding to receptors increases dramatically the likelihood of penetration, but it is not sufficient for this to occur; likewise it is not sufficient to bind to lipid bilayer. It is necessary to do both of these, and presumably to do them in a very particular way.

Figure 3 expands on this theme. Several workers who have measured the binding of diphtheria toxin to cells have recognized a number of apparently different interactive states. Cells seem to have some surface material, probably glycoproteins,[21] to which diphtheria toxin binds. We call this "receptor." The binding constant appears to be about 10^{-8} M,[22] and it is quite reversible. We can remove the bound toxin by washing, for example, or by competition with excess CRM_{197}, and the cell will not suffer.

If ammonium ions or chloroquine are present, or if the temperature is below 20°C or so, the bound toxin remains in this condition, but at temperatures above 20° it rapidly proceeds to an irreversibly bound state, as described by P. Boquet and co-

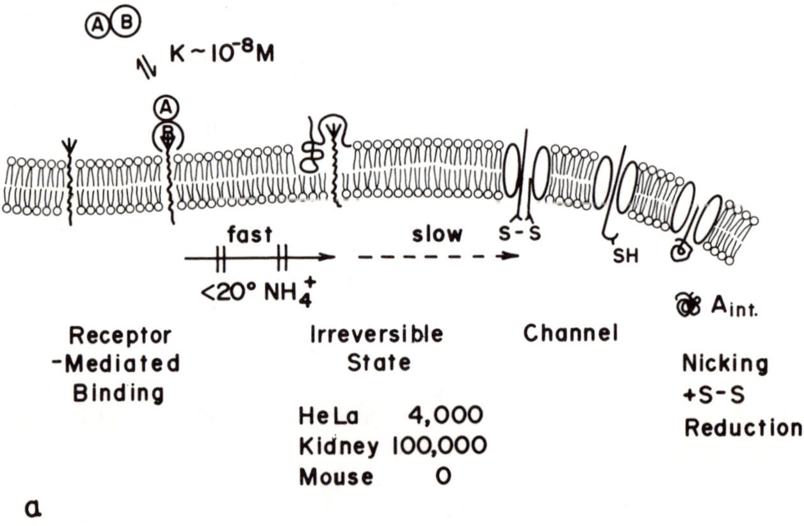

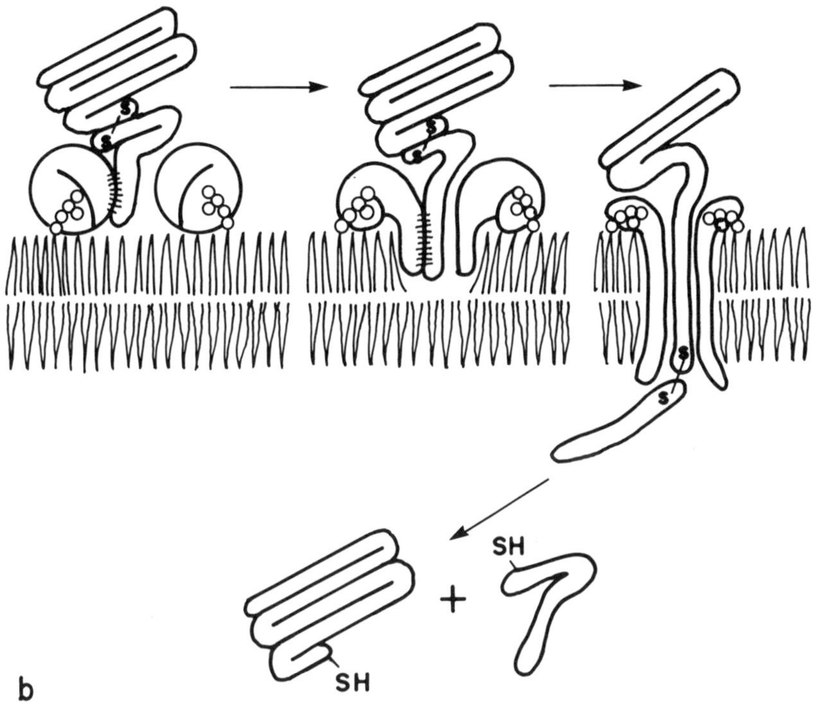

Figure 3. a) Interactions between diphtheria toxin and a plasma membrane and the suggested formation of a channel that leads fragment A into the cytosol; it is not clear to what extent the proposed channel may be lined by fragment B or by cellular proteins. b) Suggested interactions between cholera toxin and a plasma membrane, including the formation by B subunits of a channel through which the S subunit enters the cytosol.

SOURCE: Reproduced, by permission of the American Chemical Society, from D.B. Gill, "The Arrangement of Subunits in Cholera Toxin," *Biochemistry* 15 (1976): 1242-48.

workers,[23] a condition in which the toxin cannot be removed, even by extracellular antitoxin, and yet it has not yet advanced to the stage where it can interact with the cell cytoplasm. We propose that a part of the B region, while still bound to the receptor, becomes intimately associated with the thickness of the lipid bilayer.

There are some 4,000 of these irreversible binding sites in a HeLa cell, a great many more in kidney cells,[24] but none in mouse cells: a fact that presumably accounts for the insensitivity of murine cells to diphtheria toxin. We frankly do not know what happens next, and for this reason we have, at this point in Figure 3,a, changed the graphical representation of the proteins involved. My suggestion is that the toxin forms a channel through the plasma membrane, either by itself or in conjunction with existing membrane proteins, and that fragment A passes through this channel.[25] We know the intact unreduced toxin is toxic, and therefore must suppose that the S-S bonds and the peptide bond connecting A to B are cleaved during this passage, most probably at the cytoplasmic surface, so that A is liberated on the inside.

Now, since the proposed channel could not be very wide, it is easiest to suppose that fragment A travels through the channel in an unwound configuration and has to spontaneously refold in the cytoplasm before catalyzing ADP-ribosylation. It is therefore encouraging to know that fragment A does refold after denaturation; in fact it exhibits its original enzymatic activity as soon as we remove a denaturing condition, such as heat or high detergent concentration.

Another school of thought maintains that the toxin is taken up by lysosomes and released from there into the cytoplasm proper. Inhibition of intoxication by ammonium ions or chloroquine, which among other effects disturbs lysosome function, lends some credence to this idea. The difficulty remains, however, of explaining how fragment A is released from lysosomes, particularly at very low toxin concentrations, when each irreversibly bound toxin molecule has a high probability of generating a cytosolic fragment A.[26]

CHOLERA TOXIN

When we started working with cholera toxin we had no idea that the story would turn out to be very similar to that of diphtheria toxin. On the face of it the two proteins are quite different. Unlike diphtheria toxin, cholera toxin is a multisubunit protein. Five of the six subunits, the B subunits, are identical; are thought to be arranged in a regular ring; and are responsible for the binding of cholera toxin to the G_{M1} ganglioside on cell surfaces. The remaining subunit, the A subunit, is presumed to be arranged axially on one side of the ring.[27] It is usually, but not always,[28] nicked into a peptide A_1 that is enzymically active, and a shorter peptide, A_2, that is required for the positioning of A_1 with respect to the B ring.

Either nicked or intact cholera toxin is toxic,[29] but activity requires only A_1, which, coincidentally or not, is about the same size as fragment A of diphtheria toxin, and shares with the latter the property of SDS stability and heat stability. Like fragment A, it is nontoxic and requires the remainder of the molecule, both A_2 and B, for toxicity. Activity in vitro requires reduction of the single disulfide bond in A, a clear analogy to diphtheria toxin, and separation of either A or fragment A_1 from the B subunits, which is conveniently achieved by SDS treatment (Figure 2). Unlike diphtheria toxin it does not seem to be necessary to nick the peptide bond between A_1 and A_2 for activity. The best analogy is drawn between fragment B of diphtheria toxin and the $A_2$5B complex of cholera toxin as the regions required for reception and entry (Figure 3,b).

The enzymic activity of fragment A_1 was less easy to establish than that of diphtheria toxin fragment A, partly because the target protein is membrane-bound and so the reaction cannot be run in an entirely soluble system. Some years ago, however, Carolyn King and I were able to show that fragment A_1 of cholera toxin elevated the adenylate cyclase activity of broken cells with kinetics that suggested it was an enzyme.[30] Using pigeon erythrocytes we found that the adenylate cyclase of the membranes could be elevated by the toxin only in the presence of pigeon erythrocyte cytosol. Over the years we have been able to establish that several cytosolic components are necessary, namely, NAD, GTP, and a protein. The NAD is required absolutely: that is, if we remove all endogenous NAD from the system by pretreatment with a NAD glycohydrolase there is no response whatsoever to cholera toxin.[31] The GTP is also required absolutely. We can, however, replace GTP by a GTP-generating system, and in most cases there is sufficient GDP or GMP available for the GTP requirement to be satisfied by the mere addition of ATP, CTP, or UTP, relying on endogenous kinase to generate GTP. K. Enomoto has shown that when measures are taken to prevent the regeneration of GTP from GDP, ATP is ineffective and GTP or another guanine nucleotide must be specifically provided.[32]

The cytosolic protein remains an enigma. It is required in high concentrations for maximum activity, and the activity falls off very rapidly indeed when the cytosol is diluted. But there is a slight residual activity in washed ghosts. We do not know if this is an inherent low activity or whether it should be attributed to a small amount of cytosolic factor that has been absorbed by the ghosts and cannot be removed by repeated washings. One of the things we discovered about the interaction of cholera toxin with this system is that all five ingredients—toxin, membranes, cytosol, NAD, and nucleotide—have to be present at the same moment. If the four other ingredients are preincubated, for example, there is no activation of cyclase until NAD is added; the activation of cyclase commences on the addition of NAD without any observable lag (Figure 4). Thus it looks as though we are dealing with a single event. We suspected of course that the cholera toxin was catalyzing an ADP-ribosylation, just like diphtheria toxin, and the specificity of the NAD requirement supported such a notion, although it did not exclude the possibility that NAD was acting as an oxidizing agent, as it does in 200 or so other reactions.[33]

The problem with testing the ADP-ribosylation hypothesis was two-sided. First, when NAD is incubated with pigeon erythrocyte lysate, there is a substantial background incorporation, both into poly ADP-ribose and to ADP-ribosyl proteins, not all of them nuclear proteins. Second, we did not expect a toxin-specific signal to be particularly large, for we anticipated only 1,000 to 2,000 copies of adenylate cyclase in a pigeon erythrocyte. Thus we had both high noise and a low signal. We were able to solve both these problems by suitable changes of the incubation conditions. To decrease the noise we added thymidine, a potent inhibitor of poly ADP-ribose synthesis, and reduced the NAD concentration, because poly ADP-ribose synthesis increases linearly with the NAD concentration. We then had to remove the endogenous NAD, about 100 μM, from the membranes and cytosol in a manner that did not destroy the responsiveness to cholera toxin. Unfor-

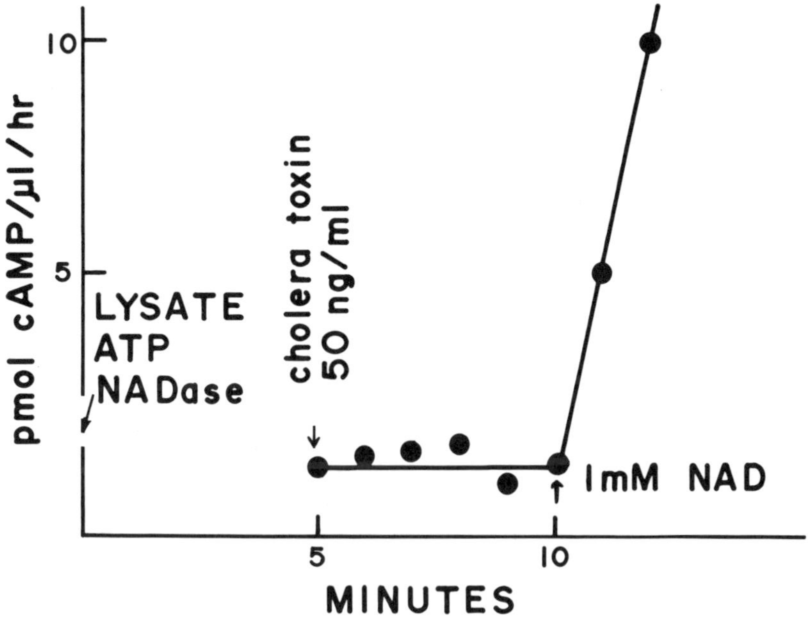

Figure 4. Initiation of cyclase activation by NAD. Endogenous NAD of a fresh lysate of pigeon erythrocytes was hydrolyzed by Neurospora NAD glycohydrolase. When cholera toxin was added five minutes later there was no response, but adenylate cyclase activation started immediately (within the discrimination of the experiment) upon the addition of NAD at ten minutes. Each sample was assayed for adenylate cyclase activity under standard conditions after washing.

tunately this entailed a 50 percent drop in cytosol concentration, and thereby a substantial reduction in reactivity, but not a total loss. To compensate we raised the total toxin concentration by 1,000-fold or more (Figure 5).

With these revised conditions we were for the first time able to show a toxin-specific incorporation of ADP-ribose into the membranes of pigeon erythrocytes under conditions that resulted in adenylate cyclase activation—and only under such conditions.[34] The prime target, a single polypeptide of 42,000 daltons (Figure 1,b) is located on the inner surface of the cytoplasmic membrane; it is a protein already described by T. Preuffer.[35] He established that there is, on the inner surface of the membrane, a guanyl nucleotide-binding protein of 42,000 daltons that interacts with

Lysate[1]	{Membranes + Cytosol}	($OD_{540} \sim 100$) →	NAD-free membranes / NAD-free cytosol	($OD \sim 50$)
ATP		5mM	ATP	5mM
NAD		5mM	$NAD^{32}P$	10μM
Toxin/SDS, DTT		1-10ng/ml	Toxin	$\sim$ 10μg/ml
			Thymidine	10mM

Figure 5. Incubation conditions for assay of cholera toxin in vitro. On the *left,* maximum sensitivity for adenylate cyclase activation is exhibited by a concentrated lysate supplemented with NAD and ATP; on the *right* are representative conditions used to demonstrate ADP-ribosylation. 1) Trace of guanine nucleotide present.

adenylate cyclase and mediates the hormonal stimulation of activity. We can readily see now why there is a GTP requirement for the cholera toxin activation. Presumably, as depicted in Figure 6, the guanyl nucleotide-binding protein (N) can be ADP-ribosylated, is a substrate for the toxin only when it has GTP bound. The role of the cytosolic factor in this interaction remains unsolved.

The numbers are interesting. Up to a point, cyclase activation and ADP-ribosylation of the guanine nucleotide protein run in parallel. The cyclase activity is maximal when 1,000 to 2,000 ADP-ribose residues are attached per cell, that is, one per beta adrenergic receptor, or, we may assume, one per functional adenylate cyclase catalytic moiety (C). There seems to be an excess

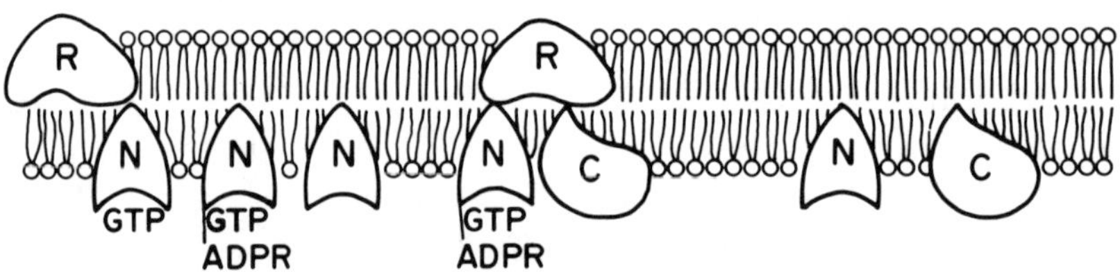

Figure 6. Suggested states of components of the adenylate cyclase system. The guanylnucleotide regulator (N) is considered to exist largely dissociated from, and in substantial excess over, the catalytic component (C). N, C, and GTP form a ternary complex that is enzymically active until the GTP is split. In the presence of cholera toxin, N with GTP bound may be ADP-ribosylated and thereby stabilized in the GTP-bound state. Thus the ADP-ribosyl N is preferentially associated with C.

of the guanyl nucleotide protein, however, and we can continue to add ADP-ribose residues to this excess without further activation of the enzyme. We suppose that: 1) the N and C components of adenylate cyclase are dissociable, a proposal that has much outside support; [36-39] 2) when the regulatory protein (N) is bound to GTP it has a higher affinity for the catalytic component; and 3) when ADP-ribosylated, thus preventing hydrolysis of the GTP, N·GTP binds to the catalytic moiety more or less permanently. Thus, upon ADP-ribosylation, a portion of the available N protein, the modified form, will associate with as much of the catalytic component of cyclase as is available. A further complication is that the N protein is not the only cholera toxin target; others exist. They are not as readily modified as the N protein, but can be labeled with extensive incubation with high toxin or NAD concentrations. The consequences of such secondary ADP-ribosylations are not yet known.

We have extended the demonstration of a principal target protein of 42,000 daltons to rat fat cells,[40] rat reticulocytes,[41] and frog erythrocytes.* Other workers have extended them to S49 lymphoma cells, turkey erythrocytes,[42] and liver.[43]

The investigations of D. Cassel and Z. Selinger support the following model to explain the activation of adenylate cyclase by cholera toxin and hormones (Figure 7).[44, 45] They have introduced evidence that there is a cycle of binding and hydrolysis of GTP attached to the regulatory protein. Only when the GTP is bound and there is a ternary complex, C·N·GTP, is the cyclase active. When the GTP hydrolyzes, the cyclase is inactive and the bound GDP, to regain activity, must now be displaced in favor of GTP. Hormones promote the displacement of the GDP and thus increase the proportion of the enzyme in the active configuration. The nonhydrolyzable guanine nucleotide analogues, such as Gpp(NH)p, activate cyclase quasi-permanently by binding to the guanine nucleotide site, forming the ternary complex, and stabilizing this complex against hydrolysis. A third way of activating the cyclase is to trap the ternary complex by ADP-ribosylation,

* D. M. Gill and L. E. Limbird: unpublished observations.

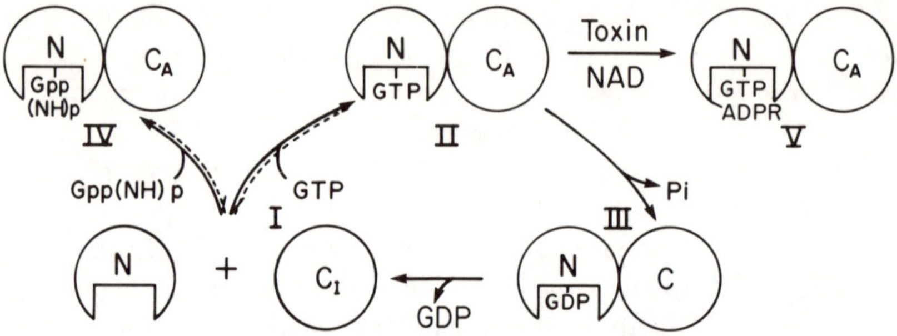

Figure 7. Control of adenylate cyclase activity. Inactive catalytic components (I) forms an active ternary complex with N and GTP (II), but reverts to inactivity (III) when the GTP is hydrolyzed. Hormones promote the replacement of GDP by GTP, and thus increase the proportion of C in an active state. The activity may also be increased by supplying GppNHp, which is resistant to hydrolysis and so forms a long-lived ternary complex (IV). Alternatively, ADP-ribosylation may stabilize bound GTP and prolong the life of active complex (V).

which, according to much evidence, reduces the cyclase-associated hydrolysis of GTP.[46,47]

A final parallel demonstrates that the analogy with diphtheria toxin may be even more remarkable. Our model essentially proposes independent cycles of the C and N components of adenylate cyclase between dissociation of GDP and the formation of a new ternary active complex with GTP ($N \cdot GTP + C \rightarrow N \cdot C \cdot GTP \rightarrow N \cdot C \cdot GDP \rightarrow N \cdot GDP + C$). We propose that N·GTP is the cholera toxin target. During protein elongation each peptide bond EF 2 binds GTP and carries it to a pretranslocation ribosome. The GTP is thereupon split. After translocation, EF 2 is released as a complex with GDP, and cycles independently of the translocated ribosome. EF 2 is the diphtheria toxin target ($EF\ 2 \cdot GTP + Rib \rightarrow EF\ 2 \cdot GTP \cdot Rib \rightarrow EF\ 2 \cdot GDP \cdot Rib \rightarrow EF\ 2 \cdot GDP + Rib$).

From here on the analogy breaks down, because: 1) EF 2 does not require the presence of GTP in order to react with diphtheria toxin;[48] and 2) it is not clear that ADP-ribosylation of EF 2 inhibits the ribosomal GTPase activity. It was originally reported that this was so,[49,50] but subsequent work has shed some doubt on this proposition.[51,52]

Figure 8 is a revision of certain aspects of the action of diphtheria and cholera toxins, demonstrating once again some of the

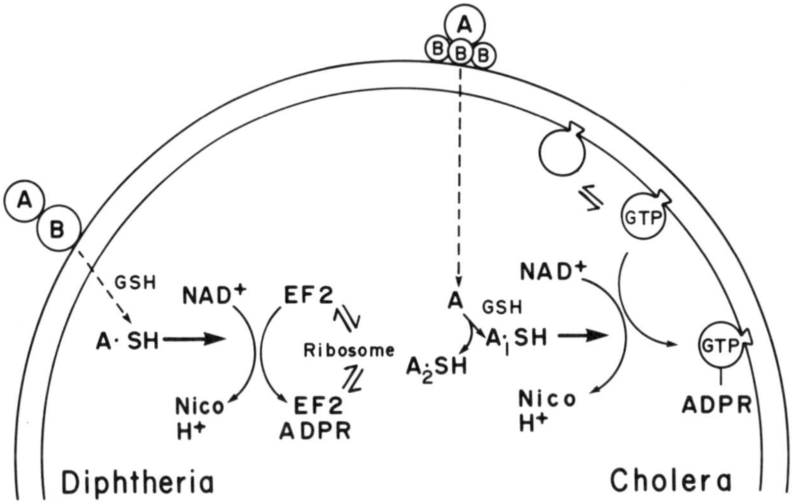

Figure 8. Summary of intracellular events on diphtheria and cholera intoxication.

striking similarities between the two. One wonders why two toxins have come to have such analogous modes of action, especially when one remembers that the unrelated exotoxin A of *Pseudomonas aeruginosa* catalyzes the same reaction as diphtheria toxin, and that the enterotoxin of *E. coli* is hard to distinguish functionally from cholera toxin.

NOTES

1. D. M. Gill, "Seven Toxic Peptides That Cross Cell Membranes," in *Bacterial Toxins and Cell Membranes,* ed. J. Jeljaszewicz and T. Wadström (London: Academic Press, 1978): 291-332.

2. R. J. Collier, "Diphtheria Toxin: Mode of Action and Structure," *Bacteriological Reviews* 39 (1975): 54-85.

3. B. H. Iglewski, P. U. Liu, and D. Kabat, "Mechanism of Action of *Pseudomonas aeruginosa* Exotoxin A: Adenosine Diphosphate Ribosylation of Mammalian Elongation Factor 2 In Vitro and In Vivo," *Infection and Immunity* 15 (1977): 138-44.

4. D. M. Gill and R. Meren, "ADP-Ribosylation of Membrane Proteins Catalyzed by Cholera Toxin: Basis of the Activation of Adenylate Cyclase," *Proceedings of the National Academy of Sciences* 75 (1978): 3050-54.

5. D. Cassel and T. Pfeuffer, "Mechanism of Cholera Toxin Action: Covalent Modification of the Guanyl Nucleotide-Binding Protein of the Adenylate Cyclase System," *Proceedings of the National Academy of Sciences* 75 (1978): 2669-73.

6. D. M. Gill and S. H. Richardson, "ADP-Ribosylation of Adenylate Cyclase

Catalyzed by the Heat-Labile Enterotoxin of *Eschericia Coli,* Comparison with Cholera Toxin," submitted to the *Journal of Infectious Diseases.*

7. S. Olsnes, "Abrin and Ricin: Structure and Mechanism of Action of Two Toxic Lectins," *Bulletin de l'Institut Pasteur* 74 (1976): 85-99.

8. S. Olsnes et al., "Ribosome Inactivation by the Toxic Lectins Anrin and Ricin. Kinetics of the Enzyme Activity of the Toxin A Chains," *European Journal of Biochemistry* 60 (1975): 281-88.

9. K. Refsnes et al., "Modeccin-A Plant Toxin Inhibiting Protein Synthesis," *Biochemical and Biophysical Research Communications* 79 (1977): 1176-83.

10. C. Fernandez-Puentes and D. Vazquez, "Effects of Some Proteins That Inactivate the Eukaryotic Ribosome," *FEBS Letters* 78 (1977): 143-46.

11. S. Ohno et al., "Purification and Characterization of Active Component and Active Fragment of Colicin E3," *Journal of Biochemistry* 82 (1977): 1045-53.

12. B. Oudega and F. K. DeGraaf, "Enzymatic Properties of Cloacin DF13 and Kinetics of Ribosome Inactivation," *Biochimica et Biophysica Acta* 425 (1976): 296-304.

13. K. Schaller and M. Nomura, "Colicin E2 Is a DNA Endonuclease," *Proceedings of the National Academy of Sciences* 73 (1976): 3989-93.

14. T. J. Moehring and J. M. Moehring, "Slection and Characterization of Cells Resistant to Diphtheria Toxin and Pseudomonas Exotoxin A: Presumptive Translational Mutants," *Cell* 11 (1977): 447-54.

15. J. F. Collins, S. Raeburn, and E. S. Maxwell, "Aminoacyltransferase II from Rat Liver. II. Some Physical and Chemical Properties of the Purified Enzyme and Its ADPR Derivative," *Journal of Biological Chemistry* 246 (1971): 1049-54.

16. D. M. Gill and D. M. Steinhaus, "Modification of Diphtheria Toxin by NAD," *Journal of Hygiene, Epidemiology, Microbiology and Immunology* 18 (1974): 316-23.

17. R. J. DeLange, R. E. Drazin, and R. J. Collier, "Amino Acid Sequence of Fragment A, an Enzymatically Active Fragment from Diphtheria Toxin," *Proceedings of the National Academy of Sciences* 73 (1976): 69-72.

18. P. Boquet et al., "Binding of Triton X-100 to Diphtheria Toxin, Crossreacting Material 45 and Their Fragments," *Proceedings of the National Academy of Sciences* 73 (1976): 4449-53.

19. P. Falmagne, P. Lambotte, and J. Dirkx, "Isolation and Characterization of the Cyanogen Bromide Peptides from the B Fragment of Diphtheria Toxin," *Biochimica et Biophysica Acta* 535 (1978): 54-65.

20. P. Boquet, "Transport of Diphtheria Toxin Fragment A Across Mammalian Cell Membranes," *Biochemical and Biophysical Research Communications* 75 (1977): 696-702.

21. R. K. Draper, D. Chin, and M. I. Simon, "Diphtheria Toxin Has the Properties of a Lectin," *Proceedings of the National Academy of Sciences* 75 (1978): 261-65.

22. T. R. Ittelson and D. M. Gill, "Diphtheria Toxin: Specific Competition for Cell Receptors," *Nature* 242 (1973): 330-32.

23. P. Boquet and A. M. Pappenheimer, "Interaction of Diphtheria Toxin with Mammalian Cell Membranes," *Journal of Biological Chemistry* 251 (1976): 5770-78.

24. J. L. Middlebrook, R. B. Dorland, and S. H. Leppla, "Association of

Diphtheria Toxin with Vero Cells: Demonstration of a Receptor," *Journal of Biological Chemistry,* in press.

25. Gill, "Seven Toxic Peptides" (See note 1).

26. Ibid.

27. D. M. Gill, "The Arrangement of Subunits in Cholera Toxin," *Biochemistry* 15 (1976): 1242-48.

28. D. M. Gill and R. S. Rappaport, "The Origin of the Enzymatically Active A_1 Fragment of Cholera Toxin," *Journal of Infectious Diseases,* in press.

29. J. J. Mekalanos, R. J. Collier, W. R. Romny, "Enzymic Activity of Cholera Toxin 2: Relationship to Proteolytic Processing and Disulfide Bond Reduction," *Journal of Biological Chemistry,* manuscript submitted for publication.

30. D. M. Gill and C. A. King, "The Mechanism of Action of Cholera Toxin in Pigeon Erythrocyte Lysates," *Journal of Biological Chemistry* 250 (1975): 6424-32.

31. D. M. Gill, "The Involvement of NAD in the Mechanism of Action of Cholera Toxin," *Proceedings of the National Academy of Sciences* 72 (1975): 2064-68.

32. K. Enomoto and D. M. Gill, "The Involvement of GTP in the Action of Cholera Toxin *In Vitro,*" *Journal of Supramolecular Structure,* in press.

33. Gill, "Involvement of NAD" (See note 31).

34. Gill and Meren, "ADP-Ribosylation" (See note 4).

35. T. Pfeuffer, "GTP-Binding Proteins in Membranes and the Control of Adenylate Cyclase Activity," *Journal of Biological Chemistry* 252 (1977): 7224-39.

36. G. L. Johnson, H. R. Kaslow, and H. R. Bourne, "Reconstitution of Cholera Toxin-Activated Adenylate Cyclase," *Journal of Supramolecular Structure* suppl. 2, abstr. 300 (1978): 122.

37. E. M. Ross and A. G. Gilman, "Resolution of Some Components of Adenylate Cyclase Necessary for Catalytic Activity," *Journal of Biological Chemistry* 252 (1977): 6966-69.

38. M. Hebdon et al., "Properties of the Interaction of Fluoride- and GppNHp-Regulatory Proteins with Adenylate Cyclase," *Proceedings of the National Academy of Sciences* 75 (1978): 3693-97.

39. P. M. Lad, A. F. Welton, and M. Rodbell, "Evidence for Distinct Guanine Nucleotide Sites in the Regulation of the Glucagon Receptor and of Adenylate Cyclase Activity," *Journal of Biological Chemistry* 252 (1977): 5942-46.

40. C. C. Malbon and D. M. Gill, "ADP-Ribosylation of Membrane Proteins and Activation of Adenylate Cyclase by Cholera Toxin in Fat Cell Ghosts: Comparing Euthyroid and Hypothyroid Rates," manuscript submitted for publication.

41. L. E. Limbird et al., manuscript submitted for publication.

42. G. L. Johnson, H. R. Kaslow, and H. R. Bourne, "Genetic Evidence That Cholera Toxin Substrates Are Regulatory Components of Adenylate Cyclase," *Journal of Biological Chemistry* 253 (1979): 7120-23.

43. S. Beckner and M. Blecher, "Cholera Toxin-Catalysed Activation of Adenylate Cyclase and ADP-Ribosylation from ^{3}H-NAD$^+$ in Cloned Rat Hepatocytes," *Pharmacologist* 20 (1978): 146.

44. D. Cassel and Z. Selinger, "Catecholamine-Stimulated GTPase Activity in Turkey Erythrocyte Membranes," *Biochimica et Biophysica Acta* 452 (1976): 538-51.

45. ———, "Mechanism of Adenylate Cyclase Activation by Cholera

Toxin: Inhibition of GTP Hydrolysis at the Regulatory Site," *Proceedings of the National Academy of Sciences* 74 (1977): 3307-11.

46. Ibid.

47. S. L. Levinson and A. J. Blume, "Altered Guanine Nucleotide Hydrolysis as Basis for Increased Adenylate Cyclase Activity after Cholera Toxin Treatment," *Journal of Biological Chemistry* 252 (1977): 3766-74.

48. D. M. Gill, A. M. Pappenheimer, Jr., and J. B. Baseman, "Studies on Transferase II Using Diphtheria Toxin," *Cold Spring Harbor Symposia on Quantitative Biology* 34 (1969): 595-602.

49. S. Raeburn et al., "Interaction of Aminoacyl Transferase II and GTP: Inhibition by Diphtheria Toxin and NAD," *Proceedings of the National Academy of Sciences* 61 (1968): 1428-34.

50. K. Kloppstech, R. Steinbeck, and F. Klink, "Characteristics of the Reaction Between Diphtheria Toxin, Pyridine Coenzymes and the GTP-splitting Transfer Factor F1," *Hoppe-Seyler's Zeitschrift für Physiologische Chemie* 350 (1969): 1377-84.

51. E. Bermek, "Formation of a Complex Involving ADP-Ribosylated Human Translocation Factor, Guanosine Nucleotide and Ribosomes, *FEBS Letters* 23 (1972): 95-99.

52. D-M Chuang and H. Weissbach, "Studies on Elongation Factor II from Calf Brain," *Archives of Biochemistry and Biophysics* 152 (1972): 114-24.

OPIATE RECEPTORS AND ENDORPHINS

Eric J. Simon

INTRODUCTION

The receptor I have been asked to discuss is probably the most recently discovered of any of the many receptors under consideration in this volume. It is therefore not surprising that the implication of the opiate receptor in human disease has not yet been established. Rather than go no further, as I probably should, at this point I plan to proceed as follows. I will begin by presenting a brief summary of the state of the art in this area of research.[1,2] After that I shall discuss those disease states and pathological conditions where there is some evidence, though frequently controversial, that appears to implicate the opiate receptor-endorphin system. I apologize in advance for the heavy inclusion of the ligands, which ought to be restricted to the receptor, but the field is too embryonic to permit me to do otherwise; the fact that little is yet known about receptor modulation by its ligands provides a suitable rationalization for this course of action.

THE DISCOVERY OF OPIATE RECEPTORS

The hypothesis that narcotic analgesics must bind to highly specific sites or receptors in the central nervous system (CNS) in order to produce their many well-known responses has been held by some investigators for several decades. There was, however, the question of why the human and animal brain should contain sites that can bind with high specificity and affinity substances derived from plants. Nevertheless, the evidence for such receptor sites was compelling. It consisted primarily of the remarkable stereospecificity, and, for certain parts of the molecules, structural specificity displayed by many of the pharmacological actions of narcotic analgesic drugs.

As thousands of analogues of morphine were synthesized in the search for the still mythical nonaddictive analgesic, it became clear that one enantiomer of a racemic mixture, usually the levorotatory one, was generally much more active than the other. Moreover, whereas some parts of the molecule could be drastically altered with relatively little change in potency, tampering with certain regions had dramatic effects. The most interesting and most thoroughly studied such region is the substituent on the tertiary nitrogen, one of the functional groups essential for narcotic analgesic activity. When a larger alkyl group, for example, an allyl or cyclopropylmethyl group, is substituted for the methyl group, analgesic potency is reduced and the drug takes on the new pharmacological activity of a potent, specific antagonist against many of the actions of morphine and related narcotics. Some drugs have both agonist and antagonist properties, while others such as naloxone and naltrexone (the N-allyl and N-cyclopropylmethyl analogues of oxymorphone) are "pure" antagonists.

These kinds of specificities are most easily explained by interaction with a binding site that exhibits complementary specificity. The search for such opiate receptors began in the 1950s and bore fruit in the early 1970s. It was easy to show binding of opiates to cell constituents,[3] but to distinguish specific from nonspecific binding proved difficult.

It was the measurement of stereospecific binding that led to success. N. A. Ingoglia and V. P. Dole were the first to apply stereospecificity to the search for receptors by injecting l- and d-methadone into the lateral ventricle of rats, but they found no difference in the rate of diffusion of the enantiomers.[4] A. Goldstein and coworkers devised a method for measuring stereospecific binding of ^{3}H-levorphanol in mouse brain homogenates; they reported that only 2 percent of the total binding was stereospecific,[5] and the properties and distribution of the binding turned out to be quite different from those of the subsequently discovered "receptors."

In 1973 Simon and coworkers,[6] C. B. Pert and S. H. Snyder,[7] and L. Terenius,[8] using modifications of the Goldstein procedure, independently and simultaneously reported the observation in animal brain homogenates of stereospecific binding of opiates that represented the major portion of the total binding. The modifications involved the use of very low concentrations of labeled drugs, made possible by high specific activity, and the washing of homogenates pelleted or filtered after incubation to remove unbound or loosely bound contaminating radioactivity. Since that time stereospecific binding studies have been done in many laboratories, and much evidence has been accumulated to suggest that these stereospecific sites are indeed receptors responsible for many of the pharmacological actions of the opiates. They have been found in man [9] and in all vertebrates so far studied, but not in invertebrates.[10]

PROPERTIES AND DISTRIBUTION OF OPIATE RECEPTORS

The properties of the opiate binding sites have been studied extensively, and their distribution in the brain and spinal cord has been mapped in considerable detail by dissection and in vitro binding measurements,[11,12] as well as by autoradiography.[13-16]

Stereospecific binding is saturable, and, at saturation, binding amounts to 20-30 pmol of opiate per gram of rat brain. Affinities range from K_D of 0.025 nM for a potent fentanyl analogue [17] to little or no affinity for drugs devoid of opiate activity. The pH optimum for binding is in the physiological range, with a fairly broad optimum between 6.5 and 8.0. The addition of salts to the incubation mixture tends to reduce binding. Sodium represents an interesting exception that will be discussed later.

The inhibition of stereospecific binding by proteolytic enzymes [18,19] and a variety of protein reagents, including sulfhydryl reagents, suggests the involvement of protein moieties in opiate binding. The role of phospholipids has yet to be established. Binding is inhibited by some, but not all, phospholipase A preparations,[20,21] but not by phospholipase C or D. Moreover, H. K. Lin and Simon have recently shown that inhibition by phospholipase A can be reversed by washing the membrane preparation with a BSA solution,[22] raising further questions about the nature of the role of phospholipids in binding.

The extensive mapping studies can be summarized here only briefly. The highest levels of opiate receptors are found in areas of the limbic system and in the regions implicated in the pathways involved in pain reception, such as the perventricular and periaqueductal gray areas, the medial thalamus, the nucleus raphe magnus, and the substantia gelatinosa of the spinal cord. It has been suggested that the limbic system receptors may be involved in opiate-induced euphoria (or dysphoria) and in the affective aspects of pain perception.

Perhaps the most convincing evidence suggesting that stereospecific binding has pharmacological relevance comes from a number of studies that show excellent correlation between pharmacological potencies and in vitro binding affinities for a large number of drugs, varying in analgesic potencies over 5-6 orders of magnitude.[23,24] Such studies have also shown that neuroleptics such as haloperidol and spiroperidol have low (K_D *ca.*10^{-5}M) but definite affinities for the opiate receptor. The significance of this finding is not understood, but it is mentioned here because of its possible relationship to schizophrenia, to be discussed later.

Recently there has been considerable interest in the question of whether multiple types of opiate receptors exist. Using classical pharmacological approaches, W. R. Martin and collaborators have suggested the existence of three types of receptors, named μ (morphine), κ (ketocyclazocine), and σ (SKF 10,047).[25,26] Results from H. W. Kosterlitz's laboratory also provide evidence for the heterogeneity of opiate receptors.[27] Thus the receptors present in the myenteric plexus of the guinea pig ileum seem to have properties distinct from those in the mouse vas deferens. These workers have also reported evidence suggesting that the brain may possess at least two families of receptors differing in their affinity for enkephalins and for exogenous opiates.

ONTOGENY OF THE OPIATE RECEPTOR

The ontogenetic development of opiate receptors in the brain of rat and guinea pig was investigated by N. J. Clendeninn and coworkers[28] and by J. J. Coyle and Pert.[29] In the rat, the rate of increase of opiate binding is greatest (three- to fourfold) between the midfetal stage and three weeks postpartum. Thereafter until adulthood (ten to twenty weeks) the rise is more gradual—about twofold. Scatchard analysis of saturation curves for naltrexone binding in rat brain from one day after birth to the age of ten weeks has shown that the increase in binding is due to an increase in number of receptors rather than to enhanced affinity.

Opiate binding to guinea pig brain homogenates has demonstrated no significant difference in either receptor number or binding affinity between late fetal life and adulthood in this species. Binding in brain homogenate from a midterm fetus is about half that observed in the late fetal stage and in adult guinea pigs. The fact that guinea pigs are born with a full complement of receptors correlates well with previous reports that the guinea pig is an animal with almost full brain development before birth, whereas in the rat a significant portion of brain development continues for at least three weeks after birth. In rat brain the percentage increase of opiate receptors from newborn to adult in various regions of the brain differs widely.[30] For example, receptor binding in the hippocampus and parietal cortex increased by 830 percent and 690 percent, respectively, but in the medulla-pons and corpus striatum by only 270 and 300 percent.

The development of the opiate receptor in both rat and guinea pig closely parallels the development of other major morphological and biochemical components of the nervous system, suggesting that these receptors are themselves important components of the CNS.

CONFORMATIONAL CHANGES IN OPIATE RECEPTORS

I mentioned that sodium is an exception to the rule that increased salt concentrations present during incubation will inhibit binding. An apparent discrepancy between results in Snyder's laboratory[31] and in Simon's laboratory[32] led to the realization that sodium can distinguish between agonists and antagonists.[33,34] Whereas agonist binding is inhibited by sodium salts, the binding of antagonists is either unaffected or, in many instances, augmented by the presence of sodium. This remarkable discrimination by a small ion between closely related molecules (for example, morphine and nalorphine, oxymorphone and naloxone) is highly specific. It is exhibited, though less effectively, by lithium, but by none of the other alkali metals. In fact no other inorganic or organic cations have been found to exhibit this effect.[35] A study of the mechanism of this sodium effect has produced evidence that sodium is an allosteric effector, the binding of which produces a conformational change in the receptor. The sodium conformer has a higher affinity for antagonists and a markedly reduced affinity for agonists.

The best evidence for an alteration in receptor conformation by sodium ions came unexpectedly from a study of the kinetics of receptor inactivation by the sulfhydryl alkylating reagent, N-ethylmaleimide (NEM) by Simon and J. Groth.[36] When a membrane fraction from rat brain was incubated with NEM for various periods, followed by inactivation or removal of unreacted NEM, there was a progressive decrease in the ability of the membranes to bind opiates stereospecifically. The rate of receptor inactivation followed pseudo-first-order kinetics consistent with the existence of one SH-group per receptor essential for binding. Protection against inactivation was achieved by the addition of low concentrations of opiates or antagonists during the preincubation with NEM, suggesting that the SH-group is located near the opiate binding site of the receptor.

Considerable protection was observed—half-time of inactivation was increased from eight to thirty minutes—when inactivation was carried out in the presence of 100 mM NaCl. Since sodium salts had no effect on the alkylation of model SH-compounds such as cysteine or glutathione, this suggested that the SH-groups were made less accessible to NEM by a conformational change in the receptor protein. The fact that this protection exhibited the same ion specificity (Na^+ protect, Li^+ protects partially, K^+, Rb^+, or Cs^+ not at all), and the same dose-response to Na^+ as the differential changes in ligand affinities, suggests that the conformational change that masks SH-groups is the same as that which results in increased affinity of antagonists and decreased affinity of agonists.

These studies illustrate that the opiate receptor can alter its shape, but the physiological function of this plasticity is not yet clear. A role in the coupling of opiate binding to subsequent physical or chemical events has been suggested, as has a role for Na^+ ions in the action of opiates.

The simple model for the allosteric effect of sodium ions on opiate receptors shown in Figure 1 encompasses the various changes in receptor properties discussed here. One additional property, the evidence for which has not been discussed, is positive cooperativity of opiate binding, which is the reason for representing the receptor as (at least) a dimer.

PROGRESS IN RECEPTOR SOLUBILIZATION AND PURIFICATION

Progress in the solubilization and purification has been even slower for the opiate receptor than for some of the other receptors under consideration. One of the major reasons for this is the extreme sensitivity of the receptor to detergents, including the nonionic variety. Binding of opiates is inhibited by very low concentrations of Triton X-100 and other, usually relatively harmless, detergents.

The most promising advance is the solubilization by Simon and colleagues of an etorphine-macromolecular complex that has properties suggesting it may be an etorphine-receptor com-

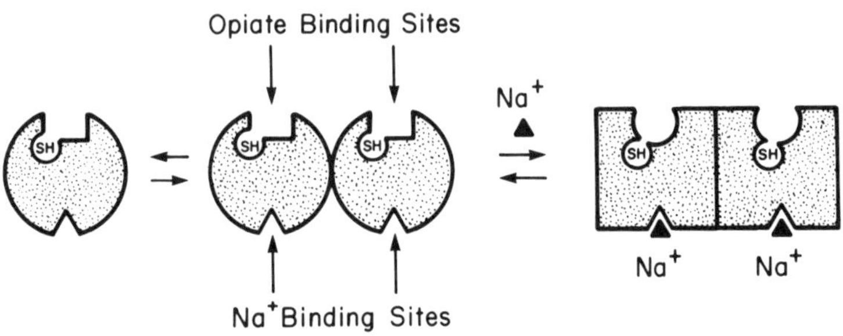

Figure 1. Model for the allosteric effect of sodium on the conformation of the opiate receptor.

plex.[37] To date the appropriate conditions have not been found under which, after dissociation of the prebound etorphine, rebinding of opiates can occur in solution. Moreover, the complex is sufficiently unstable that its purification is virtually impossible. A recent extension of this solubilization procedure by R. S. Zukin [38] may prove very useful. It involves showing that the identical procedure can be used for solubilization of an enkephalin-receptor complex. The solubilized complex was then covalently cross-linked by the use of suberimidate. This covalent enkephalin-receptor complex may lend itself to purification and to further study.

DISCOVERY OF ENDOGENOUS OPIOID PEPTIDES

The evidence that the brain of all vertebrates investigated, from the hagfish to man, contains opiate receptors led investigators to raise the question of why such receptors exist in the CNS and have survived eons of evolution. A physiological role for opiate receptors that conferred a selective advantage on the organisms seemed probable. None of the known neurotransmitters or neurohormones was found to exhibit high affinity for opiate receptors, which encouraged a number of laboratories to search for new opiate-like substances in extracts of animal brain. This search was successful first in the laboratories of J. Hughes and Kosterlitz [39] and Terenius and A. Wahlström.[40] Goldstein and his collaborators at about the same time reported opioid activity in extracts of pituitary glands.[41] Hughes utilized the in vitro bioassay for opiates, namely naloxone-reversible inhibition of electrically evoked contraction of the mouse vas deferens or the guinea pig ileum, while Terenius assayed endogenous opioid activity by measuring the ability of brain extracts and fractions to compete with labeled opiates for receptor binding.

These studies culminated in the identification of the opioid substances in extracts of pig brain by Hughes and coworkers.[42] They reported that the activity resided in two pentapeptides, Tyr-Gly-Gly-Phe-Met and Tyr-Gly-Gly-Phe-Leu, which they named methionine (Met) and leucine (Leu) enkephalin. This was confirmed by G. W. Pasternak and his associates, who found the same peptides in extracts of bovine brain.[43] Hughes's group also reported the interesting observation that the sequence of Met-enkephalin was present as amino acid residues 61-55 in the pituitary hormone β-lipotropin (βLPH). This hormone had been isolated in 1964 from pituitary glands by C. H. Li.[44] It possessed weak lipolytic activity, which was never seriously thought to be its real function. The report of Hughes and his collaborators, along with that of the Goldstein group, of the existence of opioid activity in the pituitary gland led R. Guillemin to examine the extracts of pig hypothalami and pituitary glands stored in his freezer. Two polypeptides with opioid activity were found and sequenced,[45] and they proved to have structures identical with amino acid sequences 61-76 and 61-77 of βLPH. Meanwhile, in two laboratories, potent opioid activity was found in the C-terminal fragment of βLPH (LPH 61-91) while the intact βLPH molecule was inactive.[46,47] The proliferation of endogenous peptides with opioid activity caused the author of this paper to suggest the generic term "endorphin," for endogenous morphine-like substance, which has been widely accepted. The C-terminal fragment was renamed β-endorphin by Li, while LPH 61-76 and 61-77 were named α and γ-endorphin, respectively, by Guillemin. In this discussion I shall use endorphin as the generic term for endogenous opioid peptides of which the enkephalins are a subgroup.

All the endorphins, including the enkephalins, exhibit opiate-like activity when injected intraventricularly. This activity includes analgesia, respiratory depression, and a variety of behavioral changes, including the production of a rigid catatonia. The pharmacological effects of the enkephalins are very fleeting, presumably due to their rapid destruction by peptidases. The longer-chain endorphins are more stable and produce long-lived effects. Thus analgesia due to β-endorphin, the most potent of all the endorphins so far found, can last for three to four hours. All of the responses to endorphins are readily reversed by opiate antagonists such as naloxone. There have been reports that β-endorphin and certain analogues of enkephalin can produce analgesia after systemic injection or even oral ingestion.[48,49]

Distribution of enkephalins has been studied by biochemical,[50] bioassay,[51] and immunohistochemical techniques.[52,53] The distribution in the CNS shows considerable, though not complete, correlation with the distribution of opiate receptors. Thus the globus pallidus has a very high density of enkephalin, or at least enkephalin-like immunoreactive material, while it is low in opiate receptors. Certain cortical areas dense in opiate receptors have low levels of enkephalin.

In the earlier studies of T. Hökfelt and colleagues all the immunofluorescence was found in nerve fibers and terminals, not in cell bodies.[54] In a more recent paper this group utilized colchicine,[55] which is known to arrest axonal transport. After such treatment it was possible to find immunofluorescence in cell bodies after treatment with antiserum to Met-enkephalin. More than twenty cell groups containing enkephalin have so far been observed in the brain and spinal cord, a larger number than the fifteen catecholamine cell groups known to exist in rat brain. The authors believe their results indicate these perikarya possess the machinery for enkephalin biosynthesis. This raises the interesting question of whether enkephalin is always derived from large peptide precursors, since levels of β-lipotropin and β-endorphin are very low in some of the areas that have high enkephalin levels.

Studies on the distribution of β-endorphin in the laboratories of Guillemin [56] and S. J. Watson [57] have provided convincing evidence for a distribution very different from that of the enkephalins. This has led to the suggestion that the CNS has separate enkephalinergic and endorphinergic neuronal systems. Beta-endorphin is present in the pituitary, where there is little or no enkephalin, as well as in certain regions of the brain. Brain β-endorphin seems to originate in a single set of neurons located in the periarcuate region of the hypothalamus, with axons projecting throughout the brain stem.

After this rather lengthy, but, I believe, essential introduction I will now discuss the possible involvement of the opiate receptor-endorphin system in a number of human pathologies. We will begin with the two obvious ones, pain and narcotic addiction, and then venture into the highly controversial area of the possible role of the endorphin system in mental disease.

PAIN AND ITS MODULATION

Since it was work on the opiate analgesics that led to the discovery of the endorphins and their receptors, it was natural to postulate that they may be involved in pain modulation. The fact, mentioned earlier, that all CNS regions implicated in the conduction of pain impulses have high levels of opiate receptors supports this hypothesis, as does the finding that opiates seem to inhibit selectively the firing of nociceptive neurons in the substantia gelatinosa of the dorsal spinal cord.[58] Moreover, as was also mentioned earlier, the intraventricular injection of all known opioid peptides produces analgesia. These findings did not prove that endogenous opioids are involved in the pain pathway, but were sufficiently suggestive to encourage further testing of this hypothesis.

Attempts were made to demonstrate the role of the natural opioid system in pain perception by the use of the opiate antagonist naloxone. It was postulated that, if receptor occupancy by endorphins was involved in pain modulation, the administration of an opiate antagonist should lower the threshhold or exacerbate perceived pain. Such an effect has been surprisingly difficult to demonstrate conclusively. Of a great number of attempts in animals, only those of J .J. Jacob's group were successful in demonstrating a lowering of pain threshold by naloxone[59] (decrease in latency period before rats or mice jumped off a hotplate). The results have recently been replicated in other laboratories.[60] Attempts to show exacerbation of experimental pain or lowering of pain threshhold in human volunteers have been unsuccessful.[61,62]

Better success has been achieved by studying the effect of naloxone in several types of analgesia that do not involve the use of drugs. Analgesia produced by electrical stimulation of the periaqueductal gray area of both animals[63] and humans[64] has been shown to be at least partially reversed by the administration of naloxone.

A more dramatic reversal by naloxone has been observed with analgesia resulting from electroacupuncture in mice [65] and from acupuncture in human subjects.[66] The involvement of opiate receptors and endorphins, not involving naloxone, in electroacupuncture has received additional support by the finding that the CXBK strain of mice, which is deficient in opiate receptors, shows poor analgesia in response to electroacupuncture.[67]

A recent report indicates that analgesia due to "placebo effect" is also reversed by naloxone in patients with postoperative pain after extraction of impacted third molars.[68] In this regard it is of interest that analgesia produced by hypnosis was found not to be affected by naloxone.[69]

These results, although indirect, are supportive of the idea that the endorphin system may be involved in an endogenous pain-modulation system. Such a system is likely to be of great survival value to the organisms, since it will permit it to experience pain as an important warning of tissue damage without the suffering, except in pathological states, of unbearable, disabling pain. The importance of pain to the individual is best demonstrated by a disease called congenital insensitivity to pain: individuals with this condition are unable to feel pain from either visceral or superficial tissue damage. This is a serious pathology that results in a significantly shortened life expectancy. A number of laboratories, including our own, are currently studying such patients to determine whether an abnormality in the opiate receptor-endorphin system may play a role in this inborn error.

There have indeed been two brief reports in *Lancet* proposing that naloxone may have an effect on these patients. H. Dehen and colleagues report that the flexion-reflex threshhold, which was 350 percent higher in the patient than in controls, fell dramatically by 67 percent within ten minutes of naloxone administration.[70] The amount of naloxone administered was not stated. H. Yanagida reports that tooth-pulp evoked potentials (TPEP) are absent in patients with congenital insensitivity to pain. Ten minutes after administration of naloxone (10 mg. i.v.), however, some TPEP were observed, especially after stimuli of 15 mA, the highest used. The patient, without apparent distress, was able to recognize these stimuli as pressure sensations.[71] These reports are very preliminary and the effects observed are minimal. Nevertheless, they suggest that the endorphin system could play a role in this disorder. The administration of large amounts of a long-acting antagonist should be attempted as a possibly beneficial treatment of this serious pathology.

To prove conclusively the involvement of endorphins in endogenous pain regulation requires direct demonstration of changes or turnover in endorphin level during severe acute or chronic pain and during analgesia. Terenius and Wahlström found decreases in opioid activity in the cerebrospinal fluid (CSF) of patients suffering chronic pain due to trigeminal neuralgia when compared to patients who were not in pain.[72] More recently this group also reported an increase in level of opioid activity after analgesia by electroacupuncture in patients with intractable pain.[73] These reports have the drawback that the opioid activity in the CSF represents as yet poorly characterized chromatographic fractions that seem to differ from all the known and well-characterized endorphins.

Reports on the release of well-characterized endorphins during analgesia have also appeared. D. E. Richardson and H. Akil have reported good success in relieving intractable pain in patients by electrical stimulation with electrodes implanted in the periventricular gray region.[74] The region found to be most effective, and the one at which the least side effects were seen, was near the posterior commissure, adjacent to the ventricular wall at the level of the nucleus parafascicularis. This method is adaptable for pain relief by periodic self-stimulation by the patient over a period of months or even years. The analgesia so produced was found to be partially reversible by naloxone administration. More recently Akil and colleagues have shown that stimulation analgesia is followed after fifteen to twenty minutes by a small (<twofold) but significant release of Met-enkephalin into the ventricular CSF, measured by several techniques including bioassay, receptor binding, and radioimmunoassay.[75] The same group has reported a more dramatic release of β-endorphin-like immunoreactive material under identical conditions of anal-

gesia.[76] Base-line release was undetectable (<25 fmol per ml. of CSF), while the level measured during stimulation-produced analgesia in five patients ranged from 319 to 481 fmol/ml., an increase of at least 13- to 20-fold.

Thus an impressive array of at least circumstantial evidence is accumulating suggesting the involvement of the enkephalins, and probably also brain β-endorphin, in an endogenous analgesic system.

At this point it may be appropriate to say a word about stress. Guillemin's group has shown a parallel and equimolar secretion of ACTH and β-endorphin into the blood of rats that have been severely stressed.[77] The involvement of pituitary β-endorphin in stress is further supported by the exciting discovery by R. Mains and coworkers that ACTH and β-endorphin are synthesized via a common large precursor peptide of molecular weight 31,000.[78] The question of the relationship between pituitary endorphin, evidently involved in reaction to stress, and the brain enkephalins and β-endorphin, which appear to play a role in pain modulation, is a very interesting and as yet unanswered one. The significance of the question is further enhanced by reports that certain types of stress can produce analgesia.

NARCOTIC ADDICTION

Another expected action of the endogenous opioid system is its participation in the development of narcotic addiction. The evidence for this turns out to be more difficult to obtain than that for pain modulation.

All opioid peptides will produce tolerance and physical dependence when injected repeatedly. Cross-tolerance with plant alkaloid opiates has also been shown. This does not prove that tolerance/dependence develop to endogenously produced and released endorphins, nor that these peptides and their receptors are involved in the formation of tolerance to and dependence on narcotics.

A theory that predated the discovery of the opiate receptors suggests changes in either the number or binding characteristics of opiate receptors. A change in binding affinities similar to that seen when sodium concentration is increased during in vitro binding is especially attractive, since sensitivity to agonists decreases during tolerance formation, while sensitivity to antagonists increases dramatically. W. A. Klee and R. A. Streaty examined this question in whole rat brain and found no changes in the number or affinities of opiate binding sites.[79] K. A. Bonnet and his coworkers believed this might be explicable by a "drowning out" of changes occurring in only a few brain regions.[80] But an examination of receptor number and binding affinities in the medial thalamus, periventricular gray region, and caudate nucleus gave equally negative results; whereas these three areas have high levels of receptors and/or have been implicated in various aspects of opiate action, the possibility still remains that they were not the appropriate areas to examine. It is at least equally possible, however, that detectable changes in receptors do not occur during chronic morphinization.

The possibility that a change in endorphin level might be observed during tolerance/dependence development has also received attention. A report by R. Simantov and Snyder that enkephalin levels are elevated in brains of tolerant rats [81] was refuted by more recent experiments in the same laboratory; [82] the earlier work, which had been done using a radioreceptor assay, was not supported when the much more specific radioimmunoassay was used.

A 1978 report stated that the intravenous administration of 4 mg. of human β-endorphin to human addicts led to dramatic improvement in severe abstinence syndromes; [83] there was no euphoria and little adverse effect. In a double-blind study it was found that subjects were able to distinguish morphine and β-endorphin. After endorphin treatment they felt thirsty, dizzy, sleepy, and warm, and had "a strange feeling throughout the body." All these symptoms disappeared in twenty minutes, however, whereas the beneficial effects of endorphin on the withdrawal syndrome lasted for several days. The long-lasting suppression, especially of the most severe symptoms of abstinence (vomiting, diarrhea, tremor, and restlessness), by a single dose of β-endorphin suggested to the authors the possibility that this endogenous peptide may indeed have a role in the mechanism of tolerance/dependence development to opiates.

Thus a role of the opiate receptor-endorphin system, while expected and fervently hoped for, has not yet been established. But the evidence cited is sufficiently suggestive to warrant further research in this area.

For the sake of completeness I should like to mention two recent developments of considerable interest for which the relationship to the opiate receptor is still unknown.

R. Walter's group has reported it was possible to suppress the abstinence syndrome when rats were withdrawn from chronic morphine by administration of the dipeptide Z-Pro-D-Leu.[84] There was no effect on the analgesic response to morphine. The mechanism of this phenomenon is not understood.

Based on the abundant literature that seems to implicate catecholamines in the actions of opiates, M. Gold and coworkers treated human heroin addicts with clonidine.[85] In a double-blind, placebo-controlled study, clonidine eliminated objective signs and subjective symptoms of opiate withdrawal for four to six hours in all addicts. In an open pilot study the same patients did well while taking clonidine for one week. All the patients had been addicted to opiates for six to ten years and had been on methadone for six to sixty months at the time of the study. The authors suggest that their success with clonidine indicates that abstinence may be produced by an interaction between opiate receptors and α-2 adrenergic receptors in the mediation of effects by endogenous opiates in noradrenergic areas such as the locus coeruleus.

MENTAL DISEASE

The reports that have created the greatest excitement—and also the most intense controversy—are those suggesting the involvement of the opiate receptor system in mental disease.

Widespread investigation of this question began in 1976 with side-by-side reports in *Science* by F. Bloom and coworkers[86] and Y. F. Jacquet and N. Marks.[87] Both groups reported behavioral effects from intraventricularly administered β-endorphin reminiscent of some aspects of schizophrenia. The most striking of these effects was a catatonic-like immobility, described as rigid by Bloom and as having "waxy flexibility" by Jacquet and Marks. This catatonic state was readily reversible by systemic administration of naloxone. Bloom suggested that mental illness could result from subtle derangements in endorphin metabolism. Jacquet and Marks, on the other hand, raised the question of whether β-endorphin might be an endogenous neuroleptic or an anti-psychotogen.

Soon thereafter came the report by Terenius's group that opioid peptide fractions (not characterized) present in CSF were increased in unmedicated schizophrenic patients and in manic-depressive patients in the manic phase.[88] Medication of schizophrenics and transition of manic patients to the depressed state led to a reduction in the CSF opioid activity. L.-M. Gunne and colleagues followed this up by attempts to treat schizophrenic patients with naloxone.[89] In a single-blind study (0.4 mg. naloxone i.v.) they reported a decrease in auditory hallucinations in four of the six patients treated.

Attempts to replicate this provocative result were made by several investigators, with little success. J. Volavka's group found no effect using the same dose of naloxone.[90] G. C. Davis and coworkers used 0.4 mg. of naloxone in a large number of patients, and within one hour of administration found a suggestion of improved cognition, but no change in hallucination.[91] D. S. Janowski's group [92] and A. A. Kurland and his colleagues [93] reported no effect when 1.2 mg. of naloxone was given.

The possibility was raised that the dose of naloxone administered may have been too low and the interval studied too short. H. M. Emrich's group treated hallucinating psychotic patients with 4 mg. of naloxone i.v.; [94] they found no change within half an hour of administration, but a decrease in hallucinations at a later time.

S. Watson and coworkers reported a randomized, double-blind, cross-over study on nine schizophrenic subjects who received intravenous infusions of 10 mg. of naloxone or of the vehicle only.[95] Statistical analysis of the data indicated a significant decrease in auditory hallucinations for six of the nine patients after treatment with naloxone. The authors feel that naloxone does have a beneficial effect on the hallucinatory aspect of schizophrenia, but that high doses are required and intervals longer than one hour must be studied. They also believe that only a subclass of diagnosed schizophrenic patients may respond. If these observations are sustained by further experiments it will represent an exciting step forward in the treatment of schizophrenic patients, and will imply a role of some kind for the endogenous opioid system in aspects of mental disease.

A study that seems to contradict the foregoing results involves the administration of β-endorphin to psychotic patients. N. S. Kline's group administered up to 10 mg. of β-endorphin i.v. to psychiatric patients and reported "substantial salutory effects," especially in depressed patients.[96] J. Angst in a similar study observed that intravenous infusion of 10 mg. of β-endorphin caused manic-depressive patients to switch from the depressed to the manic phase.[97] By the authors' own admission, however, both these studies were poorly controlled, mainly because of the expense of β-endorphin. But the results are sufficiently intriguing to encourage carefully controlled double-blind experiments.

Two other approaches to the question of involvement of the endorphin system in mental disease will be discussed briefly. The first began with the report by H. Wagemaker, Jr. and R. Cade that substantial improvement was observed in schizophrenic patients after sixteen weeks of hemodialysis.[98] R. M. Palmour and associates obtained some of the dialyzate and analyzed it for the presence of any unusual substance that could be a causative agent. They reported the presence of a polypeptide they have tentatively characterized at the Leu^5 analogue of β-endorphin, a peptide long searched for as the putative precursor of Leu-enkephalin.[99] Over the sixteen-week period of hemodialysis the authors reported a steady drop in the level of the opioid peptide, which seems to correlate with the time course of clinical improvement. These results, if confirmed, are of great importance. Both the existence of the peptide and its chemical structure require confirmation. If the substance proves to be an opioid peptide, its relationship to schizophrenia requires further study, because to the best of our knowledge the peptide has not been looked for in the hemodialysate of nonpsychotic patients.

The most recent approach to this problem comes from the laboratory of D. DeWied in Utrecht. Since it has been demonstrated that the peptide bond of endorphins most vulnerable to peptidase hydrolysis is the one between the N-terminal tyrosine and the glycine, this group decided to study the (des-Tyr^1) endorphins. Utilizing behavioral tests, they recently reported that (des-Tyr^1) γ-endorphin exhibits neuroleptic-like effects in rats.[100] Based on these results W. M. A. Verhoeven and colleagues administered this drug to six schizophrenic patients, all of whom were resistant to conventional neuroleptic therapy. All six patients showed improvement: three showed long-lasting improvement that survived the end of treatment, while improvement in the other three was of short duration only.[101] Clinical studies are being continued. The authors suggest that altered endorphin homeostasis may play a role in schizophrenia. They also advance the hypothesis that (des-Tyr^1) γ-endorphin may be an endogenous neuroleptic.

In summary, there are several lines of titillating evidence that suggest a role for the opiate receptor-endorphin system in some aspects of mental disease. None of the evidence provides clear-cut proof, and further work is clearly required to prove this provocative hypothesis or lay it to rest.

CONCLUDING REMARKS

The discovery of opiate receptors and their putative endogenous ligands, the endorphins, has kindled the excitement and imagination of many scientists and, through ample coverage in the news media, of the general public. Hopes have been raised that these findings may contribute to the solution of a number of human pathologies ranging from intractable pain to mental disease.

As I have indicated, there is not as yet clear-cut evidence for the involvement of the opiate receptor in any human disease, but the evidence is sufficiently suggestive to encourage much further research in many competent laboratories and hospitals.

There is an interesting difference between this area of research and those involving receptors for other hormones and neurotransmitters. In the other cases the endogenous ligand was discovered and known for some time before a receptor was postulated, searched for, and identified. The opiate field began with the identification of a drug receptor. The proof that such a receptor existed led to the search for endogenous ligands for the receptors and to the identification of a number of peptides with opioid activity. This approach is now being applied to other drug receptors, where it is believed unlikely that their existence anticipated the relatively recent development of the drug. A case in point is the discovery of specific binding sites for the benzodiazepine tranquilizer, Valium: many laboratories are currently engaged in a search for the endogenous ligand for this receptor. Does the body produce its own tranquilizing substance? Is it one of the substances we are already familiar with? Or is it a substance yet to be identified? This approach might conceivably be generalizable to other receptors for exogenous substances, such as drugs, viruses, and toxins. In those cases in which a selective advantage to the organism is not evident, a search for an endogenous ligand and a physiological role for the receptor might prove worthwhile.

It should be remembered that the opiate receptor field is only five years old, and fundamental information regarding the physiological role of the endorphins and of the receptor is still missing. A real understanding of the role of this receptor-ligand system in human disease may have to await the elucidation of its functions in normal animals and man.

NOTES

1. E. J. Simon and J. M. Hiller, "The Opiate Receptors," *Annual Reviews of Pharmacology and Toxicology* 18 (1978): 371-94.

2. L. Terenius, "Endogenous Peptides and Analgesia," *Annual Reviews of Pharmacology and Toxicology* 18 (1978): 189-204.

3. E. J. Simon and D. Van Praag, "Studies on the Intracellular Distribution and Tissue Binding of Dihydromorphine-7,8-^{3}H in the Rat," *Proceedings of the Society for Experimental Biology and Medicine* 122 (1966): 6-11.

4. N. A. Ingoglia and V. P. Dole, "Localization of *d*- and *l*-Methadone after Intraventricular Injection into Rat Brains," *Journal of Pharmacology and Experimental Therapeutics* 175 (1970): 84-87.

5. N. A. Ingoglia and V. P. Dole, "Localization of *d*- and *l*-Methadone after actions of the Morphine Congener Levorphanol in Subcellular Fractions of Mouse Brain," *Proceedings of the National Academy of Sciences* 68 (1971): 1742-47.

6. E. J. Simon, J. M. Hiller, and I. Edelman, "Stereospecific Binding of the Potent Narcotic Analgesic ^{3}H-Etorphine to Rat Brain Homogenate," *Proceedings of the National Academy of Sciences* 70 (1973): 1947-49.

7. C. B. Pert and S. H. Snyder, "Opiate Receptor: Demonstration in Nervous Tissue," *Science* 179 (1973): 1011-14.

8. L. Terenius, "Stereospecific Interaction Between Narcotic Analgesics and a Synaptic Plasma Membrane Fraction of Rat Cerebral Cortex," *Acta Pharmacologica et Toxicologica* (1973): 17-20.

9. J. M. Hiller, J. Pearson, and E. J. Simon, "Distribution of Stereospecific Binding of the Potent Narcotic Analgesic Etorphine in the Human Brain: Predominance in the Limbic System," *Research Communications in Chemical Pathology and Pharmacology* 6 (1973): 1052-62.

10. C. B. Pert, D. Aposhian, and S. H. Snyder, "Phylogenetic Distribution of Opiate Receptor Binding," *Brain Research* 75 (1974): 356-61.

11. Hiller, Pearson, and Simon, "Distribution of Stereospecific Binding" (See note 9).

12. M. J. Kuhar, C. B. Pert, and S. H. Snyder, "Regional Distribution of Opiate Receptor Binding in Monkey and Human Brain," *Nature* 245 (1973): 447-50.

13. C. B. Pert, M. J. Kuhar, and S. H. Snyder, "Autoradiographic Localization of the Opiate Receptor in Rat Brain," *Life Sciences* 16 (1975): 1849-54.

14. S. F. Atweh and M. J. Kuhar, "Autoradiographic Localization of Opiate Receptors in Rat Brain. I. Spinal Cord and Lower Medulla," *Brain Research* 124 (1977): 53-67.

15. ———, "Autoradiographic Localization of Opiate Receptors in Rat Brain. II. The Brainstem," *Brain Research* 129 (1977): 1-12.

16. ———, "Autoradiographic Localization of Opiate Receptors in Rat Brain. III. The Telencephalon," *Brain Research* 134 (1977): 393-406.

17. K. D. Stahl et al., "Receptor Affinity and Pharmacological Potency of a Series of Narcotic Analgesic, Antidiarrheal and Neuroleptic Drugs," *European Journal of Pharmacology* 46 (1977): 199-205.

18. Simon, Hiller, and Edelman, "Stereospecific Binding" (See note 6).

19. G. W. Pasternak and S. H. Snyder, "Opiate Receptor Binding: Effects of Enzymatic Treatment," *Molecular Pharmacology* 10 (1973): 183-93.

20. Simon, Hiller, and Edelman, "Stereospecific Binding" (See note 6).

21. Pasternak and Snyder, "Opiate Receptor Binding" (See note 19).

22. H. K. Lin and E. J. Simon, "Phospholipase A Inhibition of Opiate Receptor Binding Can Be Reversed by Albumin," *Nature* 271 (1978): 383-84.

23. Stahl et al., "Receptor Affinity" (See note 17).

24. R. S. Wilson et al., "Homologous N-Alkylnorketobemidones. Correlation of Receptor Binding with Analgesic Potency," *Journal of Medicinal Chemistry* 18 (1975): 240-42.

25. W. R. Martin et al., "The Effects of Morphine- and Nalorphine-Like Drugs in the Nondependent and Morphine-Dependent Chronic Spinal Dog," *Journal of Pharmacology and Experimental Therapeutics* 197 (1976): 517-32.

26. P. E. Gilbert and W. R. Martin, "The Effects of Morphine and Nalorphine-

Like Drugs in the Non-Dependent and Cyclazocine-Dependent Chronic Spinal Dog," *Journal of Pharmacology and Experimental Therapeutics* 198 (1976): 66-82.

27. J. A. H. Lord et al., "Endogenous Opioid Peptides: Multiple Agonists and Receptors," *Nature* 267 (1977): 495-99.

28. N. J. Clendeninn, M. Petraitis, and E. J. Simon, "Ontological Development of Opiate Receptors in Rodent Brain," *Brain Research* 118 (1976): 157-60.

29. J. T. Coyle and C. B. Pert, "Ontogenetic Development of ^{3}H-Naloxone Binding in Rat Brain," *Neuropharmacology* 15 (1976): 555-60.

30. Ibid.

31. Pert and Snyder, "Opiate Receptor" (See note 7).

32. Simon, Hiller, and Edelman, "Stereospecific Binding" (See note 6).

33. C. B. Pert and S. H. Snyder, "Opiate Receptor Binding of Agonists and Antagonists Affected Differentially by Sodium," *Molecular Pharmacology* 10 (1974): 868-79.

34. E. J. Simon et al., "Further Properties of Stereospecific Opiate Binding Sites in Rat Brain: On the Nature of the Sodium Effect," *Journal of Pharmacology and Experimental Therapeutics* 192 (1975): 531-37.

35. Ibid.

36. E. J. Simon and J. Groth, "Kinetics of Opiate Receptor Inactivation by Sulfhydryl Reagents: Evidence for Conformational Change in Presence of Sodium Ions," *Proceedings of the National Academy of Sciences* 72 (1975): 2404-07.

37. E. J. Simon, J. M. Hiller, and I. Edelman, "Solubilization of a Stereospecific Opiate-Macromolecular Complex from Rat Brain," *Science* 190 (1975): 389-90.

38. R. S. Zukin, "Solubilization of a Covalent ^{3}H-Enkephalin-Macromolecular Complex," *Federation Proceedings* 36 (1977): 480 (Abstr.).

39. J. Hughes, "Isolation of an Endogenous Compound from the Brain with Properties Similar to Morphine," *Brain Research* 88 (1975): 295-308.

40. L. Terenius and A. Wahlström, "Inhibitor(s) of Narcotic Receptor Binding in Brain Extracts and Cerebrospinal Fluid," *Acta Pharmacologica et Toxicologica* 35, suppl. 1 (1974): 55.

41. H. Teschemacher et al., "A Peptide-Like Substance from Pituitary That Acts Like Morphine. I. Isolation," *Life Sciences* 16 (1975): 1771-76.

42. J. Hughes et al., "Identification of Two Related Pentapeptides from the Brain with Potent Opiate Agonist Activity," *Nature* 258 (1975): 577-79.

43. G. W. Pasternak, R. Goodman, and S. H. Snyder, "An Endogenous Morphine-Like Factor in Mammalian Brain," *Life Sciences* 16 (1975): 1765-69.

44. C. H. Li, "Lipotropin: A New Active Peptide from Pituitary Glands," *Nature* 201 (1964): 924.

45. N. Ling, R. Burgus, and R. Guillemin, "Isolation, Primary Structure, and Synthesis of α-Endorphin and γ-Endorphin, Two Peptides of Hypothalamic-Hypophysial Origin with Morphinomimetic Activities," *Proceedings of the National Academy of Sciences* 73 (1976): 3942-46.

46. B. M. Cox, A. Goldstein, and C. H. Li, "Opioid Activity of a Peptide, β-Lipotropin-(61-91), Derived from β-Lipotropin," *Proceedings of the National Academy of Sciences* 73 (1976): 1821-23.

47. A. F. Bradbury et al., "C Fragment of Liptropin Has a High Affinity for Brain Opiate Receptors," *Nature* 260 (1976): 793-95.

48. C. H. Li, "β-Endorphin: A Pituitary Peptide with Potent Morphine-Like Activity," *Archives of Biochemistry and Biophysics* 183 (1977): 592-604.

49. D. Roemer et al., "A Synthetic Enkephalin Analogue with Prolonged Parenteral and Oral Analgesic Activity," *Nature* 268 (1977): 547-49.

50. R. Simantov et al., "The Regional Distribution of a Morphine-Like Factor Enkephalin in Monkey Brain," *Brain Research* 106 (1976): 189-97.

51. J. Hughes, H. W. Kosterlitz, and T. W. Smith, "The Distribution of Methionine-Enkephalin and Leucine-Enkephalin in the Brain and Peripheral Tissues," *British Journal of Pharmacology* 61 (1977): 639-47.

52. R. Elde et al., "Immunohistochemical Studies Using Antibodies to Leucine-Enkephalin: Initial Observations on the Nervous System of the Rat," *Neuroscience* 1 (1976): 349-51.

53. R. Simantov et al., "Opioid Peptide Enkephalin: Immunohistochemical Mapping in Rat Central Nervous System," *Proceedings of the National Academy of Sciences* 74 (1977): 2167-71.

54. Elde et al., "Immunohistochemical Studies" (See note 52).

55. T. Hökfelt et al., "The Distribution of Enkephalin-Immunoreactive Cell Bodies in the Rat Central Nervous System," *Neuroscience Letters* 5 (1977): 25-31.

56. J. Rossier et al., "Regional Dissociation of Beta-Endorphin and Enkephalin Contents in Rat Brain and Pituitary," *Proceedings of the National Academy of Sciences* 74 (1977): 5162-65.

57. S. J. Watson, J. D. Barchas, and C. H. Li, "β-Lipotropin: Localization of Cells and Axons in Rat Brain by Immunocytochemistry," *Proceedings of the National Academy of Sciences* 74 (1977): 5155-58.

58. A. W. Duggan, J. G. Hall, and P. M. Headley, "Suppression of Transmission of Nociceptive Impulses by Morphine: Selective Effects of Morphine Administered in the Region of the Substantia Gelatinosa," *British Journal of Pharmacology* 61 (1977): 65-76.

59. J. J. Jacob, E. C. Tremblay, and M.-C. Colombel, "Facilitation de Reactions Nociceptives par la Naloxone Chez le Souris et Chez le Rat," *Psychopharmacologia* 37 (1974): 217-23.

60. R. C. A. Frederickson, V. Burgis, and J. D. Edwards, "Hyperalgesia Induced by Naloxone Follows Diurnal Rhythm in Responsivity to Painful Stimuli," *Science* 198 (1977): 756-58.

61. A. El-Sobky, J. D. Dostrovsky, and P. D. Wall, "Lack of Effect of Naloxone on Pain Perception in Humans," *Nature* 263 (1976): 783-84.

62. P. Grevert and A. Goldstein, "Endorphins: Naloxone Fails to Alter Experimental Pain or Mood in Humans," *Science* 199 (1978): 1093-95.

63. H. Akil, D. J. Mayer, and J. C. Liebeskind, "Antagonism of Stimulation-Produced Analgesia by Naloxone, a Narcotic Antagonist," *Science* 191 (1976): 961-62.

64. Y. Hosobuchi, T. F. Adams, and R. Linchitz, "Pain Relief by Electrical Stimulation of the Central Gray Matter in Humans and Its Reversal by Naloxone," *Science* 197 (1977): 183-86.

65. B. Pomeranz and D. Chiu, "Naloxone Blockade of Acupuncture: Endorphin Implicated," *Life Sciences* 19 (1976): 1757-62.

66. D. J. Mayer, D. D. Price, and A. Rafii, "Antagonism of Acupuncture Analgesia in Man by the Narcotic Antagonist Naloxone," *Brain Research* 121 (1977): 368-72.

67. J. M. Peets and B. Pomeranz, "CXBK Mice Deficient in Opiate Receptors Show Poor Electroacupuncture Analgesia," *Nature* 273 (1978): 675-76.

68. J. D. Levine, N. C. Gordon, and H. L. Fields, "The Mechanism of Placebo Analgesia," *Lancet* (September 23, 1978): 654-57.

69. A. Goldstein and E. R. Hilgard, "Failure of the Opiate Antagonist Naloxone to Modify Hypnotic Analgesia," *Proceedings of the National Academy of Sciences* 72 (1975): 2041-43.

70. H. Dehen et al., "Congenital Insensitivity to Pain, and Endogenous Morphine-like Substances," *Lancet* (August 6, 1977): 293-94.

71. H. Yanagida, "Congenital Insensitivity and Naloxone," *Lancet* (September 2, 1978): 520-21.

72. L. Terenius and A. Wahlström, "Morphine-Like Ligand for Opiate Receptors in Human CSF," *Life Sciences* 16 (1975): 1759-64.

73. B. Sjolund, L. Terenius, and M. Erikson, "Increased Cerebral Spinal Fluid Levels of Endorphins after Electroacupuncture," *Acta Physiologica Scandinavica* 100 (1977): 382-84.

74. D. E. Richardson and H. Akil, "Pain Reduction by Electrical Brain Stimulation in Man. Part I: Acute Administration in Periaqueductal and Periventricular Sites," *Journal of Neurosurgery* 47 (1977): 178-83.

75. H. Akil et al., "Enkephalin-Like Material Elevated in Ventricular Cerebrospinal Fluid of Pain Patients after Analgetic Focal Stimulation," *Science* 201 (1978): 463-65.

76. H. Akil et al., "Appearance of β-Endorphin-Like Immunoreactivity in Human Ventricular Cerebrospinal Fluid upon Analgesic Electrical Stimulation," *Proceedings of the National Academy of Sciences* 75 (1978): 5170-72.

77. R. Guillemin et al., "β-Endorphin and Adrenocorticotropin Are Secreted Concomitantly by the Pituitary Gland," *Science* 197 (1977): 1367-69.

78. R. Mains, E. Eipper, and N. Ling, "Common Precursor to Corticotropin and Endorphins," *Proceedings of the National Academy of Sciences* 74 (1977): 3014-18.

79. W. A. Klee, and R. A. Streaty, "Narcotic Receptor Sites in Morphine-Dependent Rats," *Nature* 248 (1974): 61-63.

80. K. A. Bonnet, J. M. Hiller, and E. J. Simon, "The Effects of Chronic Opiate Treatment and Social Isolation on Opiate Receptors in Rodent Brain," in *Opiates and Endogenous Opioid Peptides, Proceedings of the International Narcotic Research Club Meeting,* ed. H. K. Kosterlitz (Amsterdam: North-Holland Publishing Co., 1976): 335-43.

81. R. Simantov and S. H. Snyder, "Elevated Levels of Enkephalin in Morphine-Dependent Rats," *Nature* 262 (1976): 505-07.

82. S. R. Childers, R. Simantov, and S. H. Snyder, "Enkephalin: Radioimmunoassay and Radioreceptor Assay in Morphine-Dependent Rats," *European Journal of Pharmacology* 46 (1977): 289-93.

83. C. Y. Su et al., "Effects of β-Endorphin on Narcotic Abstinence Syndrome in Man," *Journal of the Formosan Medical Association* 77 (1978): 133-41.

84. R. Walter et al., "Inhibition by Z-Pro-D-Leu of Development of Tolerance to and Physical Dependence on Morphine in Mice," *Proceedings of the National Academy of Sciences* 75 (1978): 4573-76.

85. M. Gold, D. E. Redmond, Jr., and H. D. Kleber, "Clonidine Blocks Acute Opiate Withdrawal Symptoms," *Lancet* (September 16, 1978): 599-601.

86. F. Bloom et al., "Endorphins: Profound Behavioral Effects in Rats Suggest New Etiological Factors in Mental Illness," *Science* 194 (1976): 630-32.

87. Y. F. Jacquet and N. Marks, "The C-Fragment of β-Lipotropin: An Endogenous Neuroleptic or Antipsychotogen?" *Science* 194 (1976): 632-34.

88. L. Terenius et al., "Increased Levels of Endorphins in Chronic Psychosis," *Neuroscience Letters* 3 (1976): 157-62.

89. L.-M. Gunne, L. Lindstrom, and L. Terenius, "Naloxone-Induced Reversal of Schizophrenic Hallucinations," *Journal of Neural Transmission* 40 (1977): 13-19.

90. J. Volavka et al., "Naloxone in Chronic Schizophrenia," *Science* 196 (1977): 1227-28.

91. G. C. Davis et al., "Intravenous Naloxone Administration in Schizophrenia and Affective Illness," *Science* 197 (1977): 74-77.

92. D. S. Janowsky et al., "Lack of Effect of Naloxone on Schizophrenic Symptoms," *American Journal of Psychiatry* 134 (1977): 926-27.

93. A. A. Kurland et al., "The Treatment of Perceptual Disturbances in Schizophrenia with Naloxone Hydrochloride," *American Journal of Psychiatry* 134 (1977): 1408-10.

94. H. M. Emrich et al., "Indication of an Antipsychotic Action of the Opiate Antagonist Naloxone," *Pharmakopsychiatrie* 10 (1977): 265-70.

95. S. Watson et al., "Effects of Naloxone in Schizophrenia: Reduction in Hallucinations in a Subpopulation of Subjects," *Science* 201 (1978): 73-76.

96. N. S. Kline et al., "β-Endorphin-Induced Changes in Schizophrenic and Depressed Patients," *Archives of General Psychiatry* 34 (1977): 1111-13.

97. J. Angst, "Preliminary Results of Treatment with β-Endorphin in Depression," in *Endorphins in Mental Health Research,* ed. E. Usdin, W. E. Bunney, Jr., and N. S. Kline (London: MacMillan Press, Ltd., 1979): 518-28.

98. H. Wagemaker, Jr. and R. Cade, "The Use of Hemodialysis in Chronic Schizophrenia," *American Journal of Psychiatry* 134 (1977): 684-85.

99. R. M. Palmour et al., "Characterization of a Peptide from the Serum of Psychotic Patients," in *Endorphins in Mental Health Research,* ed. E. Usdin, W. E. Bunney, Jr., and N. S. Kline (London: MacMillan Press, Ltd., 1979): 581-93.

100. D. DeWied et al., "Neuroleptic-Like Activity of (Des-Tyr[1]) γ-Endorphin in Rats," *Lancet* (May 13, 1978): 1046.

101. W. M. A. Verhoeven et al., "(Des-Tyr[1]) γ-Endorphin in Schizophrenia," *Lancet* (May 13, 1978): 1046-47.

RED CELL RECEPTORS FOR MALARIA PARASITES

Louis H. Miller

INTRODUCTION

The evidence for receptor specificity in parasitic protozoa is limited, although parasites with complex life cycles in vertebrate and invertebrate hosts must have specific mechanisms for the recognition and invasion of host cells. The most detailed information has derived from studies on the invasion of red blood cells (RBCs) by malaria parasites.

This paper will discuss the mechanism by which malaria merozoites gain entry into RBCs. Merozoites do not puncture a hole in the membrane so as to enter the cytoplasm of the RBC, but instead enter within an invagination of the RBC membrane. The invaginated membrane seals at its orifice and becomes the vesicle in which the parasite develops. The problem is: How does a free-floating parasite (the merozoite) with limited motility gain entry into a nonphagocytic cell, the mature RBC? We will attempt to answer this question, at least in part.

LIFE CYCLE OF MALARIA

Plasmodia are protozoa transmitted by *Anopheles* mosquitoes. Sporozoites inoculated by the mosquito become localized within parenchymal cells of the liver, where they divide repeatedly. Within a liver cell one sporozoite can develop into 30,000 merozoites, each capable of invading an RBC. Once within RBCs the parasites divide by a process called schizogony, which results in the production of merozoites that are liberated to invade other RBCs. Disease in malaria is caused by these repeated cycles of asexual development in RBCs. Some merozoites develop into gametocytes (sexual parasites) that are infectious to mosquitoes.

RECEPTORS ON RBCs FOR *PLASMODIUM KNOWLESI* MEROZOITES

The first step in invasion is recognition and attachment. It has been known for a long time that each species of malaria has a limited host range, and that some species have a preference for subpopulations of RBCs within one host, for example, reticulocyte preference of *P. vivax*.[1] Studies in vitro on invasion of RBCs from various hosts by *P. knowlesi* indicate that the parasite can only invade RBCs from those hosts susceptible to infection in vivo (Table 1).[2-4] For example, RBCs from Cebus monkeys, the only new world monkey resistant to infection, were resistant to inva-

TABLE 1. THE SUSCEPTIBILITY OF PRIMATES AND SUBPRIMATES IN VIVO AND THEIR ERYTHROCYTES IN VITRO TO INFECTION BY *Plasmodium knowlesi*.

Group	Species	In Vivo	In Vitro
Man (white)		S	S
Man (black)		S & R	
Duffy positive		NT	S
Duffy negative		NT	R
Great apes	*Pan troglodytes*	S	S
	Hylobates agilis	S	S
Old world monkeys	*Macaca fascicularis*	S	S
	Macaca mulatta	S	S
	Erythrocebus patas	S	S
	Cercopithecus aethiops	S	S
New world monkeys	*Aotus trivirgatus*	S	S
	Saimiri sciureus	S	S
	Cebus apella	TP	R
Marmosets	Saguinis geoffroyi	NT	S
	Saguinis mystax	NT	S
Prosimians	Galago crassicaudatus	R	R
	Nycticebus coucang	R	R
Subprimates		R	R

NOTES: S = susceptible; R = resistant; TP = transient parasitemia; NT = not tested.

sion in vitro.[5] The high degree of specificity suggested a surface receptor. As further evidence for this it was observed that pronase or chymotrypsin treatment of human RBCs converted these cells from susceptible to refractory,[6] suggesting that the receptor was a protein or a glycoprotein. It is of interest that no proteolytic enzyme, including pronase, has any effect on invasion of rhesus RBCs,[7] an observation that remains unexplained to date.

Finally, on direct observation of the invasion sequence, J. A. Dvorak and coworkers observed a sequence of events during invasion that was characterized by attachment, widespread deformation, and slow entry within the invaginated RBC membrane. When the merozoite came in contact with a nonreceptor cell there was no observable interaction.[8] From these observations it was concluded that a specific determinant on the RBC surface initiated the interaction.

Since a determinant on the RBC surface is responsible for invasion, it seemed possible that a blood group determinant might be involved in invasion. More than 300 blood group determinants grouped into blood group systems have been identified; the structures of a few blood group antigens are known. Each blood group system is defined by antisera that react with specific determinants and cause RBC agglutination. Individuals have been identified whose RBCs fail to agglutinate with all antisera from a blood group system, and these are called negative, or null, for that system. For example, Rh-null RBCs are not agglutinated by any antiserum against Rh determinants. Cells lacking determinants for various blood group systems were tested for susceptibility to invasion by *P. knowlesi*. All were invaded with one exception—Duffy blood group negative RBCs (Table 2).[9] The Duffy blood group system has two principal determinants, Fy^a (detected by anti-Fy^a) and Fy^b (detected by anti-Fy^b), whose presence on the RBCs is determined by the codominant alleles Fy^a and Fy^b. Cells that do not react with either anti-Fy^a or anti-Fy^b are homozygous for a third allele, Fy, and are called Duffy negative. Another antiserum, anti-Fy3, reacts with all cells carrying Fy^a or Fy^b determinants, but not with Duffy-negative RBCs [Fy(a−b−)]. As stated earlier, Duffy-negative cells are resistant to invasion by

TABLE 2. SUSCEPTIBILITY OF HUMAN RBCs TO INVASION IN VITRO BY *Plasmodium knowlesi* MEROZOITES.

Refractory	Susceptible
Fy(a — b —)	Bombay (ABO null)
All blacks	Rh null
One white Australian	S — s — U —
Two American Indians	Ko = Kell negative
	Tj(a) = P,P_1, p negative
	Kx negative
	Jk(a — b —)
	Le(a — b —)
	Lu(a — b —)

SOURCE: L. H. Miller, F. M. McAuliffe, and S. J. Mason, "Erythrocyte Receptors for Malaria Merozoites," *American Journal of Tropical Medicine and Hygiene* 26 (1977): 204-08.

P. knowlesi. The majority of black Africans have this phenotype; it is extremely rare in populations with no black admixture.[10] RBCs from blacks with either Fy^a or Fy^b determinants are fully susceptible, although many are undoubtedly heterozygous for the Duffy-negative gene.

Following the observation that Duffy-negative RBCs were refractory to invasion by *P. knowlesi,* it was suggested that this may be the genetic basis for known resistance of blacks to *P. vivax.*[11] Three studies have supported the Duffy-malaria hypothesis. First, the Duffy blood group phenotype was determined in eleven blacks who had previously been exposed to *P. vivax* infected mosquitoes.[12] The six Duffy-positive blacks developed RBC infection; the five Duffy-negative blacks were resistant. Second, H. C. Spencer and collaborators determined the relationship between the Duffy blood group phenotype and malarial infections in a village in Honduras in which 247 of 420 persons were Duffy negative.[13] *P. falciparum* occurred with equal frequency in Duffy-positive and -negative individuals. In contrast, all fourteen slides demonstrated that *P. vivax* infections occurred in Duffy-positive persons. Third, we determined the blood group phenotype of thirteen blacks who had been infected with *P. vivax* in Vietnam.[14] All were Duffy positive.

MECHANISM OF RESISTANCE TO INVASION BY *P. KNOWLESI*

The observation that Duffy-negative human RBCs are resistant to invasion poses two important questions: 1) Is resistance the result of the Duffy-negative phenotype or some other factor closely linked on the chromosome?; and 2) How does the Duffy blood group system function in the invasion sequence?

Evidence that the Duffy-negative genotype is the basis for resistance is the refractoriness to invasion of RBCs from three rare, nonblack Duffy negatives.[15] In these three nonblacks, the Duffy-negative phenotype is probably the result of mutation to Duffy negativity rather than black gene admixture. First, no other common black blood group determinants were found in these three. Second, the Duffy-negative RBCs in these three may differ from the common black Duffy-negative RBCs. Two of the three produced anti-Fy3 that binds to all Duffy-positive RBCs. Only one Duffy-negative black is known to have produced anti-Fy3. Thus for untreated human RBCs there is an absolute correlation between the presence of Fy^a and Fy^b determinants and susceptibility to invasion.

We can assume therefore that resistance to invasion is based on the absence of a Duffy-associated determinant. Proteolytic enzymes that removed the Duffy determinants, chymotrypsin, for example, reduced invasion of those cells.[16] Conversely, trypsin or neuraminidase treatment did not remove Duffy determinants and did not reduce invasion. Furthermore, coating of Fy(a+b−) RBCs with anti-Fy^a markedly reduced invasion.[17] Anti-Fy^b specifically blocked invasion of Fy(a−b+) RBCs, but only when it was added to the culture medium.*

In some situations, however, invasion occurred in the absence of Fy^a or Fy^b determinants. First, Duffy-negative cells that had been treated with trypsin or neuraminidase were invaded, although the cells still lacked the Fy^a and Fy^b determinants (Table 3).[18] Second, RBCs from New World monkeys, lacking Duffy determinants, were invaded. Third, RBCs from higher apes and Old

* L. H. Miller: unpublished data.

TABLE 3. EFFECT OF ENZYMATIC TREATMENT OF HUMAN DUFFY BLOOD GROUP-POSITIVE [FY(A+B−)] AND -NEGATIVE [FY(A−B−)] RBCs ON INVASION RATE AND ELECTROPHORETIC MOBILITY

Treatment	Invasion Rate (Percent)[1] Fy(a+b−)	Invasion Rate (Percent)[1] Fy(a−b−)	Percent Reduction of Electrophoretic Mobility
None	12.0 (4)	0.2 (4)	—
Trypsin	13.9 (4)	6.1 (4)[2]	35
Neuraminidase	18.7 (2)	1.4 (3)[2]	79

NOTES: 1) The figures in parentheses represent the number of experiments performed using each cell type. The invasion rate is the mean of the results of all experiments. 2) The Fy(a−b−) RBCs remained phenotypically Fy(a−b−) after trypsin or neuraminidase treatment.

SOURCE: S. J. Mason et al., "The Duffy Blood Group Determinants: Their Role in the Susceptibility of Human and Animal Erythrocytes to *Plasmodium knowlesi* Malaria," *British Journal of Haematology* 36 (1977): 327-35.

World monkeys were still invaded after chymotrypsin treatment, although the Duffy-b determinants were removed. It is impossible at this time to explain the conflicting data on the requirement for Fy^a and Fy^b determinants for invasion, because the chemistry of the Duffy determinants and associated structures are unknown. It may be that a structure associated with the Duffy determinant, but not Fy^a or Fy^b itself, may be the requirement. One associated determinant, Fy3, cannot be the requirement, since chymotrypsinized human Duffy-positive RBCs that still have Fy3 determinants are refractory to invasion.

What role the Duffy-associated structures play in the invasion sequence is also unclear. Observation of the interaction between Duffy-negative RBCs and *P. knowlesi* merozoites demonstrated that merozoites induced widespread deformation of these RBCs, but failed to enter within a localized invagination of the RBC.[19] The initial interaction was specific, however, since *P. knowlesi* merozoites did not interact with RBCs from prosimii or subprimates.

Two hypotheses could explain these observations. First, there may be two receptors for invasion with Duffy-associated antigen involved in the second stage of invasion, endocytosis. Alternatively, the Duffy-negative determinant may have a lower affinity receptor for the merozoite. This would allow recognition, but not hold the merozoite in place for invasion.

RECEPTORS ON RBCs FOR *P. FALCIPARUM*

Although *P. falciparum, P. vivax,* and *P. knowlesi* all invade human RBCs, *P. falciparum* appears to use different receptors. First, the host range for *P. falciparum* differs from that of *P. vivax.*[20] Second, Duffy-negative RBCs that are refractory to invasion by *P. knowlesi,* and probably *P. vivax,* have normal susceptibility to *P. falciparum.*[21] Third, there is differential susceptibility of enzyme-treated RBCs to invasion by *P. falciparum* and *P. knowlesi.*[22] Chymotrypsin markedly reduces susceptibility to *P. knowlesi;* invasion by *P. falciparum* is normal, or only minimally reduced. Conversely, trypsin or neuraminidase reduces susceptibility to *P. falciparum; P. knowlesi* invasion of these cells is normal.

ULTRASTRUCTURAL OBSERVATIONS ON THE INVASION PROCESS

It seems reasonable to assume there are specific receptors on both the parasite and host cell by which the parasite attaches to host cells. Even if we accept the idea of receptors between merozoite and RBC, we still must explain how a merozoite of limited motility can enter within the invaginated plasma membrane of a nonphagocytic cell, the mature RBC. Recent ultrastructural studies have helped us understand this process.

In 1969 R. Ladda and associates reported on the invasion of RBCs by merozoites of *P. berghei* and *P. gallinaceum,* and established that merozoites enter within an invagination of the RBC membrane rather than by penetrating it.[23] Recently it was observed that a junction formed between the apical region of the merozoite and the RBC (Figure 1). The RBC membrane to which the merozoite was attached became thickened.[24] As invasion progressed, the depression in the RBC deepened and conformed to the curvature of the merozoite. At this time the thickened, electron-dense zone on the RBC membrane was no longer ob-

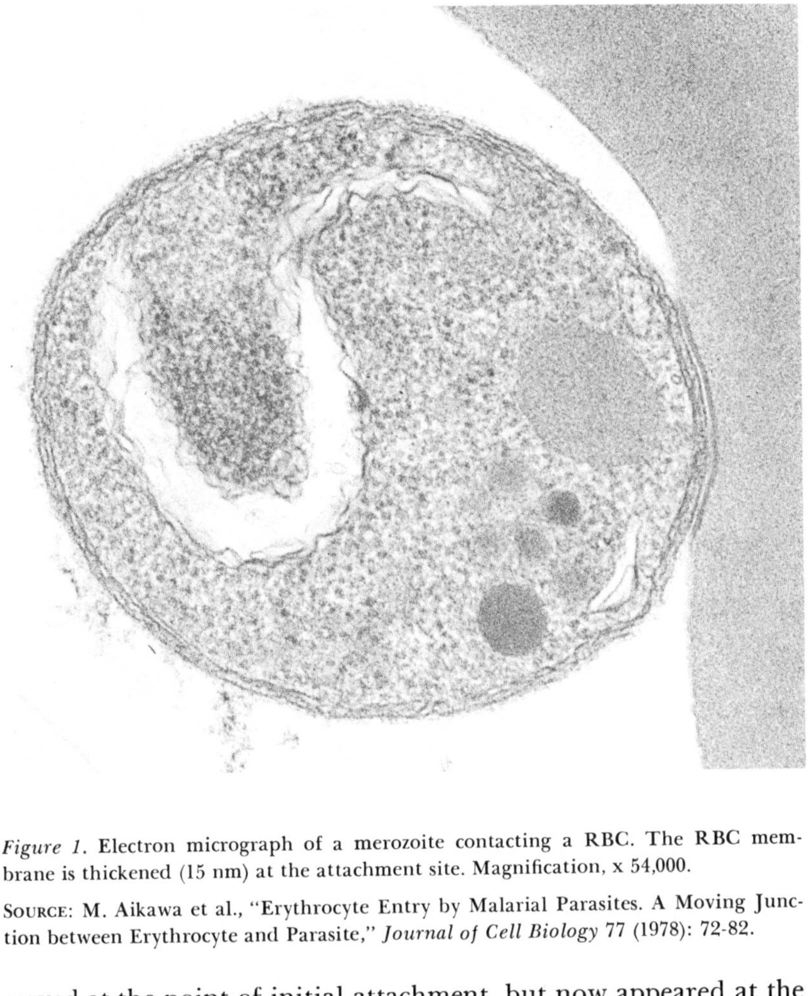

Figure 1. Electron micrograph of a merozoite contacting a RBC. The RBC membrane is thickened (15 nm) at the attachment site. Magnification, x 54,000.

SOURCE: M. Aikawa et al., "Erythrocyte Entry by Malarial Parasites. A Moving Junction between Erythrocyte and Parasite," *Journal of Cell Biology* 77 (1978): 72-82.

served at the point of initial attachment, but now appeared at the orifice of the merozoite-induced invagination (Figures 2 and 3). When entry was completed, that is, when the merozoite was within the RBC, the junction appeared to fuse at the posterior end of the merozoite, closing the orifice (Figure 4).

The extracellular merozoite was surrounded by a surface coat that was susceptible to proteolytic enzymes.[25] During invasion, no surface coat was visible on the portion of the merozoite within the RBC invagination, whereas the surface coat on the

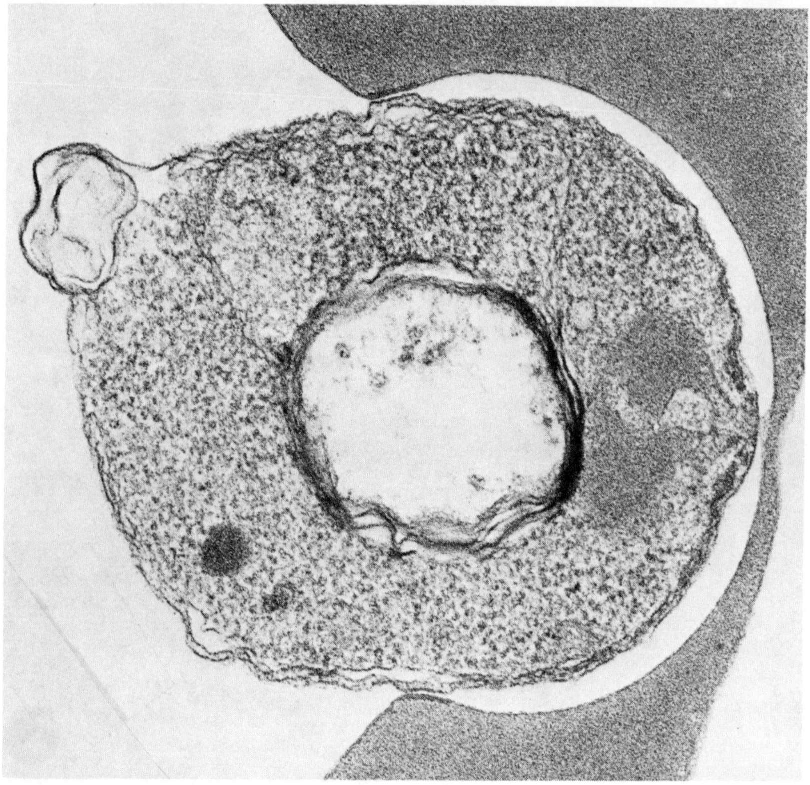

Figure 2. An advanced state of RBC entry by merozoite. The junction is formed between the thickened membrane of the RBC and the plasma membrane of the merozoite. An electron-opaque projection connects the apical end and the RBC membrane. Magnification, x 50,000.

SOURCE: M. Aikawa et al., "Erythrocyte Entry by Malarial Parasites. A Moving Junction between Erythrocyte and Parasite," *Journal of Cell Biology* 77 (1978): 72-82.

portion remaining outside the RBC appeared to be similar to that seen on free merozoites.[26-28]

Throughout invasion the apical end remained in contact with the RBC.[29-32] Material from rhoptries (paired organelles in the apical portion of the merozoite) appeared to empty through a common duct onto the RBC membrane.[33] It is likely that the contents of the rhoptries are important for junction movement and interiorization, but the chemical nature and physiological

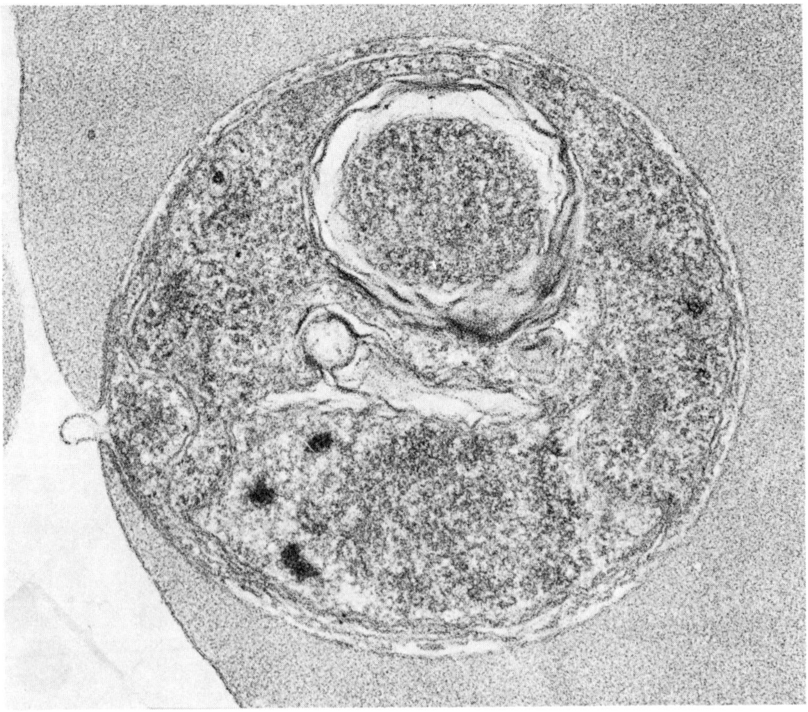

Figure 3. RBC entry by a merozoite is almost completed and a small orifice is seen at the posterior end of the merozoite. The junction is now moved to the posterior end of the merozoite. Magnification, x 54,000.

SOURCE: M. Aikawa et al., "Erythrocyte Entry by Malarial Parasites. A Moving Junction between Erythrocyte and Parasite," *Journal of Cell Biology* 77 (1978): 72-82.

function are unknown. A. Kilejian has suggested that the rhoptries and micronemes contain a histidine-rich protein that invaginates the RBC membrane.[34] The isolation and characterization of this material is of obvious importance.

ISOLATION OF THE ATTACHMENT PHASE OF INVASION

One step in the structural and functional definition of the invasion sequence was the development of a method for isolation of the attachment phase of invasion. This method used

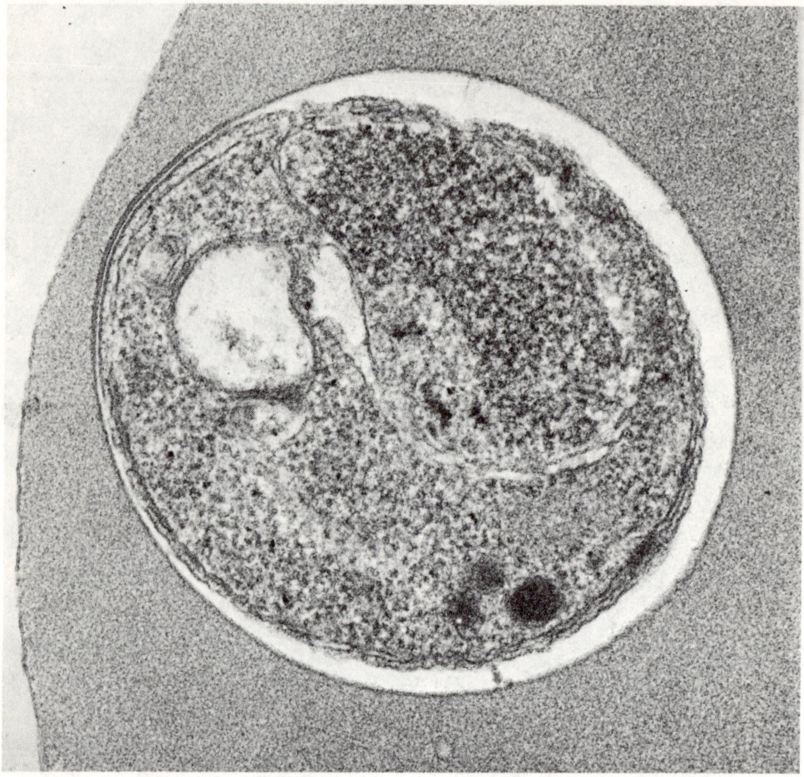

Figure 4. A merozoite is now inside a RBC. The posterior end of the merozoite, however, is still attached to the thickened RBC membrane. Magnification, x 50,000.

SOURCE: M. Aikawa et al., "Erythrocyte Entry by Malarial Parasites. A Moving Junction between Erythrocyte and Parasite," *Journal of Cell Biology* 77 (1978): 72-82.

cytochalasin-treated *P. knowlesi* merozoites that attached to but did not invade rhesus RBCs.[35] Because cytochalasin B has multiple effects on cells we did not attempt to define its mechanism of action, but instead used it operationally to isolate the early step in invasion. Characteristics of the attachment between cytochalasin-treated merozoites and RBCs can be summarized as follows.

First, attachment was specific, in that the cytochalasin-treated merozoites only attached to RBCs from species that were susceptible to invasion. Second, the assay detected differences in attachment affinity, in that attachment to rhesus RBCs was consistently greater than to human RBCs. This corresponded to the higher invasion rate of rhesus than of human RBCs. Third, certain manipulations of merozoites in vitro that markedly reduced invasion had no effect on attachment: for example, incubation of meroboites at 4°C.* Fourth, the cytochalasin-treated merozoites were attached so that the end of the merozoite containing the rhoptries (the apical end) was usually in apposition to the RBC. This orientation of attachment was the same as during invasion, that is, during invasion the apical end containing the rhoptries was oriented toward the RBC. Fifth, the RBC membrane to which the cytochalasin-treated merozoite was attached became thickened and formed a junction with the plasma membrane of the apical portion of the merozoite (Figures 5 and 6). No movement of the junction between the RBC and merozoite occurred, however. Vacuoles surrounded by a unit membrane appeared in the RBC cytoplasm near the attachment site (Figure 7).

Because *P. knowlesi* merozoites do not invade Duffy-negative human RBCs, we were surprised to find that cytochalasin-treated merozoites attached equally well to Duffy-positive and -negative RBCs. In addition, the apical orientation for attachment was the same for Duffy-positive and -negative RBCs. Ultrastructural studies, however, identified an important difference in the basis for attachment. Cytochalasin-treated merozoites formed a junction with Duffy-positive human RBCs (Figure 8) similar to that observed with rhesus RBCs. In contrast, no junction was observed between cytochalasin-treated merozoites and Duffy-negative RBCs. Instead, filaments arranged in a cylindrical fashion extended from the edge of the apical end of merozoites to the RBC. It appears that the Duffy-associated determinants are involved in junction formation, not in the initial attachment.

SUMMARY

Invasion of RBCs by malaria merozoites requires specific determinants on the surface of the RBC to which merozoites are

* J. G. Johnson and L. H. Miller: unpublished data.

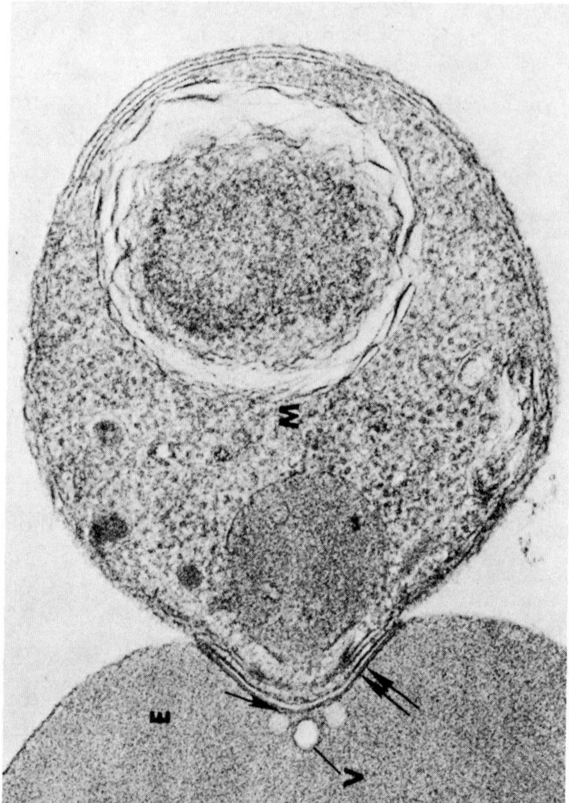

Figure 5. Electron micrograph showing the attachment between the apical end of a cytochalasin-B-treated merozoite (M) and rhesus erythrocyte (E). The erythrocyte membrane is thickened (double arrow) at the attachment site. A few vacuoles (V) are seen in erythrocyte cytoplasm, and some are in contact with the invaginated erythrocyte membrane (arrow). Magnification, x 50,000.

SOURCE: L.H. Miller et al., "Interaction between Cytochalasin-B-Treated Malarial Parasites and Red Cells: Attachment and Junction Formation," *Journal of Experimental Medicine* 149 (1979): 172-84.

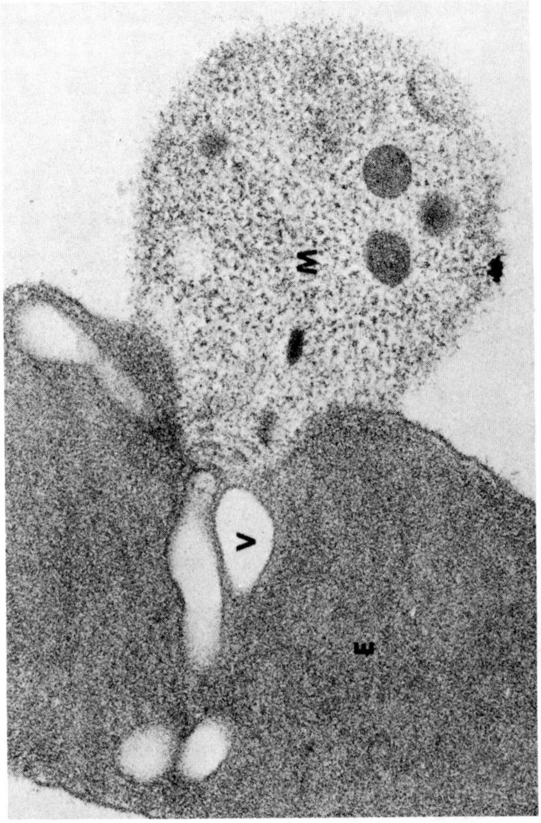

Figure 6. Electron micrograph showing attachment between a cytochalasin-B-treated merozoite (M) and a rhesus erythrocyte (E). Note several vacuoles (V) in the erythrocyte cytoplasm. Magnification, x 57,000.

SOURCE: L.H. Miller et al., "Interaction between Cytochalasin-B-Treated Malarial Parasites and Red Cells: Attachment and Junction Formation," *Journal of Experimental Medicine* 149 (1979): 172-84.

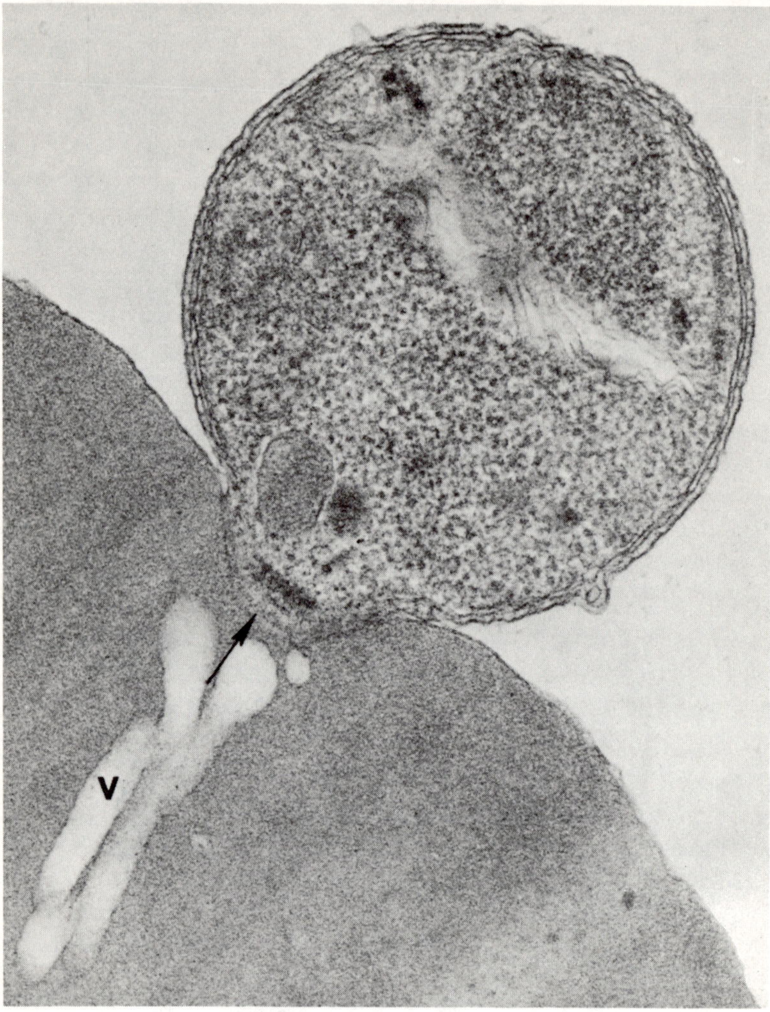

Figure 7. Electron micrograph showing the attachment between the apical end of a cytochalasin-B-treated merozoite (arrow) and a Duffy-positive human erythrocyte. Several elongated vacuoles (V) are present near the attachment site. Magnification, x 50,000.

SOURCE: L.H. Miller et al., "Interaction between Cytochalasin-B-Treated Malarial Parasites and Red Cells: Attachment and Junction Formation," *Journal of Experimental Medicine* 149 (1979): 172-84.

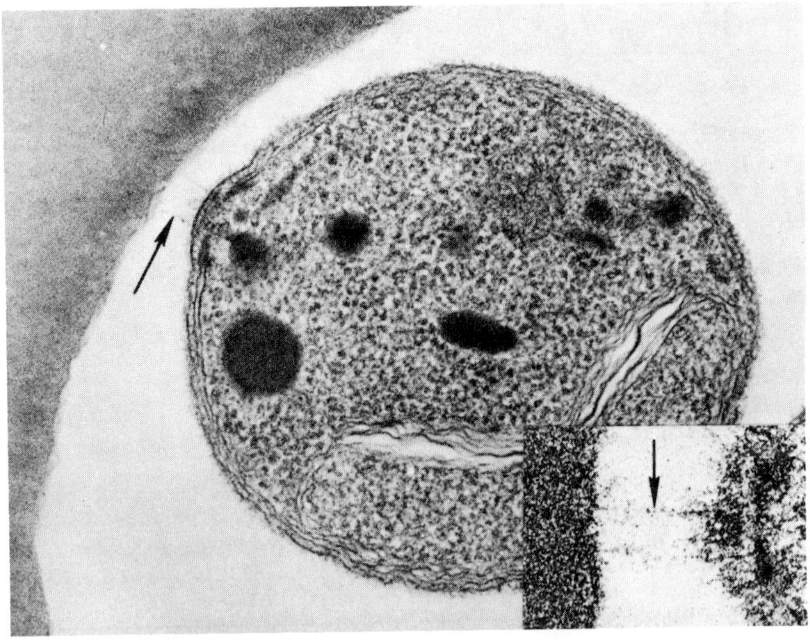

Figure 8. Electron micrograph showing a cytochalasin-B-treated merozoite connected with a Duffy-negative human erythrocyte by two fine fibrils (arrow) extending from the edge of the apical end. Magnification, x 50,000. *Inset:* two fine fibrils connecting the apical end of a merozoite and a Duffy-negative human erythrocyte. This specimen was treated with 10% albumin before preparation for EM embedding. Magnification, x 78,000.

SOURCE: L.H. Miller et al., "Interaction between Cytochalasin-B-Treated Malarial Parasites and Red Cells: Attachment and Junction Formation," *Journal of Experimental Medicine* 149 (1979): 172-84.

attached. Attachment is followed by two events: release of the contents of rhoptries at the apical end of the merozoite, and junction formation. During the invasion of human RBCs by *P. knowlesi* merozoites, junction formation appears to require Duffy-associated determinants. As the merozoite enters the invagination in the RBC surface, the junction, which is in the form of a circumferential zone of attachment between RBC and merozoite, moves so as to remain at the orifice of the invagination.

Biochemical characterization of the merozoite plasma membrane should help in identifying the receptors on the merozoite and RBC and may suggest novel ways to block invasion.

NOTES

1. L. H. Miller and R. Carter, "Innate Resistance to Malaria," *Experimental Parasitology* 40 (1976): 132-46.

2. G. A. Butcher, G. H. Mitchell, and S. Cohen, "Mechanism of Host Specificity in Malaria Infection," *Nature* 244 (1973): 40-42.

3. L. H. Miller et al., "Influence of Erythrocyte Membrane Components on Malaria Merozoite Invasion," *Journal of Experimental Medicine* 138 (1973): 1597-1601.

4. S. J. Mason et al., "The Duffy Blood-Group Determinants: Their Role in the Susceptibility of Human and Animal Erythrocytes to *Plasmodium knowlesi* Malaria," *British Journal of Haematology* 36 (1977): 327-35.

5. Ibid.

6. Miller et al., "Influence of Erythrocyte" (See note 3).

7. Ibid.

8. J. A. Dvorak et al., "Invasion of Erythrocytes by Malaria Merozoites," *Science* 187 (1975): 748-50.

9. L. H. Miller et al., "Erythrocyte Receptors for (*Plasmodium knowlesi*) Malaria: Duffy Blood-Group Determinants," *Science* 189 (1975): 561-63.

10. A. E. Mourant, A. C. Kopec, and K. Domaniewska-Sobczak, *The Distribution of the Human Blood Groups, and Other Polymorphisms,* 2nd ed. (London: Oxford University Press, 1976).

11. Miller et al., "Erythrocyte Receptors" (See note 9).

12. L. H. Miller et al., "The Resistance Factor to *Plasmodium vivax* in Blacks: the Duffy Blood-Group Genotype, *FyFy*," *New England Journal of Medicine* 295 (1976): 302-04.

13. H. C. Spencer et al., "The Duffy Blood Group and Resistance to *Plasmodium vivax* in Honduras," *American Journal of Tropical Medicine and Hygiene* 27 (1978): 664-70.

14. L. H. Miller et al., "The Duffy Blood-Group Phenotype in American Blacks Infected with *Plasmodium vivax* in Vietnam," *American Journal of Tropical Medicine and Hygiene* 27 (1978): 1069-72.

15. Mason et al., "The Duffy Blood Group" (See note 4).

16. Miller et al., "Erythrocyte Receptors" (See note 9).

17. Ibid.

18. Mason et al., "The Duffy Blood Group" (See note 4).

19. Miller et al., "Erythrocyte Receptors" (See note 9).

20. L. H. Miller, F. M. McAuliffe, and S. J. Mason," Erythrocyte Receptors for Malaria Merozoites," *American Journal of Tropical Medicine and Hygiene* 26 (1977): 204-08.

21. L. H. Miller et al., "Evidence for Differences in Erythrocyte Surface Receptors for the Malaria Parasites, *Plasmodium falciparum* and *Plasmodium knowlesi*," *Journal of Experimental Medicine* 146 (1977): 277-81.

22. Ibid.

23. R. Ladda, M. Aikawa, and H. Sprinz, "Penetration of Erythrocytes by

Merozoites of Mammalian and Avian Malaria Parasites," *Journal of Parasitology* 55 (1969): 633-44.

24. M. Aikawa et al., "Erythrocyte Entry by Malarial Parasites. A Moving Junction between Erythrocyte and Parasite," *Journal of Cell Biology* 77 (1978): 72-82.

25. L. H. Miller, M. Aikawa, and J. A. Dvorak, "Malaria (*Plasmodium knowlesi*) Merozoites: Immunity and the Surface Coat," *Journal of Immunology* 114 (1975): 1237-42.

26. Ladda, Aikawa, and Sprinz, "Penetration of Erythrocytes" (See note 23).

27. Aikawa et al., "Erythrocyte Entry" (See note 24).

28. L. H. Bannister et al., "Structure and Invasive Behavior of *Plasmodium knowlesi* Merozoites in Vitro," *Parasitology* 71 (1975): 483-91.

29. Dvorak et al., "Invasion of Erythrocytes" (See note 8).

30. Ladda, Aikawa, and Sprinz, "Penetration of Erythrocytes" (See note 23).

31. Aikawa et al., "Erythrocyte Entry" (See note 24).

32. Bannister et al., "Structure and Behavior" (See note 28).

33. Aikawa et al., "Erythrocyte Entry" (See note 24).

34. A. Kilejian, "Studies on a Histidine-Rich Protein from *Plasmodium lophurae*," in *Biochemistry of Parasites and Host-Parasite Relationships*, ed. H. Van den Bossche (Amsterdam: Elsevier North-Holland, Inc., 1976): 441-48.

35. L. H. Miller et al., "Interaction between Cytochalasin B-Treated Malarial Parasites and Red Cells: Attachment and Junction Formation," *Journal of Experimental Medicine* 149 (1979): 172-84.

PARTICIPANTS

Alexander G. Bearn, M.D.
Stanton Griffis Distinguished Medical Professor
Cornell University Medical College
New York, New York

Celso Bianco, M.D.
Associate Professor
Department of Pathology
Downstate Medical Center
State University of New York
Brooklyn, New York

Purnell W. Choppin, M.D.
Laboratory of Virology
The Rockefeller University
New York, New York

Max Cooper, M.D.
Professor of Pediatrics and Microbiology
Cellular Immunobiology Unit
Comprehensive Cancer Center
Departments of Pediatrics and Microbiology
School of Medicine
University of Alabam
Birmingham, Alabama

Ian Creese, Ph.D.
Assistant Professor
Department of Neurosciences
School of Medicine
University of California at San Diego
La Jolla, California

Richard L. Crowell, Ph.D.
Professor of Microbiology and Immunology
Department of Microbiology and Immunology
The Hahnemann Medical College and Hospital of Philadelphia
Philadelphia, Pennsylvania

Vincent P. Dole, M.D.
Laboratory of Biology and Addictive Diseases
The Rockefeller University
New York, New York

Daniel B. Drachman, M.D.
Professor of Neurology
Director, Neuromuscular Unit
School of Medicine
The Johns Hopkins University
Baltimore, Maryland

D. Michael Gill, Ph.D.
The Biological Laboratories
Harvard University
Cambridge, Massachusetts

Joseph L. Goldstein, M.D.
Paul J. Thomas Professor of Genetics and Chairman
Department of Molecular Genetics
The University of Texas Health Center at Dallas
Dallas, Texas

Calderon Howe, M.D.
Professor and Head
Department of Microbiology
School of Medicine
Louisiana State University Medical Center
New Orleans, Louisiana

Thomas J. Kindt, Ph.D.
Chief, Laboratory of Immunogenetics
National Institute of Allergy and Infectious Diseases
National Institutes of Health
Bethesda, Maryland

Richard M. Krause, M.D.
Director
National Institute of Allergy and Infectious Diseases
National Institutes of Health
Bethesda, Maryland

Louis H. Miller, M.D.
Head, Malaria Section
Laboratory of Parasitic Diseases
National Institute of Allergy and Infectious Diseases
National Institutes of Health
Bethesda, Maryland

Joel Moss, M.D., Ph.D.
Laboratory of Cellular Metabolism
National Heart, Lung, and Blood Institute
National Institutes of Health
Bethesda, Maryland

Victor Nussenzweig, M.D.
Professor
Department of Pathology
School of Medicine
New York University Medical Center
New York, New York

Lennart Philipson, M.D.
Professor
Department of Microbiology
Biomedical Center
University of Uppsala
Uppsala, Sweden

S. Michael Phillips, M.D.
Allergy and Immunology Section
Department of Medicine
School of Medicine
University of Pennsylvania
Philadelphia, Pennsylvania

Jesse Roth, M.D.
Chief, Diabetes Branch
National Institute of Arthritis and Metabolism and Digestive Diseases
National Institutes of Health
Bethesda, Maryland

Lawrence I. Rothfield, M.D.
Professor and Chairman
Department of Microbiology
School of Medicine
The University of Connecticut Health Center
Farmington, Connecticut

Andreas S. Scheid, M.D.
Laboratory of Virology
The Rockefeller University
New York, New York

William T. Schrader, M.D.
Associate Professor
Department of Cell Biology
Baylor College of Medicine
Houston, Texas

Eric J. Simon, Ph.D.
Professor of Experimental Medicine
Department of Medicine
New York University Medical Center
New York, New York

Armen H. Tashjian, Jr., M.D.
Professor of Toxicology
Harvard School of Public Health
Professor of Pharmacology
Harvard Medical School
Laboratory of Toxicology
Boston, Massachusetts

Rolf M. Zinkernagel, M.D.
Department of Immunopathology
Scripps Clinic and Research Foundation
La Jolla, California

INDEX